Food Packaging and Preservation

Food Packaging and Preservation

Edited by

M. MATHLOUTHI
Faculté des Sciences
Université de Reims Champagne-Ardenne

SPRINGER-SCIENCE+BUSINESS MEDIA, B.V.

First edition 1994

© 1994 Springer Science+Business Media Dordrecht
Originally published by Chapman & Hall in 1994

Typeset in 10/12 Times by Acorn Bookwork, Salisbury, Wiltshire

ISBN 978-0-7514-0182-0   ISBN 978-1-4615-2173-0 (eBook)
DOI 10.1007/978-1-4615-2173-0

A catalogue record for this book is available from the British Library

Library of Congress Catalog Card Number: 93-74917

∞Printed on acid-free text paper, manufactured in accordance with ANSI/
NISO Z39.48-1992 (Permanence of Paper)

Preface

This book is an updating of *Food Packaging and Preservation, Theory and Practice* published in 1986 by Elsevier Applied Science. Since that date, many things have changed in the world. Hence the name given to the first IFTEC meeting held at the Hague (NL), November 15–18, 1992 *Food Technology for a Changing World*. Is the world changing for better or worse and what can food technology improve? The keynote lecture of the IFTEC meeting dealt with hunger and the challenge it represents to food science and technology. In the preface to the 1986 book it was suggested that food packaging could solve some of the problems of crop preservation in countries where starvation is prevalent.

However, such thoughts did not solve any problems. The famine is still spreading in Africa. The unbalanced north–south situation evoked in the 1986 preface has not improved. The international market of foods and agricultural products is constantly changing and food packaging scientists can only explore new ways to help cope with this. Some of these ideas are approached in this book, particularly in chapters 9, 10 and 12.

The preservation of a packaged food and the increase in shelf-life require knowledge of the permeability of the plastic film to O_2, CO_2 and water vapour, as well as migration and flavour retention. These aspects are described with reference to the structure of the polymer itself and from the analytical point of view. New trends in food packaging such as the microwavability of packaged foods or their direct control by NMR imaging, as well as the effect of irradiation and the use of trehalose for food preservation, are discussed.

The chapters in this book were either presentations in international symposia (IFTEC, The Hague (NL)), Conditionnement alimentaire; 2 défis: Innovation et environnement, Pouzauges (France); or original contributions. Food packaging and preservation is a new multidisciplinary speciality. It deserves to be treated as science, which is what we aim to do in this book.

M. M.

Foreword

This book includes papers presented at a symposium on *Food Packaging Interactions and Packaging Disposability*, held at the International Food Technology Exposition and Conference, 15–18 November 1992, Den Haag, The Netherlands, as well as several additional contributions. Plastics are important agents for packaging foods and their use is expected to increase because: (1) they are convenient and cost-effective; (2) new or improved types are continually being developed; (3) new information regarding package/food and package/environment interactions is rapidly accruing; and (4) the problems of recycling, reuse and disposability are being addressed. With the exception of the first item, these important matters are dealt with in this book.

Several chapters relate to permeability characteristics of plastic packaging materials. Specific matters addressed are the relationship between selective diffusion properties of packaging materials and shelf-life, methods of measuring oxygen and carbon dioxide permeability of films and the relationship between light-transmitting properties of packaging materials and oxidation of dairy products.

Attention is also given to using plastic packaging materials to aid in establishing modified atmosphere conditions for cheese, fruits and vegetables; the purpose being to achieve increased shelf-lives of food products at refrigerated temperatures.

Two types of interactions between plastic packaging materials and food are considered: (1) that involving food flavors; and (2) that involving migration of packaging constituents into food.

Also considered in detail is the influence of packaging materials on the response of foods to microwave heating, to ionizing radiation and to spatial mapping of water and fat molecules by nuclear magnetic resonance.

Finally, recycling, reuse and disposability of plastic packaging materials are covered in several chapters.

Readers with an interest in food packaging will find the subject matter of this book timely, representative of some of the best work in the field of food packaging, and of considerable value.

Owen Fennema
Professor of Food Chemistry
University of Wisconsin

Contributors

A. Amorós

Universidad Politécnica de Valencia, Escuela Universitaria Ingeniería Técnica Agrícola, Orihuela, Spain

M.B.M. Audhuy-Peaudecerf

ENSEEIHT, 2 Rue Carles Camichel, 31071 Toulouse Cedex, France

A.L. Baner

Fraunhofer-Institut für Lebensmitteltechnologie und Verpackung, Schragenhofstr. 35, 8000 Munich 50, Germany

J.O. Bosset

Station Fédérale de Recherches Laitières, CH-3097 Liebefeld-Bern, Switzerland

A.J. Campbell

Department of Product and Packaging Technology, Campden Food and Drink Research Association, Chipping Campden, Gloucestershire GL55 6LD, UK

C.A.L.S. Colaço

Quadrant Research Foundation, Maris Lane, Cambridge CB2 2SY, UK

E. Eggink

Agrotechnological Research Institute (ATO-DLO) PO Box 17, 67000 AA Wageningen, The Netherlands

Z. El Makhzoumi

12 Rue du Danube, 51100 Reims, France

R. Franz

Fraunhofer-Institut für Lebensmitteltechnologie und Verpackung, Schragenhofstr. 35, 8000 Munich 50, Germany

P.U. Gallmann

Station Fédérale de Recherches Laitières, CH-3097 Liebefeld-Bern, Switzerland

N. Gontard

Université de Montpellier II, GBSA, Place E. Bataillon 34095 Montpellier, France

S. Guilbert

CIRA-SAR, 73 Rue J.F. Breton BP 5035, 34032 Montpellier, France

L.D. Hall

Herchel Smith Laboratory for Medicinal Chemistry, Cambridge University School of Clinical Medicine, University Forvie Site, Robinson Way, Cambridge CB2 2PZ, UK

G.N.M. Huijberts

Agrotechnological Research Institute (ATO-DLO) PO Box 17, 6700 AA Wageningen, The Netherlands

B. Jasse

Laboratoire de Physico-Chimie Structurale et Macromoléculaire Ecole Supérieure de Physique et Chimie Industrielles de la Ville de Paris, 10, Rue Vauquelin 75231 Paris Cedex 05, France.

S.A.E. Lefeuvre

ENSEEIHT, 2 Rue Carles Camichel, 31071 Toulouse Cedex, France

J.P. Leiris (de)

1, Avenue Général de Larminat 94000 Créteil, France

J.P.H. Linssen

Agricultural University, Department of Food Science, PO Box 8129, 6700 EV Wageningen, The Netherlands

G. Martinez

Consejo Superior de Investigaciones Científicas Centro de Edafología Y Biologíca Aplicada del Segura, Av. La Fama, 1. 30003 Murcia, Spain

M. Mathlouthi

Laboratoire de Chimie Physique Industrielle, Faculté des Sciences, Université de Reims Champagne-Ardenne B.P. 347, 51062 Reims Cedex, France

G. Ongen-Baysal

Agrotechnological Research Institute (ATO-DLO) PO Box 17, 6700 AA Wageningen, The Netherlands

O. Piringer

Fraunhofer-Institut für Lebensmitteltechnologie und Verpackung, Schragenhofstr. 35, 8000 Munich 50, Germany.

M.T. Pretel

Universidad Politécnica de Valencia, Escuela Universitaria Ingeniería Técnica Agrícola Orihuela, Spain

F. Riquelme
Consejo Superior de Investigaciones Científicas Centro de Edafología Y Biologíca Aplicada del Segura, Av. La Fama, 1. 30003 Murcia, Spain

F. Romojaro
Consejo Superior de Investigaciones Científicas Centro de Edafología Y Biologíca Aplicada del Segura, Av. La Fama, 1. 30003 Murcia, Spain

J.P. Roozen
Agricultural University, Department of Food Science, PO Box 8129 67000 EV Wageningen, The Netherlands

B. Roser
Osmotica Foods Inc., 1920 Fifth Street, Davis, California 95616, USA

M. Serrano
Universidad Politécnica de Valencia, Escuela Universitaria Ingeniería Técnica Agricola, Orihuela, Spain

A.M. Seuvre
Institut Universitaire de Technologie, Université de Bourgogne 21000 Dijon, France

R. Sieber
Station Fédérale de Recherches Laitières, CH-3097 Liebefeld-Bern, Switzerland

J. Smegen
Agrotechnological Research Institute (ATO-DLO) PO Box 17 6700 AA Wageningen, The Netherlands

J.J. Wright
Herchel Smith Laboratory for Medicinal Chemistry, Cambridge University School of Clinical Medicine, University Forvie Site, Robinson Way, Cambridge CB2 2PZ, UK

Contents

3 Food flavour and packaging interactions 48

J.P.H. LINSSEN and J.P. ROOZEN

4 Microwavability of packaged foods 62

S.A.E. LEFEUVRE and M.B.M. AUDHUY-PEAUDECERF

5 Effect of irradiation of polymeric packaging material on the formation of volatile compounds 88

Z. EL MAKHZOUMI

6 Package coating with hydrosorbent products and the shelf-life of cheeses   100

M. MATHLOUTHI, J.P. de LEIRIS and A.M. SEUVRE

7 Trehalose – a multifunctional additive for food preservation   123

C.A.L.S. COLAÇO and B. ROSER

8 Packaging of fruits and vegetables: recent results   141

F. RIQUELME, M.T. PRETEL, G. MARTÍNEZ, M. SERRANO,
A. AMORÓS and F. ROMOJARO

9 Bio-packaging: technology and properties of edible and/or biodegradable material of agricultural origin 159

N. GONTARD and S. GUILBERT

10 Bacterial poly(hydroxyalkanoates) 182

G. EGGINK, J. SMEGEN, G. ONGEN-BAYSAL
and G.N.M. HUIJBERTS

11 NMR imaging of packaged foods 197

J.J. WRIGHT and L.D. HALL

1 Permeability and structure in polymeric packaging materials

B. JASSE, A.M. SEUVRE and M. MATHLOUTHI

Abstract

The materials used in food packaging are very often common polymers. Their permeability to gases and vapours is at the origin of their barrier properties and capacity for protection of the food. The permeability coefficient, which is at thermodynamic equilibrium equal to the product of diffusivity and solubility, depends on the structure of the polymer as well as the properties of diffusing molecules. Polymer properties affecting permeability, such as free volume, crystallinity, tacticity, cross-linking, orientation and thickness, are reviewed as well as permeant characteristics, size and shape and polarity, especially for water vapour, which are described in relation to their influence on permeability. Different experimental methods of determination of permeability are also summarized.

1.1 Introduction

The preservation of a food product packed in a plastic film mainly depends on the maintaining of its original quality by protecting it against external deteriorative influences. This is achieved through the barrier properties of the packaging material. The required protection of the foodstuff may be achieved with a single layer of polymer or necessitates the use of multi-layered films including different polymers, coatings and metal foils. The barrier properties, hence the protecting capacity of a package, mainly originate from its permeability to gases and vapours that are noxious to the quality of the product. For the majority of foods, the gain or loss of moisture leads to either a physical or biological defect. A loss of water may lead to undesirable drying detrimental to both the texture of the product and the purse of the manufacturer. A gain of water may lead water activity (A_w) to approach the region of microbial spoilage above $A_w = 0.8$.

More harmful than moisture is oxygen for foods from plant or animal origin. Its fixation to the product is irreversible. It causes lipid oxidation and provokes rancidity especially when the package allows light transmittance. The other requirements for the preservation of the qualities (physical, chemical, sanitary, organoleptic) of the food are to prevent changes in taste, colour and odour and if a modified atmosphere is applied inside the package to maintain its composition in CO_2 and N_2. All these deterioration processes are time and temperature dependent. That is why the packaging very often bears a notice like 'use before' or 'recommended deadline'.

The choice of a packaging material should take into account all these constraints as well as those caused by further treatment, storage and handling of the packaged food. The polymeric materials are so varied and their combinations so diverse that one can always find an appropriate film or laminate for a given application. However, the absolute barrier does not exist. It is necessary to adapt barrier properties to the anticipated shelf-life. The physical properties of the material, its processability and its interaction with the food together with its cost should be taken into account. One of the crucial criteria of choice is the knowledge of the permeability of the polymeric film to the gases and vapours of the environment that may affect the preservation of the food. This knowledge requires elucidation of the factors which depend on the polymer structure or on the permeant properties that may influence the permeability. These features together with experimental methods of measurement of permeability are treated in this chapter.

1.2 Basic principles of permeability

Gas sorption and transfer in polymers are adequately described from kinetics and the steady state point of view by the values of solubility and diffusion. The amount of gas to permeate a membrane per surface and time units Q is defined by the relation:

$$Q = DS(p_1 - p_2)/\ell$$

where D and S are respectively the diffusivity and solubility coefficients; p_1 and p_2 are the ambient pressures on the two sides of the films of thickness ℓ. At thermodynamic equilibrium, the gas permeability P is given by

$$P = DS$$

The permeability thus depends on diffusivity and solubility. These two quantities are a function of free volume, cohesive energy, density and polymer morphology.

1.2.1 Behaviour in rubbery polymers

Gas sorption and transfer in rubbery polymers is typically described by Henry's law for cases in which the sorbed concentrations are low. Equilibrium sorption is given by the relationship:

$$C = K_d p$$

where C is the gas concentration in the polymer, K_d Henry's coefficient and p the pressure. It follows then that

$$P = K_d D$$

Some deviations from Henry's law can be observed when interactions exist between the penetrant and the polymer. In such case the Flory–Huggins expression provides a satisfactory description of penetrant solubility.[1]

1.2.2 *Behaviour in glassy polymers*

In glassy polymers below the glass transition temperature, gas solubility no longer obeys Henry's law and can be accurately described by the dual-mode sorption model. One mode follows Henry's law and corresponds to the expectations for rubbery polymers. An additional mode follows a Langmuir form and has been assigned to the uptake into the unrelaxed volume or 'microvoids' present in the glassy state.[2] The total sorption is the sum of these two contributions

$$C = K_d p + \frac{C'_H b p}{1 + b p}$$

where K_d is the Henry's law constant, p is the penetrant pressure, and C'_H and b are the Langmuir capacity constant and affinity constant, respectively.

Values of K_d for carbon dioxide (CO_2) sorption in numerous polymers have been computed by Bicerano[3] using the statistical thermodynamic Sanchez–Lacombe equation-of-state. A reasonable agreement is obtained between calculated and experimental values.

The time diffusivity constant D, corresponding to Henry's law, is given by:

$$D = D_m (1 + K)/(1 + FK)$$

where $F = D_H/D_d$ and $K = C'_H b/k_d$ and D_d and D_H refer to the mobility of the dissolved and Langmuir sorbed components, respectively.

Permeability is then given by:

$$P = K_d D[1 + FK(1 + b p_2)]$$

At very low values of p_2, the following limiting form applies

$$P = K_d D[1 + FK]$$

Applied to different polymers[4] such as polyethylene (PE), polyethylene terephthalate (PET), polystyrene (PS), polymethylmethacrylate (PMMA) and polyvinyl acetate (PVAc), this model proved to be in semi-quantitative agreement in the case of PE and PET, and to some extent in the case of PS. The agreement is only quantitative in the case of PMMA and PVAc. Similarly, Barbari et al.[5] using a partial immobilization model analysed satisfactorily the permeability of different polymers based on bisphenol A.

Recently, Weiss et al.[6] proposed an alternative to the dual-model. Their model is based on a single, continuous distribution of Langmuir isotherms expressed in terms of a distribution ρ (b) of affinity parameters b. The function ρ (b) is normalized in the expression

$$\int_0^\infty \rho(b)db = 1$$

and the concentration $C(p)$ is assumed to be expressible by the relation:

$$C(p) = C_{sat}p\int_0^\infty \frac{\rho_{(b)}b}{(1 + bp)} db$$

where C_{sat} is the concentration at saturation.

The calculated sorption isotherm for CO_2, CH_4, O_2 and N_2 in polycarbonate are in good agreement with experimental results.

On the other hand, Horas and Toso[7] proposed a model using a homogenization method and the effective medium theory to describe the diffusion of gas of C_4 and C_5 hydrocarbon isomers in ethyl cellulose. A good agreement is observed between the proposed model and the experimental concentration-dependent diffusion coefficient.

1.3   Experimental techniques

Different methods have been proposed to measure permeability coefficients. These parameters can be deduced from volumetric or gravimetric measurements or by differential methods using different types of detection.

1.3.1   Volumetric methods

Two volumetric methods are available, one at variable pressure and the other at variable volume.

1.3.1.1   *Variable pressure method.* This method became popular through Barrer's work.[8] It is frequently called the time-lag method. In this technique, the polymer film is placed in a cell in which the gas is introduced on one side of the membrane, with a pressure p_1. The change of pressure

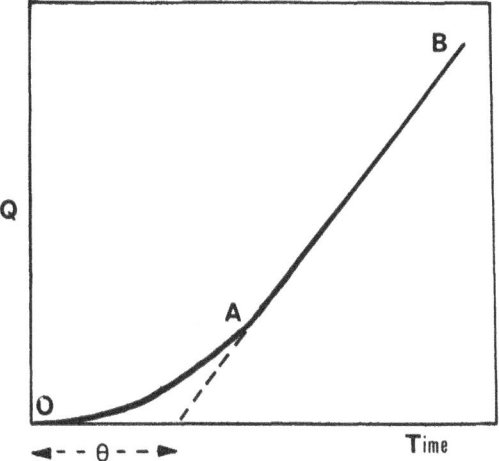

Figure 1.1 Typical permeation and time lag curve.

p_2 on the other side of the film is followed by a transducer. The experimental conditions are such that $p_1 \gg p_2 \sim 0$ so that $\Delta Q/\Delta t$ can be calculated from $\Delta p_2/\Delta t$ found from a plot of the pressure at the low-pressure side versus time (Figure 1.1). The permeability is then obtained from the relation:

$$P = \Delta Q\ell/\Delta t \, Ap$$

where ΔQ is the quantity of gas at STP which has permeated in the time interval Δt in the steady state of flow, A is the effective film area and ℓ its average thickness. Instead of a transducer, Barker et al.[9] used a mass spectrometer to measure the permeability of different PMMA samples from transient state transmission data, from which one can obtain the time lag θ. This quantity is related to the diffusivity coefficient D by $D = \ell^2/6\Theta$. From the values of P and D one may calculate the solubility coefficient $S = P/D$ and only one experiment allows the complete characterization of the behaviour of a polymer towards a given gas.[10]

1.3.1.2 *Variable volume method.* This method is based on the measure of gas or vapour flow through a polymer film using a sufficient sensitive flow meter. Because the method is dependent on the creation of a measurable volume change it has the disadvantage that film materials of very low permeability are difficult to test.

1.3.2 *Gravimetric methods*

Primarily these methods were developed to measure water vapour transmission. The weight loss of water (or weight gain of desiccant) is followed as a function of time. The method in itself is basically simple, but depending on specific conditions encountered, many variants were proposed. The simplest apparatus for obtaining water vapour transmission (or other vapours) is the procedure used by Osburn et al.[11] A plastic package containing an absorbent product is suspended from a balance beam. This device measures vapour transfer from outside to the inside of the package which results in weight increase. The use of this method with organic vapours was reported by Laine and Osburn.[12] It was also used to determine the water loss of a Nylon 6 film at different temperatures.[13] A modified version of this gravimetric technique,[14] using immersion, was applied to the determination of diffusion and solubility coefficients and permeation flow rate of organic solvents (1,1,2-trichloroethane ; 1,2-dichloroethane; methylene chloride ; benzene and toluene through a high density polyethylene film).

1.3.3 Differential methods

In these methods a stream of the gas under study is swept on one side of the polymer film, the gas crossing the film being usually carried out towards a detector by helium to produce a signal proportional to the diffusion rate. The analysis may be realized using different techniques such as a thermic conductivity detector,[15] a flame ionic detection (FID),[16] a gas chromatograph[17] or a mass spectrometer.[18,19] Absorption and desorption runs can be done as illustrated in Figure 1.2 in the case of a thermic conductivity detector. In this case the permeability P is given by:

$$P = f\sigma k \alpha S_\infty \ell / Ap$$

where f is the gas flow rate, σ the detector sensitivity, k a molecular factor,

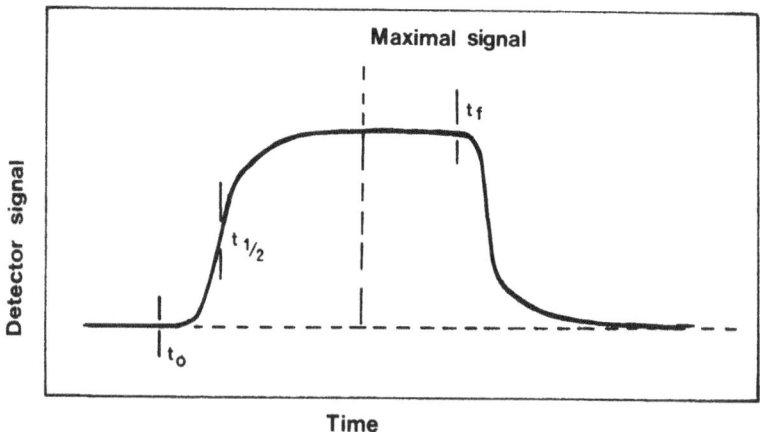

Figure 1.2 Thermal conductivity detector output as a function of time.[15]

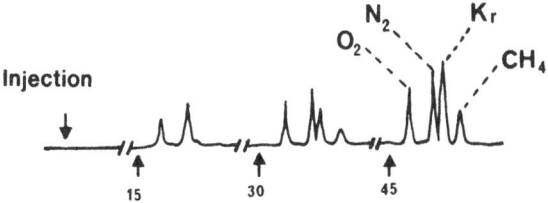

Figure 1.3 Gas chromatogram obtained in an experiment of differential measurement of permeability.[17]

α the detector attenuation, S_∞ the steady state signal intensity, ℓ the film thickness, A the film area and p the pressure of the gas.

The diffusion constant D is given by:

$$D = \frac{\ell^2}{7.2 \times t_{1/2}}$$

where $t_{1/2}$ is half time necessary to obtain the steady state.

Furthermore, the use of a gas chromatograph as well as a mass spectrometer allows one to determine the individual diffusion characteristics of each gas present in a mixture as shown in Figure 1.3.

1.3.4 *Determination of solubility*

Measurement of gas solubility can be conveniently performed using a pressure decay method. Pressure decay sorption devices can be classed either as single-volume or dual-volume types.[20] Sorption isotherms can thus be obtained as shown in Figure 1.4.[21] A compared analysis of three different methods of measurement has been given by Felder.[22]

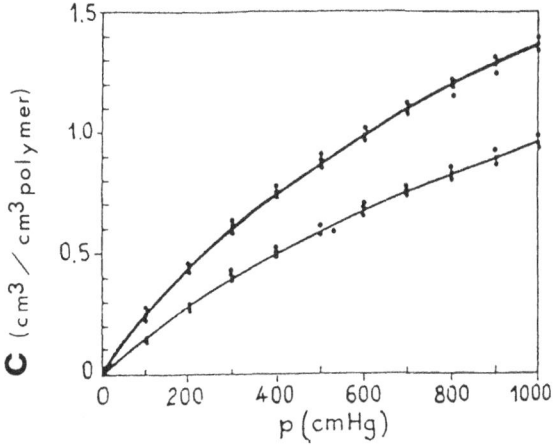

Figure 1.4 Sorption isotherm at 25°C versus pressure of an alternating copoly(vinylidene cyanide-vinyl acetate).[21] Upper curve O_2, lower curve N_2.

1.3.5 *Units of permeability*

A large number of units appears in the scientific and patent literature. Huglin and Zakaria[23] made an attempt to rationalize these units from a practical point of view and recommended the adoption of a single unit for permeability P, viz.: $(10^{-10} \text{ cm}^3.\text{cm})/(\text{cm}^2.\text{s. cmHg})$. They also proposed a conversion table for the other units encountered.

1.4 Parameters affecting permeability

1.4.1 *Polymer properties*

Polymeric materials present a wide range of structures and properties depending on their chemical structure, method of preparation and processing conditions. Furthermore a great number of structural parameters such as free volume, crystallinity, polarity, tacticity, cross-linking and grafting, orientation as well as the presence of additives or the use of polymer blends strongly affect barrier properties towards carbon dioxide, oxygen, nitrogen or water vapour. As an example, some permeability data for various polymers are given in Table 1.1.

1.4.1.1 *Influence of free volume.* The intuitive notion of free volume is widely used to explain polymer properties such as mobility or diffusion. A

Table 1.1 Permeability data for various polymers[24]

Polymer	$P((\text{cm}^3.\text{mm})/(\text{s}.\text{cm}^2\text{cmHg}))$				$\dfrac{PO_2}{PN_2}$	$\dfrac{PCO_2}{PN_2}$	Nature of polymer
	N_2, 30°C	O_2, 30°C	CO_2, 30°C	H_2O, 90%RH 25°C			
LDPE	19	55	352	800	–	–	Some crystallinity
HDPE	2.7	10.6	35	130	3.9	13	Crystalline
PP	–	23	92	680	–	–	Crystalline
Unplasticized PVC	0.4	1.2	10	1560	3.0	25	Slightly crystalline
Cellulose acetate	2.8	7.8	68	75 000	2.8	24	Glassy amorphous
PS	2.9	11	88	12 000	3.8	30	Glassy
Nylon 6	0.1	0.38	1.6	7000	3.8	16	Crystalline
PET	0.05	0.22	1.53	1300	4.4	31	Crystalline
PVdC	0.0094	0.053	0.29	14	5.6	31	Crystalline
Butyl rubber	64.5	191	1380	–	3.0	21	Rubbery
Natural rubber	80.8	233	1310	–	2.9	16	Rubbery

LDPE, low density polyethylene; HDPE, high density polyethylene; PP, polypropylene; PS, polystyrene; PET, polyethylene terephthalate; PVdC, polyvinylidene chloride.

simple way to define the free volume V_F is as follows:

$$V_F = V - V_0$$

where V is the specific volume measured experimentally and V_0 is an estimation of the specific volume at absolute zero. Lee[25] obtained a good description of the permeation of a gas through different polymers using the expression:

$$P = A_e^{-B(V-V_0)}$$

where A and B are constants characteristic of the gas.

The numerous theories taking into account free volume proposed for permeation in the literature were reviewed by Rogers.[26] Recently, Barbari,[27] using sorption and expansion data, determined the molar volumes of CO_2 and CH_4 that are inaccessible for diffusion in silicon rubber, and calculated the mobile gas concentration for CO_2 and CH_4 in glassy polycarbonate, assuming that an iso-free-volume state exists below the glass transition temperature. Then molecules which contribute to maintain the iso-free-volume in glassy state are mobile while those which do not are completely immobile.

The influence of free volume on diffusivity coefficient D and permeability P is well illustrated by a study of Maeda and Paul[28] on the antiplasticization by small molecules of a polysulfone (PSF) and poly 1,4-phenyleneoxide (PPO). As shown in Figures 1.5 and 1.6, a good agree-

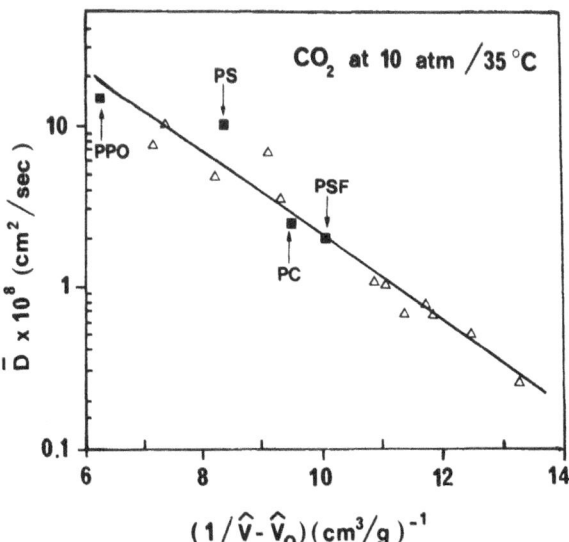

Figure 1.5 Correlation of average diffusion coefficient of CO_2 at 35°C in various polymers and mixtures of PSF and PPO with antiplasticizers (open joints) with free volume.[28] PPO: poly(phenylene oxide); PS: polystyrene; PC: polycarbonate; PSF: polysulfone.

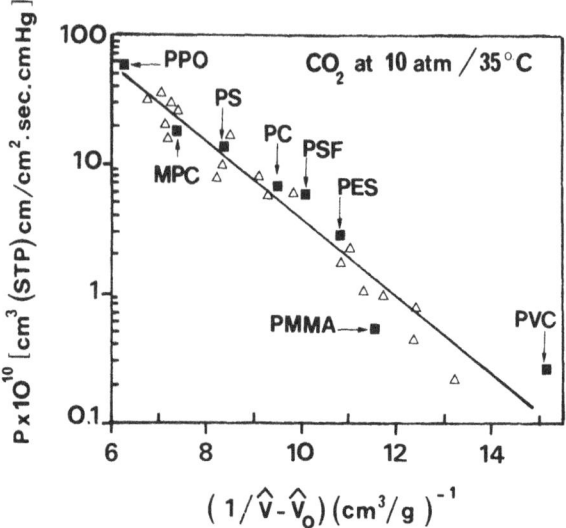

Figure 1.6 Permeability-free volume correlation for CO_2 at 35°C in various polymers and mixtures of PSF and PPO with antiplasticizers (open points).[28] MPC: tetramethyl bisphenol A polycarbonate; PMMA: polymethyl methacrylate; PVC: polyvinyl chloride.

ment is observed between free volume changes and D and P for CO_2. Both parameters decrease with a decrease in free volume. Similar results were obtained for helium and methane.

Indeed, the antiplasticizer induces an increase of the glass transition temperature and a decrease of free volume. Conversely, plasticization which lowers T_g and provokes an increase of free volume, results in an increase of permeability. Petropoulos[29] developed a model of the plasticization effect on gas permeation for the analysis and interpretation of experimental data.

1.4.1.2 *Crystallinity*. In semi-crystalline polymer, permeability depends on the degree of crystallinity, as can be shown in the results given in Table 1.2.

Usually, in such polymers, the crystalline phase presents a very low gas permeability which is considered as negligible. The solubility coefficient S,

Table 1.2 Effect of polymer crystallinity on oxygen permeability[24]

Polymer	% Crystallinity	PO_2(cm³/(mil day 100 in ² atm))
LDPE	50	480
HDPE	80	110
Nylon 66, quenched	20	8.0
Nylon 66, annealed	40	1.5

The unit of film thickness is mil (1 mil = 25μm)

i.e. the amount of gas sorbed into the polymer, directly depends on the solubility coefficient S_a and the volume fraction X_a of the amorphous phase. As far as the diffusion coefficient is concerned, Michaels *et al.*[30–32] and Yasuda and Peterlin[33] established the following relation:

$$D = D_a/\tau\delta$$

where D_a is the amorphous diffusion coefficient, τ a geometrical factor linked to the length and the section of diffusion path and δ the mobility of chain segments related to crystallites.

Therefore, permeability may be expressed as:

$$P = X_a D_a S_a/\tau\delta$$

This relation shows that permeability depends on crystallinity, free volume and orientation. As an example, Perkins[34] studied the influence of annealing under stress of different molecular weight oriented PET samples coming from fizzy drink bottles. As shown in Figure 1.7 permeability decreases with the increase of crystallinity. Weinkauf and Paul[35] observed a similar behaviour in thermotropic liquid-crystalline copolyester. However, in that case, the effect of increased crystallinity from annealing on the permeability coefficients was smaller than expected for similar changes in the crystallinity of conventional polymers. These authors using a simple two-phase model suggested that a mechanism dominated by transfer in a small volume of boundary regions could account for the bulk transfer properties in these materials.

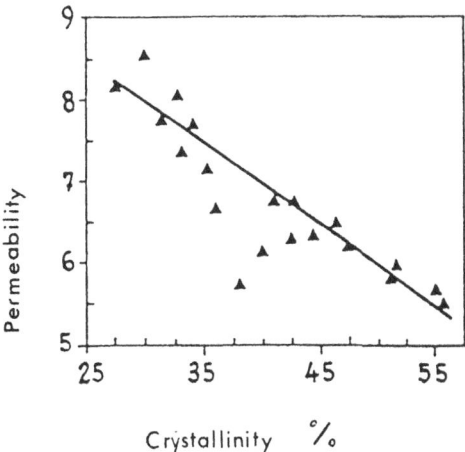

Figure 1.7 Influence of crystallinity on oxygen permeability (cm³.mil/sq.in.day.atm) of oriented PET films.[34]

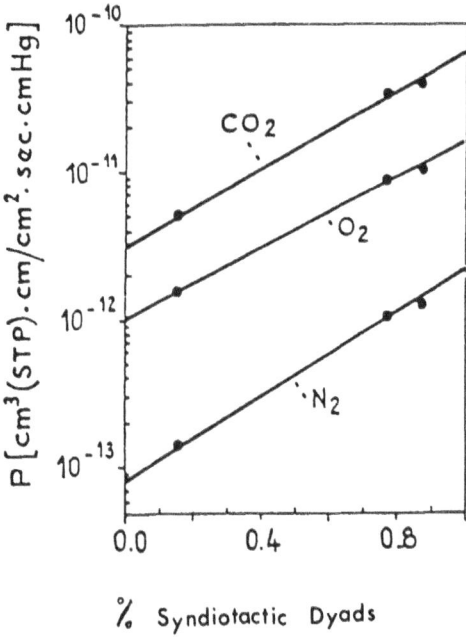

Figure 1.8 Influence of tacticity on permeability of PMMA to CO_2, O_2 and N_2.[36]

1.4.1.3 *Tacticity.* Polymerization of monomers leads to different configurational structures of the macromolecular chain depending on the type of catalyst used. When the polymer chain is obtained in the fully extended (all *trans*) planar zigzag conformation, the configuration where all the substituent groups lie on one side of the plane of the main chain is called isotactic. If substituent groups lie alternatively above and below the plane, the configuration is called syndiotactic, whereas a random sequence is said to be atactic. Min and Paul[36] studied the influence of tacticity on PMMA permeability for CO_2, O_2 and N_2. As shown in Figure 1.8, permeability increases with an increase of the percentage of syndiotactic diads according to:

$$\ln P = S \ln P_s + (1-S) \ln P_i$$

where P_s and P_i are the permeabilities of the pure syndiotactic and isotactic polymers, respectively, and S is the ratio of syndiotactic diads.

This influence of tacticity on permeability may be explained by a more dense packing of the isotactic form in the glassy state that stems in part from its lower glass transition temperature.

1.4.1.4 *Cross-linking and grafting.* The influence of these two parameters on diffusion has been the subject of few studies, mainly on cured rubbers for cross-linking.[37] Diffusion decreases with an increase of the degree of cross-linking. A similar behaviour was observed in cross-linked PE by MacDonald and Huang.[38] In PE grafted with polystyrene or polyacrylonitrile,[39] nitrogen permeability decreases with an increase of grafting amount and goes through a minimum value around 30%. On the other hand, the diffusion coefficient also decreases to reach a constant value around 20% of grafting.

1.4.1.5 *Copolymerization.* The permeability coefficient of a gas in an amorphous random copolymer system of 1 and 2 units has been shown in several cases[39-43] to obey the relationship:

$$\ln P = \phi_1 \ln P_1 + \phi_2 \ln P_2$$

where ϕ_1 is the volume fraction of component 1. As an example permeability and diffusion coefficients of different gases (N_2, Ar, O_2, H_2) increase with the increase of the hydroxybenzoic acid (HBA) amount in a thermotropic liquid, crystalline copolyesters based on HBA and 2,6 hydroxynaphthoic acid (HNA).[43]. The results were analysed on the basis of a two-phase modification of the free volume correlation suggesting that transport may likely occur in a small volume fraction of a less dense boundary phase. The films exhibited excellent barrier properties resulting mainly from very low gas solubility coefficients. On the other hand, the permeability of CO_2 through polyurethane increases with the decrease of hard segment content and the increase of soft segment length.[44]

1.4.1.6 *Orientation.* Processing always induces orientation in manufactured articles, which strongly influences mechanical properties and permeability. An asymmetrical structure is induced and the free volume distribution is changed. Transport processes would then originate in movement of molecules into voids formed by redistribution of the free volume that have a size greater than a critical value.[45] Starting from this concept, the equation for diffusion D is given by:[46]

$$D = A \exp \left[-\gamma V^*/V_m (\alpha \bot + \alpha //) \Delta T \right]$$

where A is a constant dependent on the transport species, V^* a critical volume for transport, γ a constant near unity, V_m a mean value of the molecular volume over the relevant temperature range and $\alpha = \alpha \bot + \alpha //$ the thermal expansivity. As shown in Figure 1.9, a linear relationship holds between $\ln P$ and $1/(\alpha \bot + \alpha //)$ which confirms the dependence of the permeability on orientation.

Experimentally the results published in the literature on CO_2 permeation in oriented polystyrene and polyethylene are controversial. On one hand, Wang and Porter[46] observed a reduction of permeability P with draw ratio while solubility S remains unchanged. On the other hand different

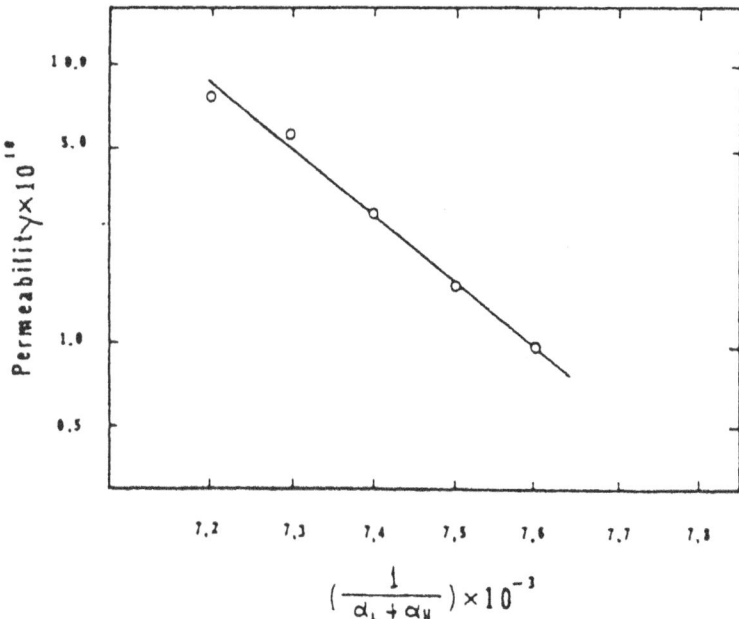

Figure 1.9 Permeability (cm^3(STP)cm/cm^2.s.cm Hg) as a function of $1/(\alpha_\perp + \alpha_{//})$ for CO_2 in oriented PS.[46]

authors observed a decrease in solubility constant S[14,47,48] as well as in permeability and diffusion constant[19,48] in function of draw ratio.

In polystyrene the decrease in permeability is assigned either to an anisotropic distribution of free volume which remains constant[46] or to a decrease in free volume.[47] It is difficult on the basis of these results to reach a conclusion. Further experiments on PS are needed to elucidate this important point. In semi-crystalline polymers such as PE, orientation induces an increase in the amount of crystallinity which also results in a decrease in permeability. However, the dramatic variations in permeability resulting from drawing were shown to be the consequence of changes of fractional free volume in the amorphous phase. A good correlation exists between solubility and diffusion coefficients of oriented samples and the estimated fractional free volume.[48] A similar behaviour was observed in polyvinyl chloride (PVC)[49] and polyvinylidene fluoride (PVF_2).[50] Table 1.3 summarizes some results on the influence of molecular orientation in different polymers on oxygen permeability.

1.4.1.7 *Polymer blends.* During the last decade, an increasing interest in polymer blends arose to improve mechanical and barrier properties of packaging materials. Usually polymers are incompatible and only few systems can be considered as compatible[51] either completely or in a narrow

Table 1.3 Effect of molecular orientation on oxygen permeability[24]

Polymer	% Elongation	PO_2(cm^3/mil day 100 in^2 atm)
PP	0	150
	300	80
PS	0	420
	300	300
PET	0	10
	500	5
Styrene–acrylonitrile	0	1.0
copolymer	300	0.9

The unit of film thickness is mil (1 mil = 25μm)

range of concentration. At room temperature, compatibility depends on phase transition conditions. In 1978, Hopfenberg and Paul[41] published the state of the art on transport phenomena in polymer blends and described the different mathematical models which apply to such systems. A linear semi-logarithmic behaviour was found to fit for permeability in case of completely miscible blends:

$$\ln P = \phi_1 \ln P_1 + \phi_2 \ln P_2$$

where ϕ_1 is the volume fraction of component 1.

A skewed curve is observed in incompatible blends. On the other hand, assuming the free volume of the blend equal to the sum of the free volumes of the components, Lee[25] obtained the following expression:

$$\ln(P/A) = [\phi_1/\ln(P_1/A) + \phi_2/\ln(P_2/A)]^{-1}$$

where A is a constant characteristic of the gas. When no volume variation occurs upon blending, this relation becomes identical to the previous linear relationship. Experimentally PMMA–SAN blends were found to obey to such a law for O_2 and CO_2 permeability.[52] Few effects on permeability were observed in these blends when phase separation occurs, as predicted by the two-phase model developed by Kraus and Rollman.[53] A similar result was obtained for CO_2 sorption in single-phase and phase-separated PS–polyvinylmethylether blends.[54] At least, modelling of transport properties of incompatible PS–polybutadiene (PB) blends showed that PS and PB phases are not topologically equivalent.[55] Simulations of sorption experiments using an unsteady-state model supports the hypothesis that the experimentally observed differences between the various volumes of the effective diffusion coefficient D_{eff} for a given blend permeant pair, can be justified only on the basis of morphological arguments.

1.4.1.8 *Film thickness.* The significance of film thickness can be summed up by stating that increased thickness reduces gas flow. One of the goals in film manufacture is to obtain as thin a film as possible with good mechanical and transport properties. However, pinholes and microvoids limit the thickness to a critical value (<10 μm). In practice a thickness of 25 μm is common. The permeability coefficient is independent of thickness. Thus, if the thickness at a barrier layer is doubled, the transmission rate of a gas or a vapour is halved. But economical considerations will limit the thickness to a definite value.

A great number of packaging materials consist in multilayered structures or laminated films. Consider the simplest case consisting of two layers of permeability P_1 and P_2 and thickness ℓ_1 and ℓ_2, respectively. Then, the total permeability P will be given by the relationship:[56]

$$\frac{1}{P} = \frac{\ell_1}{\ell P_1} + \frac{\ell_2}{\ell P_2}$$

where $\ell = \ell_1 + \ell_2$.

On the other hand, when one or both sides of the film are in contact with a liquid, a boundary resistance has to be taken into account and permeability becomes thickness dependent[57-59]. The total resistance to permeation at steady state which consists of resistances of different films and resistance of boundary layers or interfaces for a two layered film is given by:

$$\Delta C/Q = r_1 + r_2 + \ell_1/D_1 S_1 + \ell_2 D_2 S_2$$

where ΔC is the concentration difference of the fluid between the two faces of the film, Q is the permeation flux, r_1 is the boundary resistance on each side of the film and ℓ_1 the thickness of layer 1. D_1 and S_1 are the diffusion and solubility constants of layer 1.

Therefore when $\Delta C/Q$ is plotted against film thickness, two segments of straight line will result as shown in Figure 1.10a. The intrinsic permeability of each layer corresponds to the slope of each straight line and the intercept gives the boundary resistance. This model can be extended to many layers of different permeabilities, as shown in Figure 1.10b.

1.4.2 *Prediction of gas barrier properties of polymers*

We have previously seen that a relation exists between polymer structure and morphology and their barrier properties. Salame[61] assumed that the chemical segments of a polymer chain can be broken into discrete segments and assigned a numerical value in order to predict permeability. From the experimental values published on a great number of compounds, this author assigned numerical values to segments of the polymers, which were

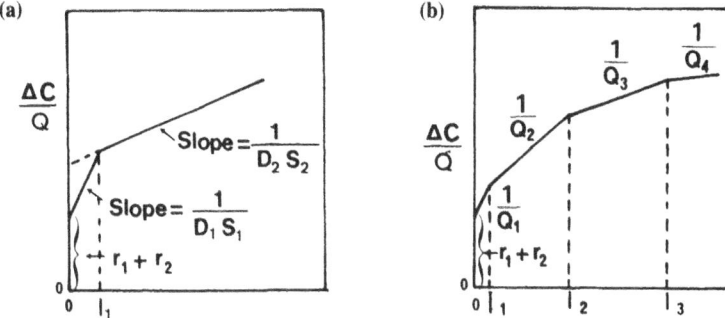

Figure 1.10 Thickness analysis for multilayer films.[60] For explanation of parts a and b see text.

called incremental 'Permachor'. When these incremental values are total-
led for the chain one gets the 'Permachor' value of the polymer (π).
Incremental 'Permachor' values for a great number of segment structures
as well as 'Permachor' values for most of the industrial polymers are listed
in ref. 60. The knowledge of π allows one to calculate permeability
according to the relationship:

$$P = Ae^{-s\pi}$$

where A and s are constants tabulated for O_2, CO_2 and N_2.

Furthermore, in the calculation of π, it is possible to take into account
side-chain substitutions and crystallinity. In addition, diffusion and solubility can also be estimated. As shown in Figure 1.11, a good agreement is
obtained between 'Permachor' values and permeability of CO_2, N_2 and O_2
in a great number of polymers.

1.4.3 *Permeant properties*

We have seen how important are such structural features as crystallinity,
free volume and tacticity in influencing the permeability of polymeric films
to gases and vapours. Such an influence also depends on the permeant
characteristics. Apart from the porous flow which depends on the microporosity of the membrane, where the pore size is preponderant and which
is not treated here, the 'diffusive solubility' flow[10] is greatly affected by the
size and shape of the penetrant molecules as well as by their polarity and
to a certain extent by temperature and pressure.

1.4.3.1 *Size and shape.* The size of the permeant molecule mainly
influences the solubility(S) and diffusion(D) coefficients. The variation of
permeability(P), which is theoretically the product of D and S, with
permeant size generally leads to a value inferior to such a product. The

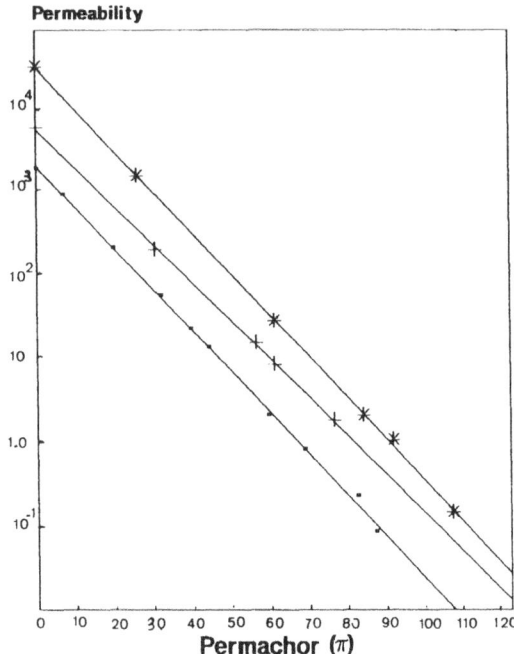

Figure 1.11 Gas permeability $((cm^3.mil)/100 \ in^2 \ day. \ atm)$ polymer structural parameter π (Permachor) correlation.[61] ■, N_2; +, O_2; *, CO_2.

increase in solubility with permeant size (diameter or molar volume) induces an increase in solubility and a decrease in diffusion coefficients. Relying on the similarity between the membrane processes and the stage processes like distillation, it was found[26] that increase in permeant size which provokes an increase in solubility requires more energy for vaporization and hence the boiling temperature increases. The decreases in diffusion coefficient with penetrant size is related to the need for a critical activation volume in the polymer proportional to that of the permeant molecule. The effect of van der Waals molar volume on diffusivity ($\log D$) is reported[62,63] for a wide range of gases and vapours in poly(vinyl chloride) (see Figure 1.12).

Similarly to the effect of the size of the molecule, the dependence of solubility in a given polymer on the shape of permeant molecules was found[26] to display a linear relationship in function of molar volume. Spherical molecules different from straight-chain paraffins in the slope and the value at zero molar volume which is greater for more spherical organic molecules. The sizes and shapes of gases and vapours seems to play a more major role in sorption and diffusion processes through a glassy polymer than a rubbery one. Comparison of the sorption isotherms at 298 K for CO_2, Ne and Ar in a glassy semicrystalline PBTP (poly(butylene tereph-

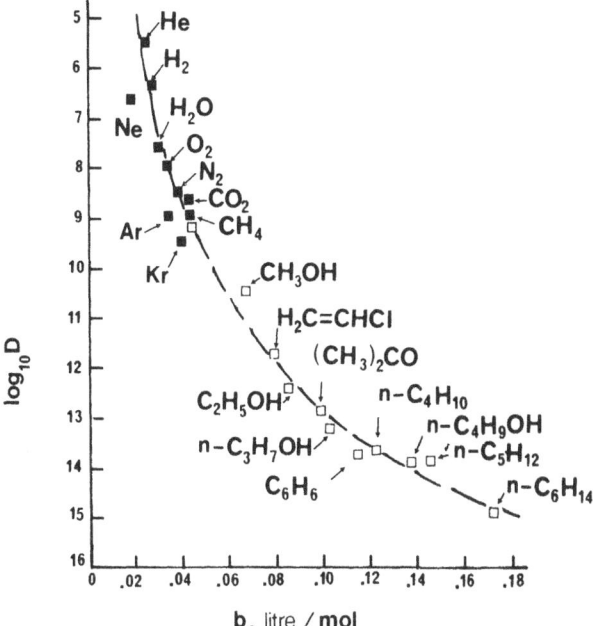

Figure 1.12 Logarithm of diffusivity (Log D) of gases and vapours in poly(vinyl chloride) versus their van der Waals volume (b) ([■] from ref. 62) ([□] from ref. 63).

thalate)) film showed[64] that this sorption follows the ascending order Ar<Ne<CO_2 although the CO_2 molecular diameter is larger than that of Ne or Ar. The linear structure of the CO_2 molecules together with the conjugated π electrons in the two (C=O) double bonds in CO_2 with the conjugated π-system in PBTP is probably at the origin of its higher solubility in non crystalline regions. Such interactions are the more important, the more the polarity is increased.

1.4.3.2 *Polarity*. Because of the interactions of polar groups in the polymeric film with polar molecules (alcohols, water, etc.), the barrier properties of these materials may be good for non-polar gas and poor for water vapour. The influence of relative humidity (RH) is due to its plasticizing effect on the polymer. When there are no polar groups in the polymer, the permeability to O_2 is not influenced by RH. For polymer containing polar groups, the package may absorb moisture[24] from the atmosphere or the liquid water in contact, and this promotes its swelling or plasticizing and reduces its barrier properties as may be seen in Table 1.4.

Likewise the permeability of polyolefins to oxygen is low because this gas only dissolves in the amorphous phase and interacts with additives. In

Table 1.4 Effect of relative humidity on O_2 transfer[24]

Polymer	Permeability (cm^3/mil day 100 in^2 atm)	
	0% RH	100% RH
Polyvinyl alcohol	0.01	25.0
Uncoated cellulose	0.13	200.0
Nylon 6	1.0	5.0
Polyvinylacetate	55.0	150.0
Acrylonitrile–styrene copolymer	1.0	1.0
Polyester	7.0	6.0
HDPE	110.0	110.0

The unit of film thickness is mil (1 mil = 25μm)

order to minimize the solubility of oxygen and the oxidation processes, and increase the polymer stability, it is recommended to use a stabilizer with organic functions similar to those on the polymer.[65]

1.4.3.3 *Temperature and pressure.* The importance of glass transition temperature T_g was underlined in the paragraph dealing with free volume. The structural changes which occur at T_g greatly affect the diffusion of gases and low molecular weight organic vapours. Small penetrant molecules like H_2 and Ar were not found to be influenced by T_g of poly(ethyl methylacrylate).[26] For gases and vapours which do not interact with the polymer, the permeability coefficient is independent of the pressure of penetrant gas.[24] In case of interaction, the permeability increases with the increase in pressure. This is due to an increase in diffusion constant D under the plasticizing effect of the permeant as well as an increase in solubility S due to the shape of the sorption isotherm.[56] The currently accepted relationship between the permeability coefficient P and temperature is:

$$P = P_o \, e^{-E_p/RT}$$

where P_o is a temperature dependent factor, E_p the activation energy for permeation, R the gas constant and T the absolute temperature.

On the other hand, a critical adsorption pressure p^* was estimated to fit with the sorption model of noble gases Ne and Ar in a PBTP film.[64] Only Henry's law of solubility takes place below p^* which is dependent mainly on the relative size of the frozen microvoids in the polymer and the noble gases atoms.

Conclusion

As permeability of polymeric packaging material depends on the sorption and diffusion of gases and vapours, it is of relevance to recall the models that apply for the prediction of these parameters in rubbery and glassy polymers. The experimental techniques used to measure permeability coefficients are generally based on volumetric, gravimetric or differential methods. These permeability coefficients are influenced by the chemical structure as well as by the method of preparation and processing conditions. Some of the structural parameters of the polymers play a major role in their effect on permeability. It is the case for free volume, crystallinity, tacticity, polarity, cross-linking and grafting as well as orientation or the presence of additives. The gases and vapours particularly important in food packaging and preservation are O_2, CO_2 and water vapour. The effect of their shape and size, interaction with the polymer and the influence of temperature and pressure on their sorption and diffusion behaviour may be helpful in the understanding of the variation of their permeation through polymeric packages.

References

1. G.K. Fleming and W.J. Koras (1986) *Macromolecules* **19**, 2285.
2. V.T. Stannett (1978) *J. Membr. Sci.* **3**, 97.
3. J. Bicerano (1992) *Computational Polym. Sci.* **2**, 177.
4. R.J. Pace and A. Datyner (1980) *J. Polym. Sci. Polym. Phys. Edn.* **18**, 1103.
5. T.A. Barbari, W.J. Koros and D.R. Paul (1988) *J. Polym. Sci. Polymer Physics* **B26**, 709.
6. G.H. Weiss, J.T. Bendler and M.F. Shlesinger (1992) *Macromolecules* **25**, 990.
7. J.A. Horas and J.P. Toso (1992) *J. Polym. Sci. Polym. Physics* **B30**, 127.
8. R.M. Barrer (1939) *Trans. Farad. Soc.* **35**, 628.
9. R.E. Barker, R.C. Tsai and R.A. Willency (1978) *J. Polym. Sci. Polym. Symp.* **63**, 109.
10. S.T. Hwang and K. Kammermeyer (1984) *Membranes in Separations*, Techniques of Chemistry Vol. 7, Wiley, New York.
11. J.O. Osburn, K. Kammermeyer and R. Laine (1971) *J. Appl. Polym. Sci.* **15**, 739.
12. R. Laine and J.O. Osburn (1971) *J. Appl. Polym. Sci.* **15**, 327.
13. J. Comyn, B.C. Cope and M.R. Werrett (1985) *Polym. commun.* **26**, 294.
14. L.N. Britton, R.B. Ashman, T.M. Aminabhavi and P.E. Cassidy (1988) *J. Chem. Ed.* **65**, 368.
15. D.G. Pye, H.H. Hoen and M. Panar (1976) *J. Appl. Polym. Sci.* **20**, 287.
16. G. Palmai and K. Olah (1984) *J. Membr. Sci.* **21**, 161.
17. M. Uchikura, H. Odani and M. Kurata (1982) *Kobunshi Ronbunshu*, **39**, 149.
18. C. de Leiris Laurent (1992) PhD Thesis, Le Havre.
19. J.A. Webb, D.I. Bower, I.M. Ward and P.T. Cardew (1993) *J. Polym. Sci. B Polymer Physics* **31**, 743.
20. W.J. Koros and D.R. Paul (1976) *J. Polym. Sci. Polym. Phys. Edn* **14**, 1903.
21. H. Hachisuka, H. Kito, Y. Tsujita, A. Takizawa and T. Kinoshita (1988) *J. Appl. Polym. Sci.* **35**, 1333.
22. R.M. Felder (1978) *J. Membr. Sci.* **3**, 15.
23. M.B. Huglin and M.B. Zakaria (1983) *Angew. Makromol. Chem.* **117**, 1.
24. R.J. Ashley (1985) In *Polymer Permeability* ed J. Comyn, Elsevier, London, p. 269.
25. W.M. Lee (1980) *Polym. Eng. Sci.* **20**, 65.
26. C.E. Rogers (1985) In *Polymer Permeability* ed J. Comyn, Elsevier, London p. 11.
27. T.A. Barbari (1993) *J. Polym. Sci. B, Polym. Phys.* **31**, 501.
28. Y. Maeda and D.R. Paul (1987) *J. Polym. Sci., Polym. Phys. Edn.* **25**, 1005.
29. J.H. Petropoulos (1992) *J. Membr. Sci.* **75**, 47.
30. A.S. Michaels, W.R. Vieth and H.J. Bixler (1964) *J. Appl. Polym. Sci.* **3**, 2735.

31. A.S. Michaels and R.E. Parker (1959) *J. Polym. Sci.* **41**, 53.
32. A.S. Michaels and H.J. Bixler (1961) *J. Polym. Sci.* **50**, 413.
33. H. Yasuda and A. Peterlin (1974) *J. Appl. Polym. Sci.* **18**, 531.
34. W. Perkins (1988) *Polymer Bull.* **19**, 397.
35. D.H. Weinkauf and D.R. Paul (1992) *J. Polym. Sci. B. Polym. Phys.* **30**, 817.
36. K.E. Min and D.R. Paul (1988) *J. Polym. Sci. Polym. Phys. Edn.* **26**, 1021.
37. V.T. Stannett (1968) In *Diffusion in Polymers* eds J. Crank and G.S. Park, Academic Press, NY, p. 41.
38. R.W. MacDonald and R.Y.M. Huang (1981) *J. Appl. Polym. Sci.* **26**, 2239.
39. P.J.F. Kanitz and R.Y.M. Huang (1971) *J. Appl. Polym. Sci.* **15**, 61.
40. A.E. Barnabeo, W. S. Creasy and L.M. Robeson (1975) *J. Polym. Sci., Polym. Chem. Edn.* **13**, 1979.
41. H.B. Hopfenberg and D.R. Paul (1974) in *Polymer Blends*, Volume 1, ed. D.R. Paul, Academic Press, NY, Chap. 10.
42. D.R. Paul (1984) *J. Membr. Sci.* **18**, 75.
43. D.H. Weinkauf and D.R. Paul (1992) *J. Polym. Sci. B. Polym. Phys.* **30**, 837.
44. H. Xiao, Z.H. Ping, J.W. Xie and T.Y. Yu (1990) *J. Appl. Polym. Sci.* **40**, 1131.
45. M.H. Cohen and D. Turnbull (1959) *J. Chem. Phys.* **31**, 1164.
46. L. H. Wang and R.S. Porter (1984) *J. Polym. Sci. B. Polym. Phys.* **22**, 1645.
47. C. Carfagna (1986) *J. Polym. Sci, B Polym. Phys.* **24**, 1805.
48. H. Sha and I.R. Harrison (1992) *J. Polym. Sci. B. Polym. Phys.* **30**, 915.
49. M.J. El-Hibri and D.R. Paul (1985) *J. Appl. Polym. Sci.* **30**, 3649.
50. M.J. El-Hibri and D.R. Paul (1986) *J. Appl. Polym. Sci.* **31**, 2533.
51. S. Krause (1978) In *Polymer Blends* Vol. 1, eds D.R. Paul and S. Newman, Academic Press, NY, p. 16.
52. J.S. Chiou and D.R. Paul (1987) *J. Appl. Polym. Sci.* **34**, 1037.
53. G. Kraus and K.W. Rollmann (1971) *Adv. Chem. Ser.* **99**, 189.
54. H. Hachisuka, T. Sato, Y. Tsujita, A. Takizawa, and T. Kinoshita (1989) *Polymer J.* **21**, 417.
55. N. Shah, J.E. Sax and J.M. Ottino (1985) *Polymer* **26**, 1239.
56. J.A. Cairns, C.R. Oswin and F.A. Paine (1974) *Packaging for Climatic Protection*, Institute of Packaging Monograph, Newnes-Butterworth London, Chapter 2.
57. S.T. Hwang, T.E. Tang and K. Kammermeyer (1971) *J. Macromol. Sci. Phys.* **B5**, 1.
58. S.T. Hwang and G.D. Strong (1973) *J. Polym. Sci. Symp.* **41**, 17.
59. H. Yasuda and C.E. Lamaze (1972) *J. Appl. Polym. Sci.* **16**, 595.
60. S.T. Hwang and K. Kammermeyer (1974) in *Polymer Science and Technology, Volume 6. Permeability of Plastics Films and Coatings to Gases, Vapors and Liquids*, ed H.B. Hopfenberg, Plenum Press, NY, p. 197.
61. M. Salame (1986) *Polym. Eng. Sci.* **26**, 1543.
62. A.R. Berens and H.B. Hopfenberg (1982) *J. Membr. Sci.*, **10**, 283.
63. B.K. Tikhomirov, H.B. Hopfenberg, V.T. Stannett and J.L. Williams (1968) *Makromol. Chem.* **118**, 117.
64. Z. Zhou and J. Springer (1993) *J. Appl. Polym. Sci.*, **47**, 7.
65. J.Y. Moisan (1985) In *Polymer Permeability*, ed J. Comyn, Elsevier, London, p. 119.

2 Alternative fatty food simulants for polymer migration testing*

A. L. BANER, R. FRANZ and O. PIRINGER

Abstract

The amounts of substances migrating from plastics into foodstuffs with high fat contents are usually higher than into aqueous foodstuffs. In most cases this is due to the higher solubility of the migrating organic compounds in fat compared to water and not due to an increase in the diffusion coefficient due to interactions between the fat and plastic as is often assumed. Ethanol and aqueous ethanol mixtures can be good fatty food simulants because they interact very little with many plastics, migrants are readily soluble in them and because they are easy to work with analytically. The utilizeable limits of aqueous ethanol and other low molecular weight solvents as food simulants are developed from the physical chemical background of partition coefficients and diffusion. The use of ethanol simulants is supported by experimental results for polyolefins and rigid polyvinyl chloride. Other low molecular weight solvents for other classes of polymers are suggested. The UNIFAC activity coefficient estimation method is applied to alternative fatty food simulants/polymer systems and estimations for sorption and partition coefficients are compared with experimental results.

2.1 Introduction

It is well known that polymeric food packaging materials are not completely inert and can transfer substances to the foods they come in contact with. Therefore, because of health and sanitation reasons, there are regulations that require the polymeric package materials to demonstrate a certain level of inertness towards the foods packaged within them. In Europe the harmonization of the European Community (EC) towards a common market starting 1 January 1993 has necessitated the introduction of a common set of laws concerning food contact materials. Similar laws concerning food contact materials can be found in other countries through-

*Paper given at the IFTEC symposium 'Food Packaging Interactions and Packaging Disposability'. The Hague, November 15–18, 1992.

out the world. One requirement of food contact law is that migration testing be carried out to determine the transfer of specific substances or the total (global) amount transferred to the food from the food contact package. This chapter will describe the physical chemical principles involved in the migration process, discuss migration testing and finally show how these principles can be used to select alternative fatty food simulants for migration tests.

2.2 Food contact material regulations in the EC and the USA

According to Article 2 of the EC framework directive 89/109/EEC,[1] articles intended to come into contact with foodstuffs should not, under normal or foreseeable conditions of use, transfer their substituents to foodstuffs in quantities which could endanger human health or bring about unacceptable changes in the foodstuffs. To fulfil this broad framework other specific directives have been written specifying the quantity of a given substituent or the total of all substituents in the food contact article which are allowed to transfer or migrate to the food (see Article 3). Substances allowed in food contact articles are published in a positive list and are assigned specific migration limits, in mg per kg food (or mg/kg polymer), determined from toxicological evaluation of the substance. Article 2 of directive 90/128/EEC is a measure of the food contact article's inertness and gives a total migration upper limit of 10 mg total of all transferred substituents per dm^2 (or 60 mg/kg food) of the food contact article. The total migration limit protects the foodstuff from unacceptable alteration and reduces the necessity for a large number of specific migration limits or specific restrictions which subsequently allows simpler and more effective control.

The basic rules for migration testing of food contact articles are given in the directive 82/711/EEC. Most foods have complex compositions which can cause considerable analytical difficulties in migration determinations. Because of these difficulties it is often necessary to use both conventional food simulant solvents and standardized test conditions which are supposed to simulate the migration behaviour of substituents under actual conditions of use in foods. Directive 85/527/EEC lists the following solvent food simulants: A) distilled water or water of similar quality; B) 3% acetic acid (w/v) in water; C) 15% ethanol (v/v) in water; D) rectified olive oil. If for technical reasons other food simulant solvents are required, a synthetic mixture of triglycerides (e.g. HB 307) or sunflower seed oil must be used in place of olive oil.

In the USA the migration of substances into food from packaging falls under the jurisdiction of the US Food and Drug Administration (FDA) which administers the Federal Food, Drug and Cosmetic Act (FD and C Act).[2] The regulations for food additives (food packaging materials are considered indirect food additives) are published in the form of positive lists in the Title 21 of the US Code of Federal Regulations. Contained in these regulations are frequently specific migration limits for substances as well as global migration limits for food contact polymer resins. Limits are also frequently placed on the package temperature and permitted food contact final use conditions. All packaging materials and their additive packages must either have prior approval for the intended end use food contact application or petition the FDA's Center of Food Safety and Applied Nutrition (CFSAN) for permission for the intended end use. In order to determine if the use of a polymer packaging material in a specific application is safe, migration tests must be carried out and the estimated daily intake (EDI) of the migrated substance determined. CFSAN has published *Recommendations for Chemistry Data for Indirect Food Additive Petitions*[3] that describe methods for measuring migration into food simulating liquids (FSL). The current FDA recommendations for food simulating solvents are 8% for aqueous and acidic foods, 8 or 50% aqueous ethanol (v/v) for alcoholic foods, and cooking oil (corn oil or HB307 are recommended) for fatty food simulants.[3] Alternative fatty food simulants allowed for specific polymers are currently 95% ethanol for polyolefins, 50% ethanol for rigid polyvinyl chloride (PVC) and polystyrene (PS) and rubber modified polystyrene (HIPS).

Specific migration measurements, and to a larger degree those of global migration, are conducted using standardized testing procedures which allow only approximate comparison of the test results with those occurring in practice. For certain foodstuffs the fatty food simulant migration values considerably exceed the values obtained for the foods themselves. to correct for these higher migration values, for example, directive 85/572/EEC specifies the use of coefficients to diminish the measured migration values in fatty food simulants and bring them closer to actual food values (see correction factor table in directive index). It should be pointed out that standardized migration test conditions are really accelerated testing conditions that are used to overcome the problem of the long contact times often needed in practice for packaged food migration measurements. These accelerated tests give estimates of the actual migration values in real foods under long contact time conditions.

It is plain to see that there is a need for migration testing particularly for new materials and additives originating from technological developments or environmental considerations and new food packaging material applications (e.g. microwave heating of food in the package).

2.3 Importance of migration testing using fatty foods

The migration into fatty foods is larger than into aqueous foods for similar time and temperature exposures. Thus migration testing of fatty foods is important in determining an upper limit of possible exposure to a migrated substance. Table 2.1 compares migration values from the literature for foods containing varying amounts of fat and for two commonly used antioxidants. The table shows the tendency for migration to increase with fat content in the food. Other factors in the foods besides fat determine the extent of migration, such as the amount of fat in the dried food mass and the type of emulsion (water-in-oil > oil-in-water).[4,5] The extent of migration also increases with increasing temperature.[5] Migration also

Table 2.1 Migration of antioxidants from PE into foods

Food ($\mu g/dm^2$)	Fat content (weight percent)	LDPE Irganox-1076 ($\mu g/dm^2$) 23°C/5 d (from ref. 4)	HDPE Irganox-1010 ($\mu g/dm^2$) 121°C/1.5 h (from ref. 5)
Water	0	2.0	19
Fat-free quark	–	2.0	–
Defatted soy meal	0.9	<0.5	–
Whole milk	3.6	9.0	–
Yoghurt	3.7	8.0	–
Beef stew	4.2	–	159
Quark 6%	6.1	234	–
Pork baby food	7.0	–	106
Chicken á la king	8.1	–	141
Corned beef hash	–	–	195
Newbury sauce	8.2	–	151
Liquid nutrient	9.6	–	21
Coffee creamer	10	17	–
Quark 10%	10.2	257	–
Soy meal	23.1	289	–
Spreadable cheese	26	10	–
Gouda cheese	27.2	780	–
Chocolate	30.5	279	–
Emmental cheese	31.9	624	–
Cream	32.3	15	–
Whipping cream	33.4	542	–
Parmesan cheese	34.8	466	–
Margarine half fat	40.1	558	–
Margarine	80.2	642	–
Mayonnaise	82.2	198	–
Butter	83.3	539	–
Corn oil	100	–	340
Olive oil	100	670	–
HB 307	100	694	–

Irganox-1010 (Ciba-Geigy Corporation): tetrakis [methylene-3-(3',5'-di-*tert*-buty-4'-hydroxyphenyl)propionate] methane.

Irganox 1076 (Ciba-Geigy Corporation): octadecyl-3-(3',5'-di-*tert*-butyl-4'-hydroxyphenyl) propionate.

HB 307: a synthetic triglyceride mixture from Unilever (Hamburg, Germany).

depends on the polymer phase and the chemical nature and molecular weight of the migrant. In the polymer phase it was observed that the migration under defined times and temperatures of Irganox-1076 from polymers into various foods increased in the order of acrylonitrile–butadiene–styrene (ABS) < impact polystyrene (HIPS) < polypropylene (PP) < high density polyethylene (HDPE) < low density polyethylene (LDPE)[6] which corresponds roughly to the order of the diffusion coefficient magnitudes of the polymers.

2.4   Advantages and disadvantages of conventional migration methods using food simulants

There are generally no basic problems measuring specific migration in aqueous food simulants. However, it is possible to find much lower migration values in distilled water in contact with a non-polar polymer compared with those found in aqueous foods due to testing problems. To avoid such problems the FDA recommends 8% (v/v) and the EEC recommends 15% (v/v) aqueous ethanol as aqueous food simulants.

With respect to specific migration measurements in olive oil (EC) or cooking oil (USA) satisfactory results can be obtained in many cases. However, to achieve sufficient analytical sensitivity in those cases requires a great deal of analytical work. In general ^{14}C radiolabelled migrants are used.[4,5]

Global migration determinations into water may in principle involve large errors because the method gravimetrically measures the amount migrated as a residual after total evaporation of water. Total evaporation of water not only leads to loss of volatile substituents but also others as well. However, in principle the volatile fraction can be measured by additional methods such as headspace gas chromatography (HSGC).

Even greater problems and analytical difficulties occur when measuring global (total) migration into oil and fat. All methods use the same principle: A material sample with known weight and surface area is placed in contact with the oil or fat food simulant under standardized conditions. After exposure the oil or fat is removed from the surface of the material and the sample is weighed again. The amount of oil or fat absorbed by the polymer is then determined by a physical or chemical method and this amount is subtracted from the mass of the second weighing. These procedures contain two significant implicit error sources: The first occurs when the polymer weighings are not carried out under standard conditions which is particularly important for polar polymers which can readily absorb or lose water. The second occurs when the amount of absorbed oil or fat is not exactly measured which can cause errors as large as the mass of the migrants.

Several methods were published for measuring the absorbed fat in polymers. The most well known methods use GC measurement of the oil[7,8] or a radiotracer technique using a ^{14}C labelled test fat mixture HB 307.[9,10] PIRA[11] has further developed the GC method which was accepted by CEN (European Committee of Normalization) as a standard test method. The main disadvantage of this GC method is that it is a very labour intensive analytical method. The main disadvantage of the radiometric method is the special equipment needed.

Finally, there is still a need for a practical method for measuring and monitoring migration from polymers into fatty foods that is less expensive and easier to perform allowing more samples to be simultaneously measured. Such a method would thus allow more effective market control. This will only be possible by using other appropriate food simulants to simulate oil or fat. n-Heptane, which was previously recommended by the FDA (U.S. Food and Drug Adminstration) as a fatty food simulant, has the great advantage of being easy to handle and analyse.[12] On the other hand, the great disadvantage of this organic solvent is its aggressiveness towards many polymers (e.g. all polyolefins and slightly polar polymers). The n-heptane rapidly penetrates into the polyolefins and behaves as an extracant for the polymer substituents thus leading to considerable over-estimation of the extent of real migration. Similar behaviour in polyolefins is exhibited by all of the low molecular weight hydrocarbons as well as other organic solvents with low polarity. Larger molecular weight alcohols (>C4) tend to swell and extract PE and PP polymers in migration studies.[13] The use of ethanol and ethanol/water mixtures have yielded migration results similar to those obtained for migration into oil and fat in polyolefins.

2.5 Physical principles of migration

2.5.1 *The importance of the partition coefficient*

The solute partition coefficient between the polymer and food contact phase, $K_{P/F}$ (units: dimensionless) is the central parameter in defining the limiting cases of migration coefficient is defined as the concentration of solute (i) in the polymer (P) at equilibrium, $c_{i,\infty}{}^P$ (units: mass/volume), divided by the concentration of the solute in the food (F) at equilibrium, $c_{i,\infty}{}^F$:

$$K_{P/F} = c_{i,\infty}{}^P / c_{i,\infty}{}^F \qquad (2.1)$$

The two limiting cases for the migration of polymer constituents and additives into foods are: (1) water which is the lower limit and (2) fat or oils which are the upper limit. These cases are determined by the chemical natures of the migrant, the polymer and the food. Polymers, by comparison with water, are not very polar in nature. Polymer constituents or additives must be similar in nature to the polymer in order to effectively remain incorporated in the polymer. Therefore, the solubility of the polymer constituents and additives is quite low in water compared to the polymer phase and the corresponding partition coefficients are relatively large. The opposite case is seen for fats or oils in contact with the polymer. The fats and oils are relatively non-polar compared to water and similar enough to the polymer in chemical nature that solubility of the polymer constituents and additives is comparable in both the polymer and fat. Thus, the partition coefficient between the polymer and oil is less than or equal to one. Table 2.2 illustrates the large partition coefficient variations that can be found when BHT (2,6-di-*tert*-butyl-4-methylphenol) is partitioned between polypropylene and various fatty and aqueous foods at 40°C.[14]

Table 2.2 Partition coefficient of BHT between PP and various foods and food simulants at 40°C[14]

Milk 30% fat	0.363
Milk 10% fat	1.01
Milk 7.5% fat	1.3
Milk 3.5% fat	9.1
Milk 1.5% fat	22.8
Milk 0.1% fat	304
Pernod (40% ethanol v/v)	117
Ethanol/water (40% v/v)	134
Orange juice	5050
Grapefruit juice	4130
Water (demineralized)	35900

Table 2.2 illustrates the two limiting cases for the partition coefficient. The high fat milk products have $K_{P/F} \leqslant 1$ and water represents the other limit with the largest partition coefficient. Actual migration measurements with BHT into the 0.1% milkfat milk show the partition coefficient limiting the total migration to about 16% of the total amount in the polymer whereas in the 30% milkfat milk a mass balance shows all the BHT migrating into the milk.[14] Correlations were observed between the partition coefficient of a substance and its solubility in the food and food simulant contact phases.[14] It was observed experimentally that the partition coefficient between polymer and aqueous ethanol food simulants is relatively insensitive to temperature between 10°C and 40°C[15] apparently due to offsetting changes in solubility in the two phases.

The ratio of the packaging volume (V_P) to the food volume (V_F) plays an important role in determining the mass of solute at equilibrium in the

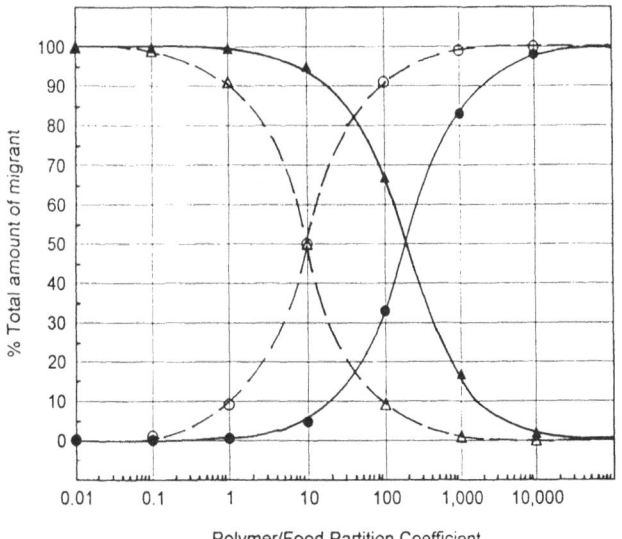

Figure 2.1 Effect of partition coefficient magnitude on solute partitioning. $\bigcirc$ = polymer, $\triangle$ = food, V_P/V_F = 0.005: solid line and solid symbols, V_P/V_F = 0.1: dashed line and open symbols.

food and polymer phases. For most food packaging applications this ratio is much less than one (normally in the range 0.1 to 0.005). Given the amount of migrant initially in the polymer, $m_{i,0}{}^P$, and the amount migrated at equilibrium into the food, $m_{i,\infty}{}^F$, it follows from the mass balance:

$$m_{i,\infty}{}^F = m_{i,0}{}^P / (1 + K_{P/F}V_P/V_L) \qquad (2.2)$$

Using equation (2.2) and assuming an initial additive concentration of 0.1% (w/w) in the polymer, Figure 2.1. shows the effect of the partition coefficient and the effect of the package to food volume ratio on how a migrant will partition at equilibrium. It is important to note in this figure that for $K_{P/F} \leqslant 1$ more than 90% of the entire amount of migrant in the polymer will be transferred into the food at equilibrium ($m_{i,\infty}{}^F \approx m_{i,0}{}^P$) for the entire practical V_P/V_F range. This situation predominates in fatty foods and defines the upper limit for potential migration. On the other hand if the migrant is much more soluble in the polymer than the food ($K_{P/F} > 1000$) then at equilibrium less than 10% of the migrant will have migrated into the food ($m_{i,\infty}{}^F \ll m_{i,0}{}^P$) over the V_P/V_F range. This second situation predominates in aqueous foods and results in much less migration than in fatty foods thus giving the lower limit to migration. The situation with large partition coefficients is very important in determining flavour sorption from aqueous foods by polymer packaging.

2.5.2 The importance of the polymer diffusion coefficient

The solute diffusion coefficient in the polymer, D_P (units: cm^2/sec), is the second most important physical parameter describing the migration from a polymer into a food. The diffusion coefficient is a function of temperature, T, and when the concentration of the migrant in the polymer is high enough D_P can increase with the migrant concentration. For most polyolefins the additive concentrates are of the magnitude of 0.1% (w/w) or less, a concentration for which the diffusion coefficient effects are not found. Plasticized PVC (polyvinyl chloride) can have plasticizer concentrations up to 40% (w/w) and thus the diffusion coefficient can be concentration dependent.[16] Swelling of the polymer by sorption of solvents or other agents can increase the diffusion coefficients of the migrants as well.

Contact time (t) and temperature (T) should be considered as the most important variables with respect to the actual mass transfer during contact with fatty foods because practically the entire amount of migrant can be transferred from the polymer into the food at equilibrium. The amount of material migrated at time t, $m_{i,t}{}^F$, is described by the following equation:

$$m_{i,t}{}^F \approx A\, 2\, c_{i,0}{}^P\, (D_P\, t/\pi)^{1/2} \approx k^x\, c_{i,0}{}^P\, t^{1/2}\, \exp(-E/RT) \qquad (2.3)$$

where A is the contact surface area between the polymer and food, E is the initial activation energy of the migration process determined from measuring $m_{i,t}{}^F$ as a function of $1/T$, R is the gas constant, k^x is a proportionality constant which is dependent on the contact area and the system measured and is independent of $c_{i,0}{}^P$, time, and temperature. Equation (2.3) shows the importance of time and temperature considerations in testing migration into fatty foods.

2.5.3 Considerations in the selection of appropriate fatty food simulating solvents

To understand the fundamentals of migration when using different food simulants (FS) it is important to describe mass transfer processes in the opposite direction, from the food simulant phase into the polymer. For the amount, $m_{FS,t}{}^P$, migrating from the food into the polymer in time t an analogous equation to equation (2.3) is:

$$m_{FS,t}{}^P = A\, 2\, C_{FS,0}{}^F\, K_{P/F}\, (D_P\, t)^{1/2} \qquad (2.4)$$

where $C_{FS,0}{}^F$ is the initial concentration of the migrant in the food. In contrast to equation (2.3), the partition coefficient plays an important role in equation (2.4) where the amount of material migrating into the polymer is proportional to $K_{P/F}$.

To better understand the influence of food simulants on migration the following description is given: Let x_l be the average migration distance of a migrant (i) during time t in the polymer with diffusion coefficient D_i. Likewise, let x_{FS} be the average distance along which the food simulant migration front moves during the same time t with a diffusion coefficient, D_{FS}, in the polymer. From the theory of diffusion the ratio of x_i/x_{FS} is:

$$x_i/x_{FS} = (D_i/D_{FS})^{\frac{1}{2}} \tag{2.5}$$

When $x_i \ll x_{FS}$, the migrant (i) in the polymer is 'overrun' by the food simulant molecules before it can migrate out of the polymer. In this case, which is also dependent on the solubility of the food simulant in the polymer (FS $K_{P/F}$ value), accelerated migration behaviour of i will be observed. When $x_i \gg x_{FS}$, the migration of i from the polymer is not affected by the migration of the food simulant into the polymer.

From equations (2.1)–(2.5) the following conclusions concerning the selection of a proper food simulant can be made.[17]

1. If notable interactions exist between the fat or oil and the polymer, for example with polyolefins, two limiting cases must be taken into account:

   Case 1a: $D_i \gg D_{FS}$, according to equation (2.5) then $x_i \gg x_{FS}$ and the migration out of the polymer is not affected by fat absorption. Since diffusion coefficients decrease with increasing molecular weight (note that molecular structure must also be considered) the migration behaviour remains unaffected by the absorption of the fat or oil for all substances whose molecular weight is smaller than that of the fat triglycerides. The molecular weights of triglycerides in question lie between 600 and 1000 which means that this case holds true for substances with molecular weights up to the lower triglyceride molecular weight limit.
   Case 1b: When $D_i \ll D_{FS}$ then the migration from the polymer is controlled by the fat or oil, which means that it is increased.

2. Food simulants having low molecular weights, e.g. heptane, iso-octane and ethanol, have $D_{FS} \gg D_i$, which practically applies for most cases of polymer additives. Here again there are two limiting cases:

   Case 2a: If the partition coefficient, $K_{P/F}$, of the food simulant is close to one ($K_{P/F} \approx 1$, i.e. the food simulant is readily soluble in the polymer) then according to equation (2.4) the amount of absorbed food simulant in the polymer, $m_{FS,t}^P$, is so large that swelling of the polymer occurs resulting in almost total extraction of the migrant from the polymer because $V_L \gg V_P$.
   Case 2b: For the case $K_{P/F} \ll 1$ the absorption of FS by the polymer is so small that the migrant's diffusion coefficient, D_i, and subsequently the migration of i is not substantially affected. This, for instance, is the case with polar ethanol food simulants in contact with non-polar polyolefins at room temperature.

When one seeks a low molecular weight food simulant alternative to edible oils or synthetic triglyceride mixtures it is necessary to test whether it belongs to case 2a or 2b. In case 2a, because there is swelling which causes almost total extraction, a time factor must be taken into consideration. Therefore, it is necessary to carry out a series of measurements to determine the contact time which according to equation (2.2) yields the same migration value as with oil or actual foods. In this case, temperature can also be used as a possible variable if necessary. In case 2b, the consideration of a time factor is not necessary as long as the migration measurement temperature used for the food simulant is the same as that used for oil.

The selection of an appropriate fatty food simulant must then have the following requirements:

1. Migration similar to that into fatty foods
   (a) simulant does not swell or extract polymer
   (b) partition coefficient similar to that for fatty food, $K_{P/F} \approx 1$
2. Migration into simulant must not greatly exceed that of fat
3. Ease of analytical analysis
   (a) less sample preparations than for foods themselves
   (b) improved accuracy

2.6 Comparison of results from migration measurements into fats and oils with those in alternative fatty food simulants

There are a number of published studies from various laboratories which allow comparisons between migration in oil and fat and migration in heptane, iso-octane, ethanol and ethanol/water mixtures. With the exception of a comprehensive study made with iso-octane the published data covers only specific migration cases.

2.6.1 *Alternative fatty food simulants for polyolefins*

2.6.1.1 *Aqueous ethanol as fatty food simulant.* Polyolefins are probably the most studied and best characterized group of food contact polymer materials because of their widespread use in food packaging. In the past twenty years extensive efforts have been made to characterize the migration behaviour of various additives out of the polyolefins into various food simulants. These studies may be used to illustrate why aqueous ethanol solutions are used as fatty food simulants in migration testing. It is well known that fats and oils can interact (migrate into) with polyolefins. The most commonly used polyolefin additives have molecular weights around 700 (except Irganox-1010 or 1330 for example) or below, therefore, additive migration falls mainly under case 1a type of migration where D_i

Table 2.3 LDPE diffusion coefficients in contact with various solvents at 23°C[18]

Solvent	Heptanol D_i (cm²/sec) × 10⁻⁹	Undecane D_i (cm²/sec) × 10⁻⁹
Water	6.1	–
Methanol	5.5	9.0
Ethanol	5.3	–
Acetone	6.8	13
Benzene	630	–
n-Hexane	610	530

D_i is calculated from permeability and solubility coefficients: $D = P/S$

$\gg D_{FS}$. This means that the diffusion of additive in the polymer controls the migration process. Thus the selection requirements for a fatty food simulant are that it should not affect the additive polymer diffusion coefficient and should have $K_{P/F} \leq 1$. Table 2.3 illustrates the results of effect of several low molecular weight food simulant solvents on the diffusion of heptanol and undecane through LDPE.[18] In Table 2.3 it may be observed that acetone, benzene and n-hexane which cause LDPE swelling induce larger diffusion coefficients of the solutes than when LDPE is in contact with water which has no swelling effect. The diffusion coefficients for the solutes in LDPE in contact with ethanol and methanol are comparable to that in water and thus would be good candidates for food simulants. Table 2.4 lists estimated diffusion coefficients for radiolabelled polyolefin additives migrating into various food simulants from LDPE, HDPE and PP at 49°C.

The important points to recognize in Table 2.4 are that within the expected uncertainty of the experimental data, the diffusion coefficients for additive migration into 95% ethanol is comparable to those for migration into corn oil. This is an example of case 2b for low molecular weight food simulants. The effect of the oil and ethanol absorption in the polymer can be seen from the increased diffusion coefficients of 1010 compared to activated carbon which has no interaction (see case 1a). Table 2.4 also illustrates the decrease in diffusion with the increase in molecular weight and the effect of the polymer on diffusion.

The migration of radiolabelled antioxidants with different molecular weights from polyolefins into olive oil, synthetic fat HB 307 and 95% ethanol were measured.[21] The migration conditions were 1 or 10 days at 40°C. Table 2.5 summarizes the results of this study using migration factors, $F_{e/f}$ and $F_{e/ov}$, which are the ratios of the amount migrated into 95% ethanol to that migrated into fat and olive oil respectively. Severely polyolefin films were studied including low density polethylene (LDPE),

Table 2.4 Diffusion coefficients[5,19,20] of additives into food simulants from polyolefins at 49°C

| Polymer | Additive | MW | D_i (cm²/sec) × 10⁻¹² | | | |
			Activated carbon	50% ethanol	95% ethanol	Corn oil
LDPE	BHT	220	5000	–	–	1500
	1076	530	45	–	–	
	1010	1178	8.7	13	130	
HDPE	BHT		23	–	–	–
	1076		–	–	15	22
	1010		–	–	0.0058	0.12
PP	1076		–	–	290	250
	1010		–	–	65	15

Diffusion data from Arrhenius equations.

two different high density polethylenes (HDPE), and two polypropylene films (PP). The antioxidants studied were BHT, Irganox PS800 (Dilauryl-thio-diproprionate or 3,3'-thio-di(laurylpropionate)), Irganox 1010 and Chimassorb 81 (2-hydroxy-4-n-octoxy-benxophenone).

Most of the ratios in Table 2.5 are close to one which means that migration into 95% ethanol is practically the same as into oil or fat. The poorest comparison between ethanol and fat and oils was found for Irganox 1010, the antioxidant with the highest molecular weight (mol. wt = 1178) which has a ratio of 0.5 for HDPE and PP. This result supports the assumptions in case 1b where only substances with high molecular weights show increased migration due to interaction between polyolefins and fats or oils. In addition, these assumptions are further supported by the smaller ratios for the synthetic fat, HB 307 (average mol. wt = 630) compared to olive oil (average mol. wt = 908). The underlying principle is the smaller the triglyceride's molecular weight the better the penetration into the polymer (due to a corresponding higher diffusion coefficient) and the greater the influence on the additive's migration out of the polymer.

In another study an interesting result was found comparing the migration of BHT from HDPE at 40°C into different food simulants.[22] After one day of contact with n-heptane, 90% of the BHT in the polymer had migrated, whereas only 20% had migrated into HB 307 and even somewhat less into corn oil under the same contact time and test conditions. About 15% of the BHT migrated into diethylene glycol, propylene glycol, and 50% ethanol/water. However, for other alcohols with longer aliphatic chains BHT migration increased up to 50% for *n*-octanol and lauryl alcohol. It was also found in the same publication that the same extent of migration is observed for Irganox 1010 migrating from LDPE into 100% ethanol compared with corn oil at 49°C. The same Irganox migration from ethylene

Table 2.5 Comparison of the results obtained from migration tests with ethanol 95% by volume (e), test fat HB 307 (f) and olive oil (ov)[21]

Polyolefin class type (Trade name)	^{14}C-labelled additive (0.5% by wt)	Migration factor		
		$F_{e/f}$	$F_{e/ov}$	$F_{ov/f}$
LDPE (Lupolen 1800 H)	BHT	1.09	1.09	1.01
	Irganox PS 800	1.03	1.06	0.98
	Irganox 1010[a]	0.82	0.90	0.91
	Chimassorb 81	1.06	1.07	0.99
HDPE (Vestolen A 6016)	BHT	0.79	0.86	0.92
	Irganox PS 800	0.82	0.90	0.91
	Irganox 1010	0.32	0.38	0.87
	Chimassorb 81	0.97	1.01	0.80
(Vestolen A 3515)	Irganox 1010	0.58	0.66	0.87
PP (Novolen 1100 HX)	BHT	1.12	1.24	0.90
	Irganox PS 800	1.29	1.32	0.98
	Irganox 1010	0.58	0.64	0.91
	Chimassorb 81	1.25	1.32	0.95
(Novolen 1300 H)	Irganox 1010	0.42	0.50	0.84

$F_{e/f}$ = migration factor ratio ethanol to synthetic fat HB307
$F_{e/ov}$ = migration factor ratio ethanol to olive oil
$F_{ov/f}$ = migration factor olive oil to synthetic fat HB307

[a]Concentration of additive 0.2% by wt.

vinyl alcohol (EVA) into 100% ethanol was found comparable with that of corn oil at 49°C.

The results for specific migration studies using ethanol as a food simulant can be extended to global migration measurements as well. A preliminary study[23] comparing global migration from a PP film into ethanol with that into olive oil for two side contact at 40°C for 10 days showed 0.58 mg/dm² migrating into 100% ethanol versus 0.62 mg/dm² into olive oil using the GC method with an accepted method accuracy of ±3 mg/dm.

In another study, the migration of Irganox 1010 and Irganox 1076 from LDPE, HDPE, and PP were measured over a temperature range from 49° to 135°C into corn oil, 95% ethanol and water/ethanol mixtures[5]. The migration temperature dependence followed the Arrhenius form of equation (2.2). Migration values into 95% ethanol food simulant were practically identical to or slightly lower than those into oil.

Very similar results were found in another migration study of Irganox 1010 from HDPE and PP into ethanol and an unspecified edible oil.[13] Migration values into 95% ethanol were 1 to 1.5 times those into the oil and values into 50% ethanol/water were 0.5 to 0.7 times those into the oil. On the other hand, much higher migration values were found using a series of more hydrophobic alcohols (butanol to octanol).

2.6.1.2 *Iso-octane as fatty food simulant.* The specific migration of various additives from different polymers into iso-octane was compared to those into olive oil.[24] Due to the strong interaction between iso-octane and the polymer (see case 2a) a contact time of 2 hours at 40°C was used to obtain comparable mass transfer into olive oil over 10 days at 40°C. The ratio of migration into iso-octane compared to that into olive oil was found to show a much greater variation (from 0.1 to 16) compared to the ratios listed in Table 2.5 for ethanol to olive oil (from 0.32 to 1.32). It is not surprising to see such large variations for iso-octane because migration is dependent on the iso-octane migration front penetrating into the polymer (x_j) (note that the migration front itself is dependent on the nature of the polymer) as well as the large differences in the applied contact testing times.

A study was made comparing the global migration of polymer components from PVC, polyurethane, a polyether/polyamide block copolymer and a silicon rubber into olive oil at 10 days at 40°C and iso-octane at 2 hours at 40°C.[25] The global migration results showed that these migration conditions are not comparable and thus iso-octane is not a suitable substitute for olive oil.

A comprehensive comparison study of global migration into iso-octane over 2 days at 20°C with olive oil over 10 days at 40°C from 130 commercial polymer samples (PP, PE, ionomer, PVC, PS, VAC, paper products and several laminates) was carried out.[26] Iso-octane and olive oil had similar results concerning the tendencies of the food simulants to provide migration data below the global migration limit in 95% of the tested samples. Nonetheless, like the specific migration study with iso-octane, there are large variations between the iso-octane and olive oil migration measurements. From the reasoning in case 2a, migration measurements in iso-octane contain in principle uncertainties due to extraction of the polymer. Due to the short contact time applied this method seems to be useful as a practical global migration screening method.

2.6.2 *Alternative fatty food simulants for rigid PVC*

The alternative fatty food simulants for rigid PVC were not investigated as extensively as the polyolefins. However, the literature contains several examples of cases where alternative fatty food simulants for oils have been used. Fatty esters, methyl palmitate,[27] methyl oleate[28] and methyl stearate[29] have recently been used for modelling fatty food and package interactions. Other researchers[30,31] have favoured aqueous ethanol mixtures as fatty food simulants. Table 2.6 shows migration data for organo-tin stabilizers into various food simulants.

In the first column of Table 2.6 the migration of dioctyl-tin from PVC was found to be more than one order of magnitude larger than 100%

Table 2.6 Organo-tin stabilizer migration data out of rigid PVC into food Simulants

Simulant	A	B
	Weight % migrated 49°C/10 days	µg/dm² 45°C/10 days
100% Ethanol	9%	–
Heptane	1%	9.55
Iso-octane	–	2.5
8% Aqueous ethanol	0.5%	–
Sunflower oil	–	2.3
Corn oil	0.07%	–
50% Ethanol	0.07%	1.28
Water	0.06%	–
3% Acetic acid	0.05%	0.22

A = Radiolabelled Thermolite 831 di(n-octyl) tin S,S′ bis (iso-octyl mercaptoacetate) 1–3% (w/w) in PVC.[30]

B = Mixture of mono and di-n-octyl tin stabilizer 2.7% (w/w) in PVC.[31]

ethanol than into corn oil. The different results for PVC compared with those for the non-polar polyolefins may be explained by a stronger interaction between the more polar PVC with ethanol than with oil as well as by the likelihood that dioctyl-tin has a more favourable partition coefficient between ethanol and PVC than between oil and PVC. By adding 50% water to ethanol the migration into oil is simulated very well. This observation is supported by a second study (in the second column of Table 2.6) and 50% aqueous ethanol is an FDA recommended fatty food simulant for rigid PVC. The results in the second column suggest that iso-octane may make a satisfactory alternative fatty food simulant although this observation is not supported by global migration results.[25,26]

2.6.3 Alternative fatty food simulants for polystyrene

A review of the literature[32–34] shows no completely satisfactory fatty food solvents for PS. Table 2.7 summarizes the measured diffusion coefficients for styrene and BHT from various migration studies using polystyrene materials. Looking at the diffusion coefficient data for styrene in GPPS (Table 2.7) it appears that the aqueous ethanol solutions (30 and 50%) would not make good fatty food simulants for PS because the polymer swells and the styrene diffusion coefficient increases. Furthermore, the partition coefficient of 50% ethanol is perhaps far enough removed from one where less styrene would partition into the aqueous ethanol than into the oil, which has a partition coefficient of approximately one. It was observed[34] that there are partitioning effects for the migration of BHT into

Table 2.7 Diffusion coefficients in polystyrene

Polymer migrant		D (cm^2/sec) $\times$ 10^{-13} $(K_{P/FS})$							
		3% Acetic		% Aqueous ethanol (v/v)			Oil		
		acid	Water	8%	30%	50%	Corn	Sun	Edible
GPPS Styrene	40°C[a]	3.0	3.0	3.0	15	50	2.8	5.0	–
		(2380)	(3030)	(1560)	(526)	(140)	–	–	–
	49°C	–	–	2.6	–	–	–	–	2.6
HIPS Styrene[b]	49°C	–	–	1.5	–	–	–	–	1.5
BHT[c]	49°C	0.03	–	–	–	–	–	–	–

GPPS = general purpose polystyrene.
HIPS = high impact polystyrene.
[a]styrene 800 μg/ml in polymer[32].
[b]styrene 200–1000 μg/ml in polymer[33].
[c]BHT 1200 μg/ml in polymer[34].

HIPS (the equilibrium migration into 50% ethanol was one quarter that into oil). Tests with activated carbon as a food simulating material have shown that the rate limiting step in the migration into aqueous food simulants (water, 8% ethanol, 3% acetic acid) is the diffusion of BHT through the polymer.[34]

2.6.4 Alternative fatty food simulants for other polymers

Very little work was done on alternative fatty food simulants for classes of polymers other than polyolefins, PS and PVC. Most migration measurements are carried out using fats (natural or synthetic – HB 307), edible oils (corn, olive, sunflower), silicone oil, or a paraffin oil. It has been reported that ethanol simulants are not good simulants for polyesters because they tend to undergo transesterification and hydrolysis reactions with the polymer and its oligomers.[35] n-Propanol does not appear to be heavily absorbed in PET and as a result may be a potential fatty food simulant. Likewise, the polymer swells with ethanol and methanol but less with n-propanol. Hydrocarbons such as n-heptane may be appropriate fatty food simulants for PET and nylon because of the extreme differences in polarity between the polymers and n-alkanes and consequently low sorption. Other possible fatty food simulants may be the fatty esters.

2.7 Estimation methods for alternative fatty food simulants

When experimental data are not available it may be desirable to use estimation methods to estimate the suitability of an alternative fatty food simulant with a particular polymer system. To estimate the applicability of a particular solvent as a fatty food simulant for migration testing two parameters need to be estimated: the amount of simulant absorbed in the polymer and the partition coefficients of potential migrants between polymer and simulant.

2.7.1 *Estimation of simulant sorption in polymers*

The amount of simulant sorbed by the polymer is important because sorbed simulants can act as plasticizers for the polymer increasing the diffusion coefficient of potential migrants. The change in the diffusion coefficient due to sorbed simulant should not be significantly different from that found for a polymer in contact with fat or oil. Polymer solvents[36] can be disregarded as alternative simulants because of their excessive interaction with the polymer. Correlation methods using the Regular Solution Theory solubility parameters[36] work to some degree but do not give quantitative predictions and are not always qualitative[37]. Semi-empirical activity coefficient estimations methods using molecule group contributions can be used for widely varying classes of substances. Recent studies have shown the UNIFAC method to be one of the most widely applicable methods for estimating activity coefficients in polymers.[15,38] It is assumed that a pure simulant dissolves in the polymer but no polymer dissolves in the simulant. The solubility is defined as the concentration in the polymer corresponding to a unit of solute activity. Because the slopes of the activity versus absorbed weight fraction curves are often quite small as the activity approaches one the solubility is defined[38] as the predicted weight fraction at an activity of 0.99. Table 2.8 shows the predicted percentage absorbed simulation weight fraction at activity = 0.99 in various food contact polymers using the UNIFAC method. The simulants are listed in order of approximate decreasing polarity.

Referring to Table 2.8, it should be mentioned that UNIFAC generally estimates well the qualitative trends but quantitatively overestimates the amount absorbed. There are exceptions, for example the absorption of water in nylon, where UNIFAC incorrectly predicts practically no sorption. UNIFAC correctly shows the expected sorption trends for polar polymers which absorb polar simulants more than the non-polar ones. For non-polar polymers the reverse behaviour is observed. The estimations of small amounts of simulants sorbed in PET is consistent with experimental observations. The UNIFAC method is perhaps better used as a way to estimate the order of magnitude of the interactions between alternative simulants and polymers. Attempts using UNIFAC to determine the amount of sorbed water and ethanol from aqueous ethanol mixtures greatly underestimated the amount of sorbed water while the amount of sorbed ethanol was greatly overestimated.

Table 2.8 UNIFAC estimated simulant absorption in polymers at 25°C

Simulant		LDPE	PP	Absorbed weight % PS	PET	N6	PVC
Water		–	–	–	0.047	–	–
	Exp.[a]	(0.0062)	(0.0071)	(0.048)	(0.8)	(4.0)	(1.5)
Methanol		0.64	0.66	0.94	0.069	>90	1.5
	Exp.[b]				(2.6[c])		
Ethanol		5.5	2.0	1.7	0.46	73	2.9
	Exp.[b]	(0.43[c])	(1.1)	(3.6)	(2.7[c])	(9.1[d])	(2.4)
Propanol		75	4.9	2.4	0.43	83	5.0
	Exp.[b]	(4.8)	(2.1)	(4.5)	(0.64)	(37)	(1.5)
Octanol		85	84	6.4	0.20	86	38
Hexylamine		82	80	67	0.28*	23*	71
Hexaldehyde		81	81	77	0.43	78*	86
Acetone		6.6	25	29	0.37	28	71
Methyl-iso-propylketone		5.5	80	68	0.44	83	83
Propylacetate		86	66	79	1.2	45	83
Methylpalmitate		86	79	59	0.062	2.4	80
Di-iso-propylether		80	81	58	0.077	2.5	30
Toluene		84	88	87	0.58	47	85
Hexane		79	81	30	0.045	1.5	37
Heptane		80	81	30	0.041	1.4	48
						(3.7[d])	

LDPE density = 0.918 (g/ml)
PP density = 0.91 (g/ml)
PS density = 1.05 (g/ml)
PET density = 1.39 (g/ml)
N6 (Nylon 6) density = 1.13 (g/ml)
PVC (hard) density = 1.39 (g/ml)
– estimated value is essentially zero
() experimental values
[a]experimental measurement at 25°C[39]
[b]experimental measurement at 25±2°C, contact with saturated vapour[39]
[c]experimental measurement at 50±1°C, contact with saturated vapour (from Ref. 37)
[d]experimental measurement at 23±.5°C, contact with liquid
*UNIFAC is missing required interaction parameter, some group interactions are set to zero.

2.7.2 Estimation of polymer/simulant partition coefficients

The estimation of potential migrant partition coefficients between polymers and food simulants using activity coefficients can be made using equation (2.6).

$$K_{P/F} = c_{i,\infty}^{P}/c_{i,\infty}^{F} = \gamma_i^{F} V_F/\gamma_i^{P} V_P \qquad (2.6)$$

Where γ_i is the activity coefficient of the migrant based on its molar fraction in the food or polymer and V is the molar volume (ml/mol). It is assumed in equation (2.6) that the migrant concentrations in the polymer and food are very dilute ($x_i < 1 \times 10^{-4}$). UNIFAC predicts activity coefficients of migrants in polymers on a weight fraction basis ($w_i = g_i/$

Σg_i) and in liquids on a mole fraction basis thus giving a variant of equation (2.6):

$$K_{P/F} = c_{i,\infty}{}^P/C_{i,\infty}{}^F = \gamma_i{}^F V_F \, den_P / \gamma_{i,w}{}^P \, MW_i \qquad (2.7)$$

Where $\gamma_{i,w}{}^P$ is the weight fraction activity coefficient, den_P is the polymer density and MW_i is the migrant molecular weight. Figure 2.2 compares UNIFAC estimated partition coefficient for BHT between LDPE and aqueous ethanol with experimental data from the literature. The estimated partition coefficients assume there is no interaction between the LDPE and the aqueous ethanol phase, the weight fraction of BHT in the polymer is 0.01%, and the corresponding mole fraction concentration in the liquid phase is 0.0001. In the dilute mole fraction range (1×10^{-3} to 1×10^{-5}) of BHT in the liquid phase the calculated $K_{P/F}$ changes very little (-1.9%). However, the estimated weight fraction activity coefficient of BHT in LDPE is dependent on the amount of sorbed ethanol, and $K_{P/F}$ increases 360% as the percentage of ethanol in LDPE increases from 0 to 1%. This range corresponds to the amount of ethanol that might be absorbed as the food simulant goes from 100% water to 100% ethanol.

Looking at Figure 2.2, UNIFAC estimates the BHT $K_{P/F}$ within a factor of 10 for the simulants with ethanol concentrations greater than 50% and is poorest for 100% water. Taking into account the possible absorption of ethanol into the polymer shifts the UNIFAC estimations higher for the ethanol containing simulants. Figure 2.2 also illustrates the negligible effect of temperature on the partition coefficient. The 100% ethanol

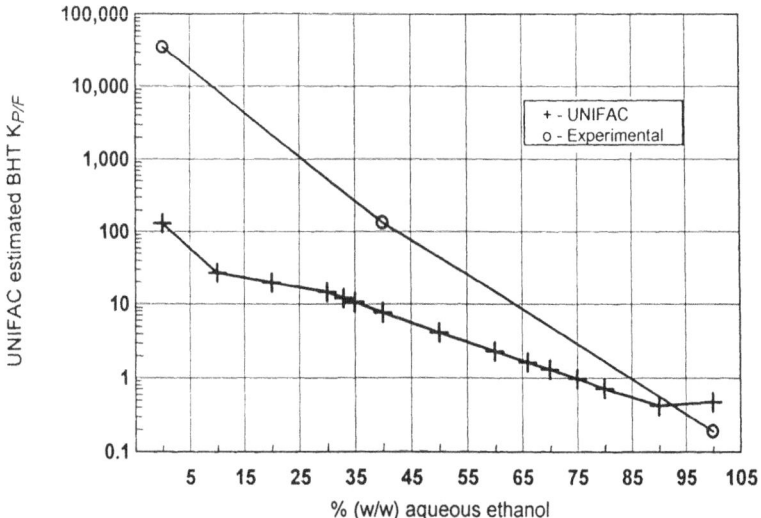

Figure 2.2 UNIFAC estimated $K_{P/F}$ versus % aqueous ethanol. Assume no interaction of ethanol with LDPE polymer at 25°C.

Table 2.9 UNIFAC estimated BHT $K_{P/F}$ between polymers and simulants at 25°C

Simulant	LDPE	PP	$K_{P/F}$ PS	PET	N6*	PVC*
Water*	130	130	86	470	1900	1600
Methanol	1.1	1.1	0.72	4.0	16	13
Ethanol	0.47	0.46	0.72	1.7	6.9	5.7
50% w/w ethanol water	4.2	4.1	2.7	15	60	50
10% w/w ethanol/water	27	27	18	98	390	320
Propanol	0.33	0.32	0.22	1.2	4.7	3.9
Octanol	0.25	0.25	0.17	0.91	3.6	3.0
Acetone*	0.23	0.23	0.15	0.84	3.4	2.8
Methylpalmitate*	0.24	0.23	0.16	0.41	3.5	2.8
Toluene	0.004	0.004	0.002	0.013	0.005	0.45
Hexane	0.54	0.53	0.35	2.0	7.8	6.4
Heptane	0.53	0.52	0.35	1.9	7.7	6.4

LDPE density = 0.918 (g/ml)
PP density = 0.91 (g/ml)
PS density = 1.05 (g/ml)
PET density = 1.39 (g/ml)
N6 (Nylon 6) density = 1.13 (g/ml)
PVC (rigid) density = 1.39 (g/ml)
* UNIFAC is missing required interaction parameter, some group interactions are set to zero.

experimental $K_{P/F}$ was measured at 23°C[40] and the 40% and 0% points were measured at 40°C.[14]

Table 2.9 compares UNIFAC estimated $K_{P/F}$ for potential alternative food simulants in several polymer systems using BHT as an example. With respect to the UNIFAC BHT partition coefficients estimations in Table 2.9, it appears that any of the pure organic solvents could be used equally well. The water and aqueous containing solvent $K_{P/F}$'s are far enough in magnitude away from $K_{P/F} = 1$ that some partitioning effects are likely to occur. This makes it necessary to evaluate aqueous simulants on a case by case basis. Table 2.9 serves to further illustrate some of the advantages and weak points of the UNIFAC method. As noted, some of the interaction parameters between functional groups that are needed are not available. This introduces even larger errors into the estimations.

2.8   High temperature fatty food simulants

The use of microwaveable and ovenable packaging has created a special need for food simulants which can withstand high temperatures of up to 150°C for 30 minutes (the current EC recommendation). Fats and oils (both edible and silicone) can withstand such high temperatures; however, aqueous and organic simulants need special high pressure cells to prevent their evaporation.[5,13] In addition the problem may be more complex as some solvent simulants may no longer be good fatty food simulants because they may swell the polymer at high temperatures. A temperature resistant organic polymer with a large surface area, Tenax (poly(p-2,6-diphenylenphenyloxide)), was used as a sorption medium for the measurement (at 150°C for 30 minutes) of global migration of PET, PP, polycarbonate (PC) and polysulfone (PSU) microwavable packages and containers.[41] Following heating the Tenax is extracted with diethyl ether and the extractant analysed by GC or dried and determined gravimetrically. It has been observed that polymer additives with molecular weights up to 600 can be measured. PET trimers are not well measured because of their extremely low volatility even at high temperatures. The Tenax method is based on the assumption that the rate of migration depends on the polymer diffusion coefficient and that all substances able to migrate have some volatility so they can be transferred to and be absorbed by a solid medium with a high surface area in contact with the polymer.

2.9   Conclusions and necessary future investigations

There exists in the literature a large amount of data on specific migration of various additives and monomers from food contact polymers. The literature used here and the literature cited within these works represent only a small part of the published data.[6,22,24,42,43] Most of the measurements described in this literature deal with specific migration into foods and into food simulants as specified in EC and FDA directives.

Two important conclusions can be made about migration into edible oils and synthetic triglycerides:

1. Fats and oils are the foods having the largest migration values.
2. With the exception of volatile migrants, which can be relatively easily measured using HSGC, the majority of migration measurements require a great deal of analytical work and also have high variability and insufficient sensitivity.

The use of food simulants with molecular weights lower than oil and fat would greatly reduce the analytical work. Meanwhile the sensitivity is improved and the variability of the measurements reduced. These substitute food simulants will be most useful in the cases where they have analytical advantages, for example, for all migrants that cannot be measured using HSGC.

For solvents having strong interactions with polymers (case 2a) which yield high migration values, the measurements should be made for each polymer and the results compared with migration measurements into oil or food to determine the corresponding contact times and temperatures.

With respect to time saving screening tests the application of such solvents is justified as demonstrated for iso-octane.[26]

Ethanol and ethanol/water mixtures containing a majority of ethanol have in general no altering interactions with most polymers (see case 2b). The most successful and best researched application of ethanol as a fatty food simulant is with the polyolefins. Swelling in slightly polar polymers, such as PVC, can be avoided by adding more water to the ethanol mixture. Methanol may be a better fatty food simulant for PS than aqueous ethanol because it is more polar and consequently less absorbed which prevents changes in the diffusion coefficient. High polar polymers, such as polyamides and polyesters, hydrocarbons like n-heptane, iso-octane or fatty esters are likely to be better food simulants than ethanol and ethanol/water. The time factor is not so critical for simulants with no altering interactions and can be regarded as identical to that applied for food and oil provided the same measurement temperatures are used. One of the problems with using aqueous ethanol mixtures is that as the ethanol content decreases to concentrations less than 50% the partition coefficient may start to play an important role in the extent of migration. Estimation methods such as UNIFAC may be used for first approximations of the extent of polymer/ alternative fat simulant interactions and for partition coefficients of potential migrants between polymers and alternative fat simulants. UNIFAC and other estimation methods are currently either not widely applicable or accurate enough to be used in place of experimental measurements.

More work is needed to find suitable alternative fatty food simulants for PS, polyesters and nylons. With the exception of studies using iso-octane almost all comparisons between alternative food simulants and oils or fats are related to specific migration measurements. More studies could be undertaken to compare global migration results using alternative simulants with those using fats and oils. High temperature global migration testing also remains an area for further research on alternative food simulants. The results show that it is possible in many cases to substitute alternative solvents or solvent mixtures for fats and oils in migration testing.

References

1. Rossi, L. (1991) In: *Food Packaging: A Burden or an Achievement?* ADRIAC, Reims, France, 10.1–10.11.
2. Schwartz, P.S. (1988) *Food Additives and Contaminants* 5, supplement No. 1, 537–541.
3. *Anon* (1988) *Recommendation for Chemistry Data for Indirect Food Additive Petitions*, version 1.1. Division of Food Chemistry & Technology, Center for Food Safety & Applied Nutrition, Food and Drug Administration, Department of Health and Human Services, Washington, D.C.
4. Koch, J. and Figge, K. (1978) *Fette Seifen Anstrichmittell*, **80**(4), 158–161.
5. Goydan, R., Schwope, A.D., Reid, R.C. and Cramer, G. (1990) *Food Additives and Contaminants* 7, 323–337.

6. Bieber, W.-D., Freytag, W., Figge, K. and vom Bruck, C.G. (1984) *Fd Chem. Toxic.* **22**, 737–742.
7. Rossi, L., Sampaola, A. and Gramiccioni, L. (1972) *Annals Istituto Superiore di Sanità* **8**, 432–439.
8. Rossi, L. (1981) *J Assoc. Official Analytical Chemists.* **64**, 697–703.
9. Figge, K. (1973) *Deutsche Lebensmittel-Rundschau.* **69**, 253–257.
10. Figge, K. (1973) *Food and Cosmetics Toxicology* **11**, 963–974.
11. Tice, P.A. (1988) *Food Additives and Contaminants* **5**, supplement no. 1, 373–380.
12. Anon (1976) *FDA Guidelines for Chemistry and Technology Requirements of Indirect Food Additive Petitions.* Division of Food Chemistry & Technology, Center for Food Safety & Applied Nutrition, Food and Drug Administration, Department of Health and Human Services, Washington, DC.
13. Lickly, T.D., Bell, C.D. and Lehr, K.M. (1990) *Food Additives and Contaminants* **7**, 805–814.
14. Keinhorst, A. and Niebergall, H. (1986) *Deutsche Lebensmittel-Rundschau* **82**, 254–256.
15. Baner, A.L. (1992) *Partition Coefficients of Aroma Compounds between Polyolefins and Aqueous Ethanol and their Estimation Using UNIFAC and GCFEOS.* Ph.D. Thesis Fraunhofer-Institut ILV, Munich, Germany and Michigan State University, E. Lansing, Michigan.
16. Vergnaud, J.M. (1991) *Liquid Transport Processes in Polymeric Materials.* Prentice Hall, Englewood Cliffs, New Jersey.
17. Piringer, O. (1990) *Deutsche Lebensmittel-Rundschau.* **86**, 35–39 and 152.
18. Becker, K., Koszinowski, J. and Piringer, O. (1983) *Deutsche Lebensmittel-Rundschau* **79**, 257–266.
19. Schwope, A.D. and Reid, R.C. (1988) *Food Additives and Contaminants* **5**, supplement no. 1, 445–454.
20. Schwope, A.D., Till, D.E., Ehnholt, D.J., Sidman, K.R., Whelan, R.H., Schwartz, P.S. and Reid, R.C. (1987) *Fd. Chem. Toxic.* **25**, 317–326.
21. Figge, K. and Hilpert, H.A. (1991) *Deutsche Lebensmittel-Rundschau* **87**, 1–4.
22. Schwope, A.D., Till, D.E., Ehnholt, D.J., Sidman, K.R., Whelan, R.H., Schwartz, P.S. and Reid, R.C. (1987) *CRC Critical Reviews in Toxicology.* **18**(3), 215–243.
23. Bücherl, T., Franz, R., Lee, T. and Knezevic, G. (1991) In: *Annual Report 1991.* Fraunhofer-Institut für Lebensmitteltechnologie und Verpackung, Munich, Germany pp. 51–53.
24. Freytag, W., Figge, K. and Bieber, W.-D. (1984) *Deutsche Lebensmittel-Rundschau* **80**, 333–335.
25. Gramiccioni, L., Di Prospero, P., Milana, M.R., Di Marzio, S. and Marcello, I. (1986) *Fd. Chem. Toxic.* **1**, 23–26.
26. de Kruijf, N. and Rijk, M.A.H. (1988) *Food Additives and Contaminants* **5**, supplement no. 1, 467–483.
27. Feigenbaum, A.E., Ducruet, V.J., Delpal, S., Wolff, N., Gabel, J. and Wittmann, J.C. (1991) *J. Agric. Food Chem.* **39**, 1927–1932.
28. Riquet, A.M., Sandray, V., Akermann, O. and Feigenbaum, A. (1991) *Sci. Aliments.* **11**, 337–355.
29. Lum Wan, J.A., Chatwin, P.C., Katan, L. and Prosser, E. (1991) *Predictive Mathematical Models of Migration; Working Party on Chemical Contaminants from Food Contact Materials, 1991 Project Review.* Food Research Institute, Norwich, UK.
30. Schwope, A.D., Till, D.E., Ehnholt, D.J., Sidman, K.R., Whelan, R.H., Schwartz, P.S., and Reid, R.C. (1986) *Deutsche Lebensmittel-Rundschau* **82**, 277–282.
31. Matos, C.M., Kroll, J., Hoppe, H., Romminger, K. and Woggon, H. (1987) *Z. Gesamte Hyg.* **33**, 258–260.
32. Till, D.E., Enhold, D.J., Reid, R.C., Schwartz, P.S., Schwope, A.D., Sidman, K.R. and Whelan, R.H. (1982) *Ind. Eng. Chem. Fundam.* **21**, 161–168.
33. Murphy, P.G., MacDonald, A. and Lickly, T.D. (1992) *Fd. Chem. Toxic.* **30**, 225–232.
34. Schwope, A.D., Till, D.E., Ehnholt, D.J., Sidman, K.R., Whelan, R., Schwartz, P.S., and Reid, R.C. (1987) *I&EC Research.* **26**, 1668–1670.
35. Begley, T.H. and Hollifield, H.C. (1990) *J. Agric. Food Chem.* **38**, 145–148.
36. Brandrup, J. and Immergut, E.H. (eds.) (1989) *Polymer Handbook*, John Wiley & Sons, New York.

37. Thalmann, W.R. (1990) *Packaging Science and Technology* **3**, 67–82.
38. Goydan, R., Reid, R.C. and Tseng, H. (1989) *Ind. Eng. Chem. Res.* **28**, 445–454.
39. Gerlowski, L.E. (1990) In: *Barrier Polymers and Structures*, ed. Koros, W.J. American Chemical Society, Washington, DC, pp. 177–191.
40. Koszinowski, J. (1986) *J. Applied Polymer Sci.* **31**, 2711–2720.
41. Fuchs, M., Kluge, S., Rüter, M., Wolff, E. and Piringer, O. (1991) *Deutsche Lebensmittel-Rundschau.* **87**, 273–276 and 311–316.
42. Van Battum, D., Rijk, M.A., Verspoor, R. and Rossi, L. (1982) *Food and Chemical Toxicology.* **20**, 955–959.
43. Figge, K. (1988) In: *Migration bei Kunstoff-verpackungen*, eds Hauschild, G. and Spingler, E. Wissenschaftliche Verlagsgesellschaft mbH, Stuttgart, pp. 33–92.

3   Food flavour and packaging interactions*

J.P.H. LINSSEN and J.P. ROOZEN

Abstract

Interactions between packaging materials and foodstuffs can affect food quality, e.g. on flavour aspects. There are three main phenomena: migration, permeation and absorption.

Several flavour compounds of an artificially flavoured drink yoghurt were absorbed by its high density polyethylene bottle. Selective absorption can disturb the delicate balance of flavour compounds and the product does not reach the consumer as was intended by the manufacturer.

Flavour change of mineral water packed in low density polyethylene lined aluminium/cardboard packages was studied by combined gas chromatography and sniffing port analysis. Mainly aromatic hydrocarbons and some carbonyls appear to be responsible for the main sensory descriptors: metallic, musty, astringent, sickly, synthetic and glue-like.

Taste recognition threshold concentrations of styrene were determined in oil in water emulsions with different amounts of fat. Thresholds increased linearly with the amounts of fat in the emulsions. Moreover, the concentrations of styrene in the head spaces of the emulsions were similar at their threshold levels.

3.1   Introduction

Food packaging interactions can be divided into three main phenomena: migration, permeation and absorption. These phenomena can occur separately or simultaneously and can affect food quality, e.g. food flavour aspects (Figure 3.1). Migration of packaging components, like residual monomers, additives or polymerization aids, can cause an undesirable contamination of food. Migration of such components may adversely affect the quality of food, e.g. alteration of flavour. Permeation of gas and vapour, particularly of oxygen, water vapour and aroma components, is of considerable interest. Particularly in combination with light, a high rate of oxygen permeation can cause oxidation problems. A high water vapour

*Paper given at the IFTEC symposium 'Food Packaging Interactions and Disposability.' The Hague, November 15–18, 1992.

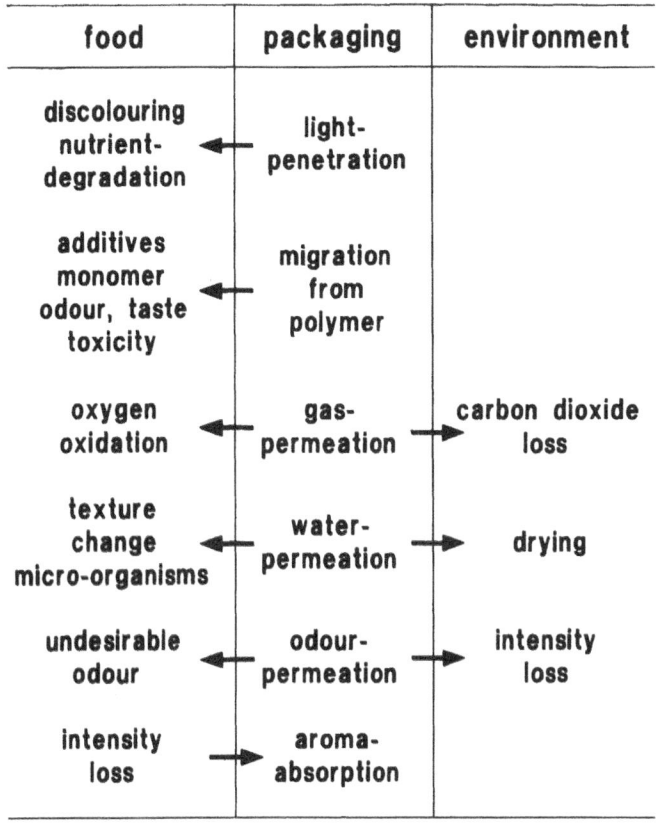

food	packaging	environment
discolouring nutrient-degradation	light-penetration	
additives monomer odour, taste toxicity	migration from polymer	
oxygen oxidation	gas-permeation	carbon dioxide loss
texture change micro-organisms	water-permeation	drying
undesirable odour	odour-permeation	intensity loss
intensity loss	aroma-absorption	

Figure 3.1 Interactions between food and packaging material.

permeation results in physical or physico-chemical alterations as wetting and drying, and can promote microbiological spoilage. These reactions can lead to an indirect alteration of flavour, while direct flavour alteration can be caused by loss of flavour components from the food or by acquiring aspecific odours from the environment through the package.[1,2] Absorption phenomena can also influence the quality of the food. Absorption of aroma components by the package can cause a direct loss of flavour components from food, as absorbed flavour compounds can no longer contribute to the flavour of a foodstuff. Moreover, selective absorption of flavour components can result in an imbalance of the aroma and the product does not reach the consumer as was intended by the manufacturer.

3.2 Absorption of flavour compounds by high density polyethylene

Absorption of flavour compounds into plastic packaging materials is a well known phenomena. The concentration of limonene is substantially reduced in an aseptic brick style package after only a few days' storage.[3,4] Also significant losses of limonene, neral, geranial, octanal and decanal from orange juice stored in PE lined packages were reported.[5] In model studies it was found that sorption of flavour compounds increased with increasing carbon chain length.[6]

In the present study absorption of volatile compounds from an artificially flavoured drink yoghurt (flavour: peach/apricot/pear) packed in a high density polyethylene bottle was investigated under practical conditions of use.

The yoghurt contains less than 0.1% fat and is a blend of 89.7% yoghurt (made from defatted milk), 8.9% sugars, 1.0% fruit juice(s), 0.4% stabilizer, artificial flavouring and colouring agents. Samples of the drink yoghurt, carefully rinsed emptied bottles and control bottles were prepared by dynamic headspace sampling.[7]

Figure 3.2 shows the chromatograms of the drink yoghurt and the emptied bottle. It indicates that flavour compounds with higher retention times and higher molecular weights were likely to be absorbed by the packaging material. Besides flavour compounds, the chromatogram of the bottle (Figure 3.2B) shows n-alkanes, branched alkanes and siloxane, which were found by analysis of the volatiles of the control bottles too. Such compounds were previously identified as volatiles from polyethylene, which probably arise from a C_{12} mineral oil fraction used as a solvent in the production of polyethylene.[8,9] Interestingly, these compounds remained in the plastic and are not likely to migrate into an aqueous product like drink yoghurt (absent in Figure 3.2A). Other compounds present in the chromatograms of Figure 3.2, such as styrene, benzaldehyde and acetophenone, were reported to be always present, when one uses dynamic headspace techniques with Tenax as adsorbent.[10,11]

The volatiles listed in Figure 3.2 were reported by Maarse and Visscher[12] as volatiles of the fruits labelled on the drink yoghurt. However, some common flavour compounds, e.g. lactones, were not found, probably because of the very low content of fruit juices added to the drink yoghurt. Figure 3.2 also shows that certain volatiles preferentially remain in the drink yoghurts while others are easily absorbed by the packaging material. Compounds with short carbon chains and sulphides remained in the aqueous phase of the drink yoghurt while compounds with medium carbon chain length (up to 8 carbon atoms) were found in the drink yoghurt as well as in the PE packaging material. Compounds with longer carbon chains and high branched chains, and compounds with a more complex structure seem to be absorbed by the packaging material. The contribution of these compounds to the flavour of the products might be reduced.

It was found that esters were very susceptible to absorption into PE packaging materials.[6] These findings were confirmed under the practical circumstances of the present case-study: mainly the higher molecular weight esters were absorbed by the HDPE bottles. As a consequence of differential sorption of volatiles, the balance of flavour compounds in the packed yoghurts might be disturbed. Volatiles added to the product and subsequently absorbed by the packaging material will not reach the consumers as was intended by the manufacturer, because these compounds are unable to contribute to the flavour of a foodstuff.

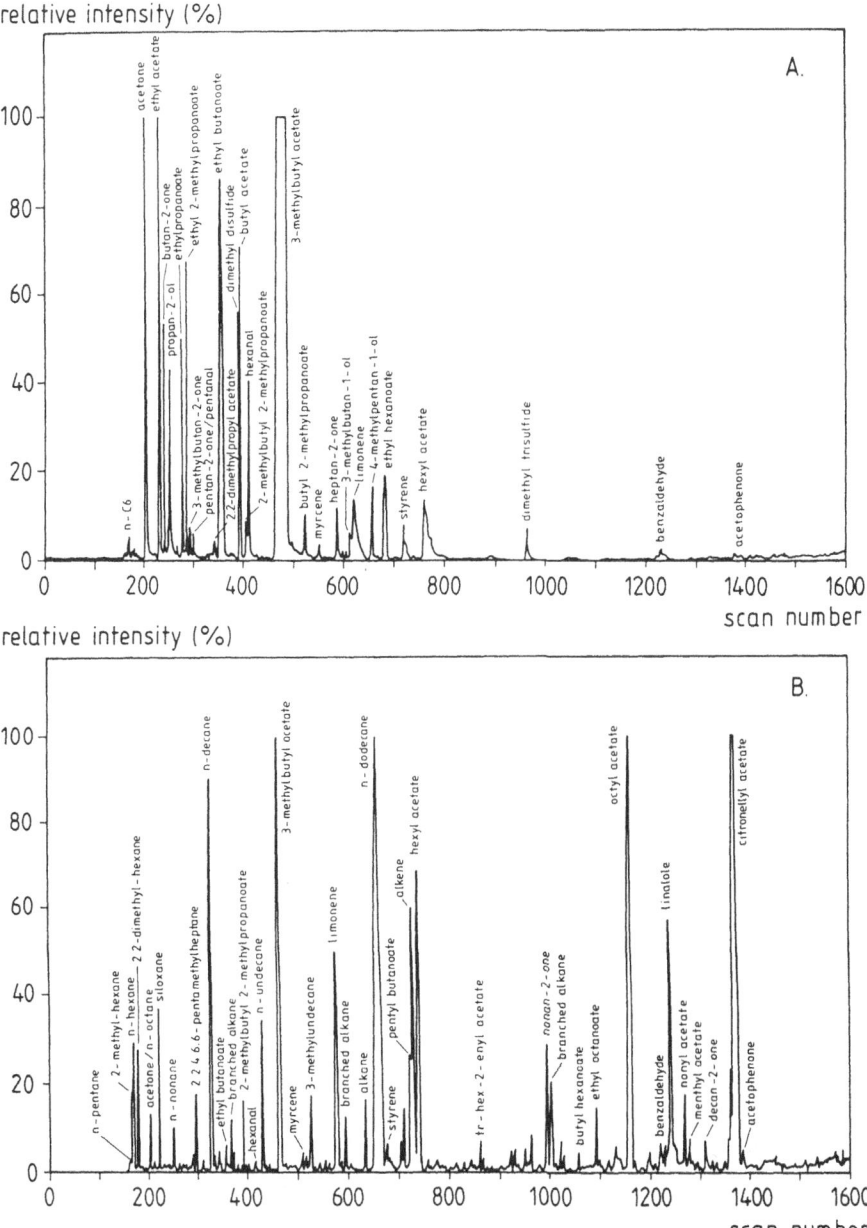

Figure 3.2 Reconstructed total ion current chromatograms of volatile compounds from drink yoghurt (A) and its HDPE bottle (B) obtained by dynamic headspace technique.

3.3 Flavour changes of mineral water packed in low density polyethylene laminated packages

Low density polyethylene (LDPE)-packaging material is used in mono- and multilayer systems as well as in lined cardboard and aluminium packages for packaging of foodstuffs. LDPE contains several low molecular weight components, which are able to migrate into the packed product and can cause a change of flavour. Such low molecular weight components can be formed during the LDPE polymerization process, in which oxygen and peroxides are added to initiate radical reactions. Consequently, it is possible to have carbonyl and carboxyl groups present in LDPE.[13] Additional oxidation products can be formed during high temperature processing in the presence of oxygen, e.g. extrusion coating.[14,15] Potts et al.[16] reported that extrusion conditions of LDPE affects the flavour of the packed product: higher melt temperatures were responsible for increased off-taste intensity of drinking water in contact with LDPE. Berg[9] reported that products such as spring water, milk and fruit juices frequently have a taint after being stored in bottles of LDPE or LDPE-lined cartons, probably caused by volatile compounds like saturated and unsaturated hydrocarbons, aromatic hydrocarbons and aromatic hydrocarbons with an unsaturated side chain. Some of them (C_3- and C_4-alkyl benzenes) were described as having an 'intense plastic off-flavour'. Also C_2–C_5 carbonyl and carboxyl compounds were reported as a reason for an off-flavour.[17,18,19] The taint was described as candle-grease, musty, rancid, soapy, pungent and acrid. In a former study the taint of water packed in LDPE-lined aluminium test pouches was described as musty, sickly, astringent, synthetic, metallic and dry.[20]

The aim of the present study was to describe and identify volatile compounds in commercial mineral water packed in LDPE-lined aluminium/cardboard packages. Techniques used were combined gas chromatography and sniffing port analysis (GC-sniffing) and gas chromatography combined with mass spectrometry (GC–MS).

The study was carried out on commercial mineral water packed in 2 L packages of LDPE-lined aluminium cardboard (test sample) and reference mineral water packed in a glass bottle as control sample. The test sample was incubated at 40°C for 24 hours. This sample and the control were used for sensory flavour intensity studies and for preparation of dynamic headspace samples for analysis by GC-sniffing and GCMS. A panel of 11 selected and trained assessors was used for the flavour intensity study. Samples were tasted with and without using nose clips. For GC-sniffing a panel of 10 selected and trained assessors was used. The latter panel was split up into five pairs because the GC was equipped with two sniffing ports, one at each side of the instrument.[21,22] The attributes: metallic, synthetic, dry, astringent, musty and sickly, found as descriptors for a LDPE taint,[20] were used in the flavour intensity study. The use of nose-

Table 3.1 Mean score ($\pm$ SD) of intensity (%) obtained for each descriptor on a visual analogue scale of 180 mm with (N) and without (WN) the use of nose-clips ($n = 11$)

| Descriptors | Mineral water | | | |
| | Reference | | 40°C/24 h* | |
	N	WN	N	WN
Metallic	28 $\pm$ 24	34 $\pm$ 25	23 $\pm$ 21	51 $\pm$ 23
Synthetic	20 $\pm$ 18	39 $\pm$ 32	18 $\pm$ 22	79 $\pm$ 21
Dry	41 $\pm$ 28	34 $\pm$ 22	35 $\pm$ 31	55 $\pm$ 26
Astringent	34 $\pm$ 25	31 $\pm$ 18	28 $\pm$ 25	48 $\pm$ 22
Musty	13 $\pm$ 10	30 $\pm$ 23	18 $\pm$ 16	58 $\pm$ 23
Sickly	21 $\pm$ 18	34 $\pm$ 24	23 $\pm$ 18	58 $\pm$ 20

*Temperature/time conditions for incubation of the commercial mineral water sample.

clips by the assessors in the flavour intensity studies diminishes significantly ($p < 0.05$) the flavour intensities of the commercial sample incubated at 40°C for 24 hours for each descriptor except for the descriptor 'dry'. The latter appears to be a kind of mouthfeel (Table 3.1).

These findings indicate that most of the descriptors are related to volatile substances. Therefore, GC-sniffing is appropriate for further studies. A preliminary analysis by GC-sniffing of the volatile compounds from water packed in LDPE lined aluminium/cardboard packages resulted in a list of 14 descriptors (Table 3.2). Besides 'other' these descriptors had to be used by the assessors at the sniffing port.

Figure 3.3A presents the average FID chromatograms ($n = 5$) of control (REF) and commercial LDPE-packed water sample, which was incubated

Table 3.2 Odour descriptors used for sniffing port analysis

Dutch expression	English translation*
Champignon geur	Mushroom-like
Cacao	Cocoa-like
Fris	Fresh
Fruitig	Fruity
Gras	Green
Kunstmatig	Artificial
Lijm	Glue-like
Metaal	Metallic
Muf	Musty
Plastic	Plastic
Prikkelend	Astringent
Wee	Sickly
Zoetig	Sweet
Zuurtjes	Candy-like

*The English translation may not be an exact synonym of the terms used by the Dutch assessors.

at 40°C. Reconstructed chromatograms of the components detected at the same time, are shown in Figure 3.3B. GC-sniffing of dummy samples revealed that detection of a smell at the sniffing port by less than five out of 10 assessors can be considered as noise.

Table 3.3 presents the compounds identified by GC–MS, and the odour described by the assessors at the sniffing port for the mineral water sample which was incubated at 40°C. Glue-like is often mentioned as a new descriptor, while synthetic (plastic), metallic, musty, astringent and sickly were also generated in the previous study.[20] The test pouches used were made of aluminium and LDPE coated at a relatively high temperature of 310°C, so they are not directly comparable with a commercial LDPE-lined package. Moreover, the pouches were immediately folded and sealed

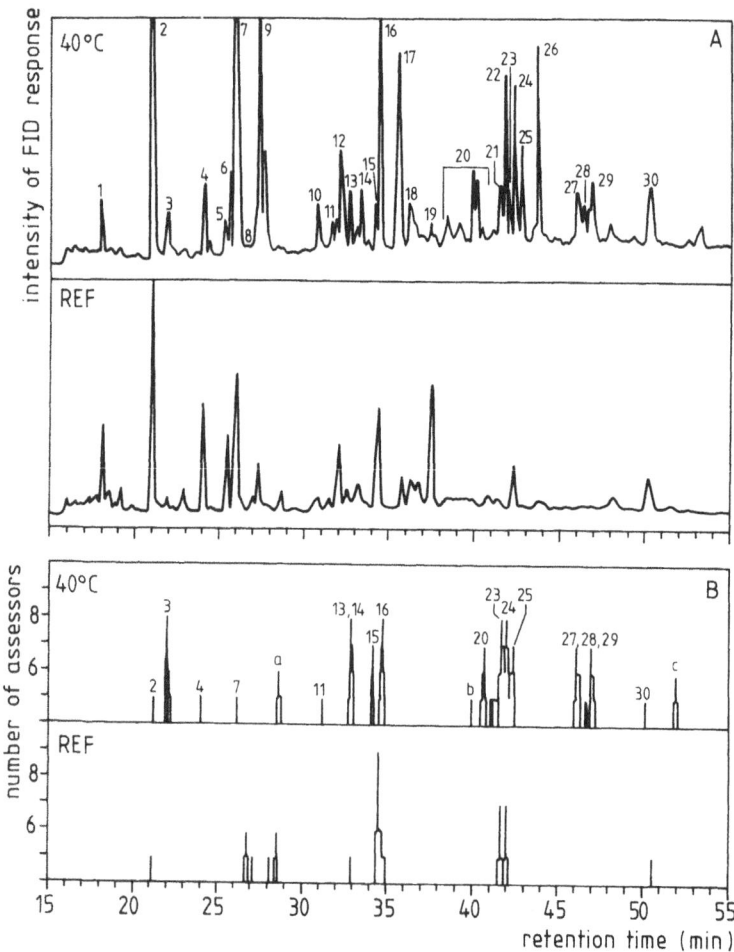

Figure 3.3 Reconstructed chromatograms of dynamic headspace samples. Average of 5 replicates (A) and the number of assessors out of 10, who had a smell impression at the sniffing port at the same time (B). Corresponding responses have the same symbol. REF: control; 40°C: LDPE packed water samples incubated at 40°C.

after LDPE lining to prevent evaporation of volatile compounds. In that case, carbonyls were identified to be responsible for the taint described. However, as can be seen in Table 3.3, mainly aromatic hydrocarbons such as toluene, xylenes, n-propylbenzene, C_3- and C_4-alkyl benzenes, and some unknowns, were found to be the compounds corresponding to the descriptors given by the assessors. Berg[9] reported similar compounds in LDPE granulate and indicated that C_3- and C_4-alkyl benzenes have an intense plastic smell. In Figure 3.3A differences between the incubated samples and control mineral water were notably found for three groups of compounds: isomers of C_3-alkyl benzenes with retention times 32–35 min (peaks 12–16) and two groups of unknown isomers with

Table 3.3 Volatile compounds of mineral water incubated at 40°C in LDPE lined aluminium/cardboard packages and their odour description

Peak no*	Compound	Odour description
1	Pentanal	
2	Toluene	Plastic, astringent, glue-like, 'other'
3	Hexanal	Green, artificial, mushroom-like, fruity, musty, 'other'
4	4-OH-4Me-2-pentanone	musty, fruity, 'other'
5	Isopropylacetate	
6	Ethylbenzene	
7	p-, m-Xylene	Plastic, musty, astringent, artificial, metallic
a		mushroom-like, 'other'
8	Heptanal	
9	o-Xylene	
10	α-Pinene	
11	n-Propylbenzene	Fruity, sweet, 'other'
12	C_3-Alkyl benzene	
13	C_3-Alkyl benzene	Plastic, astringent, glue-like
14	C_3-Alkyl benzene	Musty, astringent, metallic
15	Octanal	Sickly, musty, candy-like, 'other'
16	C_3-Alkyl benzene	Astringent, fresh, sweet, candy-like, fruity, fresh
b		green, mushroom-like, sweet
17	2,2,4,6-6-Pentamethylheptane	
18	C_3-Alkyl benzene	
19	Limonene	
20	C_4-Alkyl benzene + branched alkanes	Artificial, fresh, mushroom-like, musty, metallic, sweet, sickly
21	2-Nonanone	
22	Unknown	
23	Nonanal	Astringent, cocoa-like, mushroom-like, 'other'
24	Unknown	Plastic, glue-like, metallic, artificial, musty
25	Unknown	Plastic, glue-like, metallic
26	Unknown	
27	Unknown	Plastic, astringent, sickly, metallic, artificial
28	Unknown	
29	Unknown	Plastic, glue-like, metallic, musty, artificial
30	Decanal	Artificial, plastic, metallic, plastic, sweet, glue-like, 'other'
c		

*Peak no corresponds to the peak symbols in Figure 3.3.

retention times 42–44 min (peaks 22–26) and 46–47 min (peaks 27–29). At the same retention times sniffing port assessors agreed upon the detection of volatile compounds from the LDPE packed water samples (Figure 3.3B). The absence of a descriptor for 2,2,4,6,6,-pentamethylheptane is remarkable, because this compound has been suggested for the taint sometimes found in LDPE-packed products.[8] In Table 3.3 pentanal, hexanal, heptanal and octanal were found to be present in the LDPE-packed mineral water sample, which was incubated at 40°C. The unknown compounds are (probably branched) carbonyl compounds. The latter are hardly present in the control sample, but appear as higher peaks in the FID chromatograms of the LDPE-packed mineral water samples after incubation. The results of the study indicate that the main descriptors for volatile compounds (separated by gas chromatography) in commercial mineral water packed in LDPE lined aluminium/cardboard and incubated at 40°C, are: plastic, astringent, musty, sickly, glue-like and metallic. Isomers of C_3-alkyl benzenes and isomers of unknown carbonyls appear to be responsible for these descriptors. Storage at higher temperatures for a longer time can imply flavour deterioration.

3.4   Taste recognition threshold concentrations of styrene in oil in water emulsions

Polystyrene (PS) is used as a polymer for food packaging. Examples are yoghurt and dessert packaging, foamed trays for meat and crystal clear trays for salads and vegetables. PS contains traces of residual styrene monomer. Styrene monomer is able to migrate into foodstuffs and can impart an off-flavour to the packed product. Several authors found styrene in food products present in very low amounts.[23,24,25] Also, an off-flavour in chocolate and lemon cookies packed in PS trays was detected by a sensory panel.[26] The intensity of off-flavour depended on the level of residual styrene in the packaging material, type of food matrix and contact time.

Threshold values are important parameters for off-flavour perception and it is desirable to know more about the mechanism of release of off-flavours from a food product. Many food products are of the emulsion type and therefore emulsions are used as a model for flavour release.[27,28,29] In this model the concentration of the flavour compound in the aqueous phase was supposed to be crucial. It was also suggested that partitioning with the vapour phase has to be taken into account.[30] The following model for emulsions was proposed:

$$K_{ve} = \frac{K_{vd} \cdot K_{dc}}{1 + (K_{dc} - 1)\, \Phi_d} \tag{3.1}$$

in which K_{ve}, K_{vd} and K_{dc} are the equilibrium partition coefficients between vapour- and emulsion phase, between vapour- and dispersed phase and between dispersed- and continuous phase, respectively. Φ_d represents the volume fraction of the dispersed phase. In case of an O/W-emulsion the enumerator of equation (3.1) can be replaced by the equilibrium partition coefficient between vapour phase and water ($K_{vw} = C_v/C_w$), which results in:

$$K_{ve} = \frac{K_{vw}}{1 + (K_{ow} - 1)\,\Phi_d} \qquad (3.2)$$

in which K_{ow} ($= C_o/C_w$) is the equilibrium partition coefficient between the oily and aqueous phase in the O/W-emulsion. Furthermore C_e is defined as:

$$C_e = (1 - \Phi_d)\,C_w + \Phi_d C_o \qquad (3.3)$$

in which C_e, C_o and C_w represent the concentration in the emulsion phase, the oily phase and the aqueous phase, respectively.

This study deals with the taste recognition threshold concentrations of styrene in a model of O/W-emulsions (3–30% fat). Partition coefficients between vapour phase and emulsions and between vapour phase and water have been determined. C_w can be calculated from equations (3.2) and (3.3). The validity of equation (3.1) for the release of styrene from O/W-emulsions was verified. The concentrations of styrene were calculated for the vapour phases of emulsions and water with styrene contents at their taste recognition threshold concentrations (TRTCs). Samples of O/W-emulsions were spiked with aliquot amounts of styrene and presented as 15 mL samples in closed glass bottles in series of nine solutions to the sensory panel (Table 3.4). Panels consisted of 48–53 untrained assessors for tasting the O/W-emulsions.[31]

The TRTC of styrene was defined as the concentration of styrene for which 50% of the answers of the assessors were positive in the recognition test. Figure 3.4 shows the linear regression of the amount of oil in an O/W-emulsion versus 50% TRTC value. The linear regression equation calculated is: 50% TRTC (ppm) = 0.068 (% oil)+0.035 (r = 0.99). Additionally, the 50% TRTC value for water[32] is presented in Figure 3.4.

The results imply that the TRTC is higher in products with higher fat content, and so making styrene less noticeable in high fat products. The increase of the TRTC is linearly with the fat content of the O/W-emulsions. However, migration of styrene from polystyrene is more likely into products with a higher fat content.[33,34] These two diverging influences hinder accurate prediction of the probability of off-flavour development.

Table 3.5 gives the partition coefficients of styrene between vapour phase and emulsions with different amounts of fat (K_{ve}). C_e measured and C_e calculated from Figure 3.4 represent the TRTC values of styrene for each emulsion. Using equations (3.2) and (3.3), and the measured value of K_{vw} ($= 0.027 \pm 0.001(n = 3)$), C_w is calculated at the TRTC values.

Table 3.4 Concentrations (ppm) of styrene in standards and test sample series used for determining taste recognition threshold concentrations in O/W-emulsions with different amounts of fat

| | O/W-Emulsions | | | | | |
| | Fat content | | | | | |
	3	10	15	20	25	30
Sample series	0	0	0	0	0	0
	0.06	0.2	0.25	0.25	0.25	1
	0.09	0.4	0.5	0.5	0.5	2.5
	0.12	0.8	0.75	1	1	5
	0.15	1.4	1	1.5	2	7.5
	0.2	2.2	2	2.5	3	10
	0.5	3.2	3	3.5	5	15
	0.7	4.4	5	5	10	20
	1	5.6	10	10	20	30
Standard	4	25	40	40	40	40

Also the concentrations of styrene in the vapour phases (C_v) are given for emulsions containing styrene at their TRTC values. Partition coefficients could be important for estimating the chances of an off-flavour. Table 3.5 shows that the concentrations calculated for styrene in the aqueous phases of emulsions (C_w) are similar at about 15 ppb at their TRTC levels. This is in agreement with the assumption of McNulty and Karel[27] that the aqueous phase of an emulsion or probably also an emulsion type of

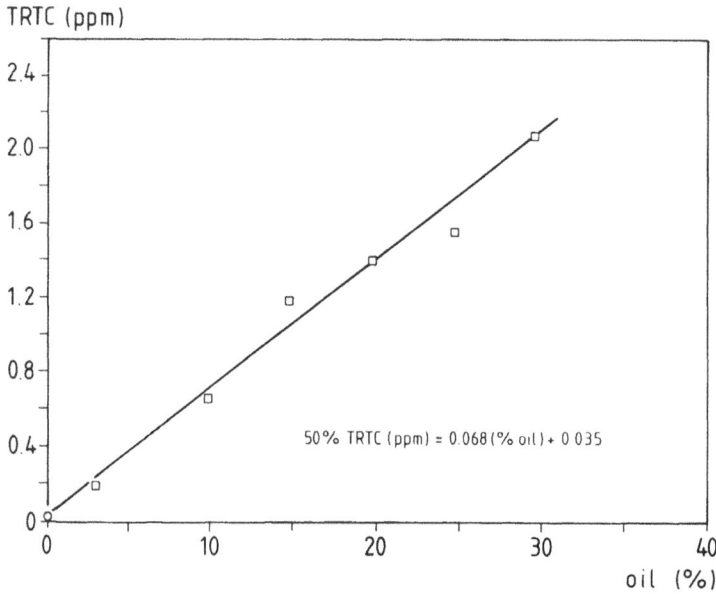

Figure 3.4 Recognition threshold concentrations of styrene in water (○) and O/W–emulsions (□) with different amounts of fat.

Table 3.5 Partition coefficients between vapour phase and emulsions (K_{ve}). Concentrations of styrene in the aqueous phase of the O/W-emulsions (C_w) and in the vapour phase of emulsions (C_v). All styrene contents are at their TRTC values (C_e)

Fraction oil	K_{ve}[a] $\times 10^{-3}$	C_e[b] ppb	C_w[c] ppb	C_e[d] ppb	C_w[e] ppb	C_v[f] ppb
0.03	1.70 ± 0.10	239	15	196	12	0.41
0.10	0.55 ± 0.02	715	15	654	13	0.39
0.15	0.39 ± 0.02	1055	15	1181	17	0.41
0.20	0.29 ± 0.01	1395	15	1396	15	0.40
0.25	0.24 ± 0.02	1735	15	1559	14	0.42
0.30	0.21 ± 0.02	2075	16	2078	16	0.44

[a] Values are means of 6 replicates $\pm$ SD.
[b] Calculated from the curve in Figure 3.4.
[c] Calculated from equations (3.2) and (3.3) using C_e calculated.
[d] C_e measured.
[e] Calculated from equations (3.2) and (3.3) using C_e measured.
[f] Calculated from $K_{ve} = C_v/C_e$, using C_e calculated.

foodstuff determines the flavour perception. Moreover, the uniform concentration of styrene in the aqueous phase of the different emulsions is close to the 50% TRTC value of 22 ppb found for water.[32] Therefore, a fixed concentration of styrene is needed in the aqueous phase of an emulsion for recognizing styrene in O/W-emulsions. As can be seen from Table 3.5, only a minor part of the styrene dissolved in the emulsion is present in the aqueous phase; the major part is hidden in the oily phase. Because of the very good solubility of styrene in the fat fractions, the TRTCs of the emulsions will increase with increasing fat contents. More styrene is needed then to reach the equilibrium at which the concentration of styrene is about 15 ppb in the aqueous phase. It was also found that the concentrations of styrene in the vapour phases (C_v) of the emulsions at their TRTCs are similar at about 0.41 ppb (Table 3.5). This can be explained by the following equilibrium:

$$C_o \rightleftharpoons C_w \rightleftharpoons C_v \qquad (3.4)$$

As already discussed, C_w is constant at the TRTCs of styrene in O/W-emulsions. Water is the continuous phase in an O/W-emulsion and thus in a closed system an equilibrium exists between continuous phase and vapour phase. So, C_v is also constant at the TRTCs of styrene. However, C_v calculated for water is higher than the constant value of 0.41 ppb, namely 0.59 ppb. The method of tasting allows styrene to evaporate very quickly from the water sample and a significant part of it has already escaped before reaching the mouth. Therefore, higher concentrations of styrene could be necessary in water for meeting its concentration in the mouth vapour phase at TRTC level.

The present findings could be important for practical situations. If the concentration of styrene in the vapour phase exceeds 0.41 ppb one could expect an off-flavour of styrene in O/W-emulsions and probably emulsion type of foods. Analysis of the styrene content in the vapour phase can be predictive then for an off-flavour caused by styrene. At the TRTCs a number of variables in equation (3.2) are known: C_v is at a constant level of 0.41 ppb, K_{ow} is 462 and K_{vw} is 0.027. Using these data in equation (3.2) and $K_{ve} = C_v/C_e$ results in a simple linear relationship between TRTC (C_e) and the dispersed (oily) phase in an O/W-emulsion: C_e (ppm) = 0.070 (% oil) +0.015. A calculation of the TRTCs for yoghurts, containing 0.1%, 1.5% and 3% fat, estimates at 22, 120 and 225 ppb, respectively. These estimates are in good agreement with the experimentally reported values: 36, 91 and 171 ppb, respectively.[31] Moreover, the latter relationship is similar to the linear regression equation of Figure 3.4, which validates equation (3.1) for this particular type of system.

References

1. Niebergall, H. and Kutski, R. (1982) Modelluntersuchungen zur Migration von Inhaltstoffe aus Verpackungsmaterialiën in Lebensmittel. *Deutsche Lebensmittel-Rundschau*, **78**, 82–87.
2. Koszinowski, J. and Piringer, O. (1987) Food/product compatibility and migration. *J. of Plastic Film and Sheeting*, **3**, 96–111.
3. Mannheim, C.M., Miltz, J. and Letzer, A. (1987) Interaction between polyethylene laminated cartons and aseptically packed citrus juices. *J. Food Sci.*, **52**, 737–740.
4. Hirose, K., Harte, B.R., Giacin, J.R., Miltz, J. and Stine, C. (1988) Sorption of d-limonene by sealant films and effect on mechanical properties. In: *Food and Packaging Interactions*, ACS Symposium Series 365, ed. J.H. Hotchkiss, Washington, 28–41.
5. Dürr, P., Schobinger, U. and Waldvogel, R. (1981) Aroma quality of orange juice after filling and storage in soft packages and glass bottles. *Alimenta*, **20**, 91–93.
6. Shimoda, M., Ikegama, T. and Osajima, Y. (1988) Sorption of flavour compounds in aqueous solution into polyethylene film. *J. Sci. Food Agric.*, **42**, 157–163.
7. Linssen, J.P.H., Verheul, A. and Roozen, J.P. (1992) Absorption of flavour compounds by packaging material: Drink yoghurt in polyethylene bottles. *Int. Dairy Journal*, **2**, 33–40.
8. Vom Bruck, C.G. and Hammerschmidt, W. (1977) Ermittlung der Fremdgeschmacksschwelle in Lebensmittel und ihre Bedeutung fur die Auswahl von Verpackungsmaterialiën. *Verpackungsrundschau*, **1**, 1–4.
9. Berg, N. (1980) Sensoric and instrumental analysis of off-flavour giving compounds from polyethylene. *Proc. 3rd Intern. Symp. on Migration*, Unilever Forschungsgesellschaft, Hamburg, pp. 266–277.
10. Schaeffer, H.J. (1989) Gas chromatographic analysis of traces of light hydrocarbons – a review of different systems in practice. *J.H.R.C.*, **12**, 69–81.
11. Lewis, M.J. and Williams, A.A. (1980) Potential artefacts from using porous polymer from collecting aroma compounds. *J. Sci. Food. Agric.*, **31**, 1017–1026.
12. Maarse, H. and Visscher, C.A. (1989) Volatile Compounds in Food: Qualitative and Quantitative Data, CIVO Zeist.
13. Shorten, D.W. (1982) Pololefins for food packaging. *Food Chem.*, **8**, 109–119.
14. Hoff, A. and Jacobsson, S. (1981) Thermo-oxidative degradation of low density polyethylene close to industrial processing conditions. *J. Appl. Polym. Sci.*, **26**, 3409–3423.
15. Hoff, A., Jacobsson, S., Pfaffli, P., Zitting, A. and Frostling, H. (1982) Degradation products of plastics. *Scand. J. Work Environ. Health*, **8**, suppl. 2, 3–27.

16. Potts, M.W., Baker, S.L., Hanssen, M. and Hughes, M.M. (1990) Relative taste performance of plastics in food packaging. *J. Plastic Film & Sheeting*, **6**, 31–43.
17. Bojkow, E. (1982) Zum Problem der geschmacklichen Beeinflussung von Molkereiprodukten durch Packmittel. Mitteilung 3, *Osterreichische Milchwirtschaft*, **37**, wiss. Beilage 2, 9–20.
18. Koszinowski, J. and Piringer, O. (1983) Die Bedeutung von Oxidationsprodukten ungesättigter Kohlenwasserstoffe für die sensorischen Eigenschaften von Lebensmittelverpackungen. *Deutsche Lebensm. Rundsch.*, **79**, 179–183.
19. Fernandes, M.H., Gilbert, S.G., Paik, S.W. and Stier, E.F. (1986) Study of the degradation products formed during extrusion lamination of an ionomer. *J. Food Sci.*, **51**, 722–725.
20. Linssen, J.P.H., Janssens, J.L.G.M., Roozen, J.P. and Posthumus, M.A. (1991) Sensory descriptors for a taint in water packed in test pouches made of polyethylene lined aluminium. *J. Plastic Film & Shtg*, **7**, 294–305.
21. Acree, T.E., Barnard, J. and Cunningham, D.G. (1984) A procedure for the sensory evaluation of gas chromatographic effluents. *Food Chem.*, **14**, 273–286.
22. Linssen, J.P.H., Janssens, J.L.G.M., Roozen, J.P. and Posthumus, M.A. (1993) Combined gas chromatography and sniffing port analysis of volatile compounds of mineral water in polyethylene laminated packages. *Food Chem.*, **46**, 367–371.
23. Withey, R. and Collins, P.G. (1978) Styrene monomer in foods: a limited Canadian survey. *Bull. Environ. Contam. Toxicol.*, **19**, 86–94.
24. Gilbert, J. and Startin, J.R. (1983) A survey of styrene monomer levels in foods and plastic packaging by coupled mass spectrometry automatic headspace gas chromatography. *J. Sci. Food Agric.*, **34**, 647–652.
25. Flanjak, J. and Sharrad, J. (1984) Quantitative analysis oif styrene monomer in foods. A limited East Australian survey. *J. Sci. Food Agric.*, **35**, 457–462.
26. Passy, N. (1983) Off-flavour from packaging materials in food products. Some case studies. In: *Instrumental Analysis of Foods*, eds Charambolous, G. and Inglett, G., Academic Press, New York, pp. 413–421.
27. McNulty, P.B. and Karel, M. (1973a) Factors affecting flavour release and uptake in O/W-emulsions. I. Release and uptake models. *J. Food Technol.*, **8**, 309–318.
28. McNulty, P.B. and Karel, M. (1973b) Factors affecting flavour release and uptake in O/W-emulsions. II. Stirred cell studies. *J. Food Technol.*, **8**, 309–331.
29. McNulty, P.B. and Karel, M. (1973c) Factors affecting flavour release and uptake in O/W-emulsions. III. Scale-up model and emulsion studies. *J. Food Technol.*, **8**, 415–427.
30. Overbosch, P., Afterof, W.G.M. and Haring, P.G.W. (1991) Flavor release in the mouth. *Food Rev. Intern.*, **7**, 137–184.
31. Linssen, J.P.H., Janssens, J.L.G.M., Reitsma, J.C.E., Bredie, W.L.P. and Roozen, J.P. (1993) Taste recognition threshold concentrations of styrene in oil in water emulsions and yoghurts. *J. Sci. Food Agric.*, **61**, 457–462.
32. Linssen, J.P.H., Legger-Huysman, A. and Roozen, J.P. (1990) Threshold concentrations of migrants from food packages: styrene and ethylbenzene. In: *Flavour Science and Technology*, eds Bessière, Y. and Thomas, A.F., John Wiley & Sons, Chichester, pp. 359–362.
33. Linssen, J.P.H. and Reitsma, J.C.E. Migration of styrene monomer from polystyrene packaging material into food simulants. In: *Proc. 7th World Conference on Packaging*, Utrecht, The Netherlands, pp. 0.3.1.–0.3.6.
34. Ramshaw, E.H. (1984) Off-flavour in packaged foods. *CSIRO Food Res.*, Q 44, 83–88.

4 Microwavability of packaged foods*

S.A.E. LEFEUVRE and
M.B.M. AUDHUY-PEAUDECERF

Abstract

In a microwave oven, the electromagnetic energy produced by a magnetron propagates into the load where it is converted into heat. The propagation is modified by the walls of the oven, the packaging and the heterogeneity of the load. The result is a standing wave which produces hot and cold points.

Among the properties of the packaging material its ability to produce heat through microwave excitation needs to be pointed out in order to get the desired profile of temperature. The microwave design of a packaging requires a basic knowledge of the properties of waves and heat.

A brief introduction to electromagnetic waves simulated by the famous Maxwell equations introduces the concept of field of influence, shows some examples of distribution of this field and explains physically how a magnetron works. The interaction between the wave and any kind of material, food or packaging is described and the process of absorption by the free carriers (conduction) and the molecules (dipolar absorption) explained. The thermal behaviour of a material is given. It explains how the thermal energy is stored and the meaning of specific heat, and how the thermal energy is exchanged by the surface of the material under black body radiation.

The classical model of heat diffusion and energy is recalled and the main properties of microwave heating in a commercial oven, the heterogeneity in space and time and the influence on the power and the frequency of the coupling of the load with the magnetron are described.

It is shown how the packaging material can act as a thermal oven in a microwave one. The different materials used for packaging are described and their thermal and electrical properties compared. The materials are classified into transparent, absorbers, susceptors and reflectors. The influence of the shape and the size of the load on the field distribution and consequently on the temperature profiles are pointed out. Some examples of different types of packaging materials designed to produce a given

*Partially given as a lecture at the Colloque International 'Conditionnement alimentaire, 2 défis: Innovation et Environnement' Pouzauges, France, 7–8 October, 1992.

temperature profile (for instance to get browning or to heat a pizza) are given as well as some physical properties useful for designing a microwave packaging.

4.1 Introduction

The microwave oven does not heat like a conventional one. The physical basis of its action is electricity. The energy transmitted by the generator is radiated by waves, caught by the food itself and transformed into internal heat. One of the consequences of the propagation is the existence of stationary waves in the ovens and of hot and cold points produced by these standing waves.

Specification of the packaging materials not only requires the knowledge of their thermal and chemical properties but also of their electrical properties. The behaviour in the oven may be conceived as a mean to get the desired heat profile. Understanding of the part the package may play requires that the physical properties of the microwave heating are known.

4.2 Electromagnetic waves: theory

It has long been known that electrical charges and currents produce forces on each other. The word itself, electromagnetism, reminds us of the earliest experiments. 'Electro' comes from a Greek origin and means amber (which is able to attract small pieces of paper when it is rubbed). 'Magnetism' comes from Magnesia, a town in Minor Asia around which an ore able to attract iron was found. To model these phenomena, it was thought that charges and currents do not exist alone but together with their fields of influence, electric E and magnetic H. To be more precise, E is associated with the charge and H with the current. Similarly a variation of charge is a current and a variation of E produces H. And also, a variation of H produces E exactly as a variation of current produces charges.

Behaviour of the fields, as a function of the charges and currents in the source, is given by resolution of the so called Maxwell's equations.[1] The transformation of the charges into current in the source gives rise, in the surrounding space, to the transformation $E \Longleftrightarrow H$ which shows that the electromagnetic propagation is similar to the mechanical one. As a consequence, the electromagnetic waves tend to fill all of a given space.

Maxwell's equations are complicated. This is due to the understanding of the physical meaning of the vectorial operators[1] $\partial \wedge$ and ∂ used in these equations and but also to the understanding of the word 'variation'. In fact, this given set describes all the electrical phenomena and machines including those operating at the industrial frequencies. The difference with the microwave oven is due to the very high frequencies which are used. The very fast variations versus time are linked with very sensitive variations with space. Hot and cold points do exist at low frequencies but we do not usually have an experimental knowledge of them.

A phenomenological approach of an electromagnetic wave may be found in a mechanical analogy. Let us consider a ship moving on the surface of the sea.

The movement will produce a wave which carries all the energy spent by the engine. This wave is a mechanical wave similar to the electromagnetic one; it is the long distance field of influence of the ship. A small boat will be sensitive to the wave and the wave is, even for it, more important than the ship. It may or may not catch the mechanical energy. This is similar to the electrical charge carriers in food (which moreover will convert this mechanical energy into heat).

The wavelength is the distance between two near peaks. The behaviour of the boat is a function of its size as compared to the wavelength λ. The wavelength is the effective mechanical unit of length. It is the same in microwave. In the air of an oven, the unit of length is approximately 12 cm. Scheme 4.1 shows the wave associated with punctual source located in the centre.

In an oven, the size of the food product has approximately the same order of magnitude as the wavelength. The diffraction produced by the

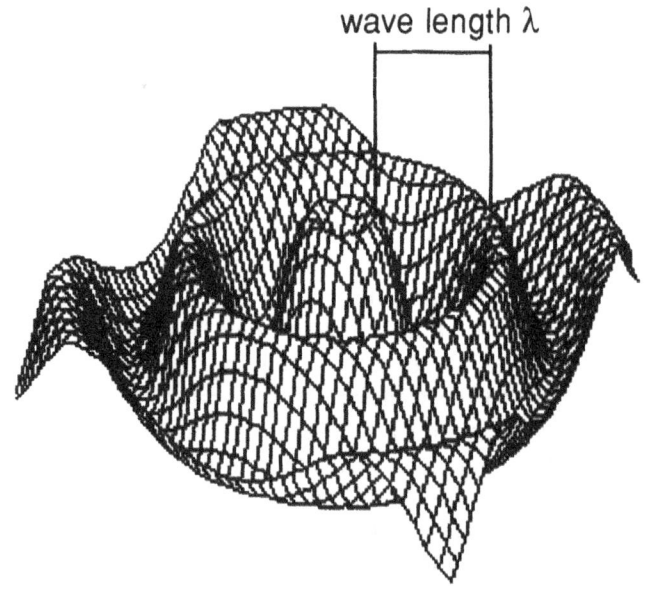

Scheme 4.1 Wave associated with a central punctual source.

load makes it difficult to predict the field configuration in the foodstuff, and, as a consequence, the temperature. The wavelength λ in vacuum or air is given by the relation:

$\lambda = 3 \times 10^{10}$ cm/s divided by the frequency of the generator.

Current values are shown in Table 4.1.

Table 4.1 Correspondence between frequencies and wavelengths

Frequency (MHz)	13.56	27.12	433.92	915	2450
λ (cm)	221.24	110.62	69.14	32.79	12.25

4.2.1 *The wave equation in the empty oven*

Maxwell's equations may provide us with a relation, for each of E or H taken apart, which is known as the wave equation. For instance, for electrical field E, this equation is:

$$\Delta E + \omega^2 \mu_0 \varepsilon_0 E = -j\omega\mu_0 J$$
Boundary conditions

where $\omega = 2\pi f$ is the pulsation, $\mu_0 = 4\pi 10^{-7}$ SI, $\varepsilon_0 \mu_0 c^2 = 1$ ($c = 3 \times 10^8$ m/s) and J the current.

The boundary conditions, at the interface between two media, express the continuity of the fields. If one medium is a perfect metal (i.e. the wall of an oven or an aluminium package), these conditions express the reflection of the wave by the metal which acts as a mirror.

The wave equation has to be understood as follows:

(1) $\Delta E + \omega^2 \mu_0 \varepsilon_0 E = 0$ expresses the electromagnetic property of air.

(2) Boundary conditions   expresses the shape of the oven.

(1) + (2)   gives the electromagnetic property of the empty oven. There is a large number of eigen modes.

(3) $= -j\omega\mu_0 J$ is the source of the waves. J is the current on the antenna of the magnetron or, for convenience, a fictive current in the slot in a wall of the oven. J selects a small set of modes along all the possibilities of the oven.

Scheme 4.2 shows one of these eigen modes (or simply modes), in a rectangular oven. The coordinates x,y are in the horizontal plane. The drawing shows the amplitude of the electric field of influence at certain height z.

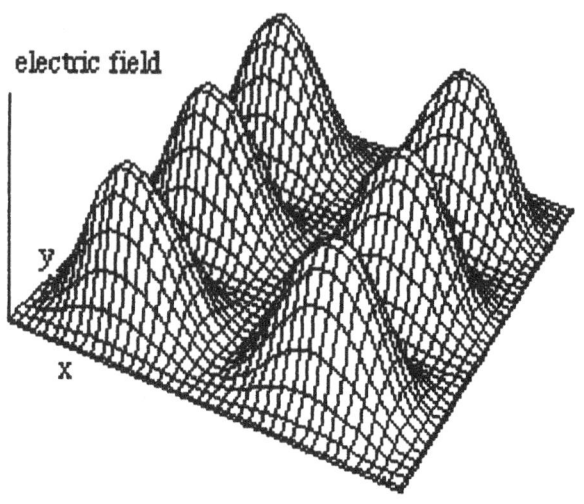

Scheme 4.2 Distribution of electrical field.

It is clear that an electric charge located at a peak will be more influenced than one located in a trough. These standing waves will produce hot and cold points in the food.

4.2.2 Electromagnetic energy

The electromagnetic energy is a kind of potential mechanical energy. It cannot be used by itself but needs electrical charges to convert it into mechanical energy. In a food product, this mechanical energy will be converted by the shocks into thermal energy.

Maxwell's equations may be used to describe the behaviour of the electromagnetic energy. For instance, it may be demonstrated that in an empty oven, there is no outside leakage and because there is no load the energy is only stored.

There is an exchange of energy from the electrical to the magnetic form and vice versa. This exchange is made at a specific frequency, an eigen frequency, characteristic of the eigen mode. If the frequency of the magnetron is close to an eigen frequency, the amplitude of the field may be very high.

Working of an empty oven should be avoided. The reasons for that are the fact that high electrical fields help arcs to occur and that the generator continuously produces active energy which is radiated back to it and produces damages. The manufacturer provides some extra loads to prevent having a pure reactive oven. When the oven is loaded by a lossy material, this produces a balance between the produced and the absorbed active energies.

4.2.3 *Production of waves by a magnetron*

The magnetron is a vacuum valve in which the electrons, emitted by the cathode, turn around under the action of a continuous electric field produced by the power supply and of a continuous magnetic field. The movement of these free carriers produces the electromagnetic radiation which carries away the main part of the mechanical energy delivered by the power supply.

Scheme 4.3 explains the main lines of the operation and of the energy transfer. When rotating, the electrons are bunched by the anode which is similar to a periodical slow wave structure. This rotating ring looks like the rotor of a low frequency generator. The frequency of rotation may be very high because the free carriers are in vacuum. When rotating, the ring induces a current in the antenna and, in turn, this current radiates the energy into the oven. Usually a waveguide is used between the magnetron and the oven, to shape the fields more carefully.

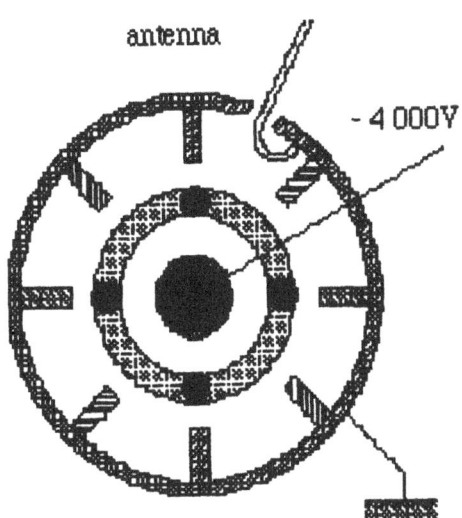

Scheme 4.3 Section of a magnetron.

4.3 Energy absorption in natural media

From the electrical point of view at the microwave frequencies, natural media are compounds of free charge carriers, bounded charge in molecules and neutral particles. These charges move under the influence of the electrical field of the wave (the influence of the magnetic field is neglected because the velocity v of the charge is low).

4.3.1 *Free carriers*

Let n be the density of carriers the charge of which is $(-e)$, and the velocity v. When moving under the action of E, they produce a current given by:

$$J = n(-e)v$$

The action of E is expressed by the equation of mechanics which is:

$$(-e)E = m\tau^{-1}v$$

where τ is the relaxation time for free carriers, time between two successive shocks which transfer to the surroundings the kinetic energy of the carrier. The relaxation time τ is short as compared to the period of the electromagnetic waves. From the electrical point of view, the movement defines the conductivity σ:

$$J = \sigma E, \sigma = n\, e^2\, \tau/m, -e = -1.60219\ 10^{-19}\ C$$

and the power P transmitted by shocks, which is the kinetic energy $n(1/2\, mv^2)$ divided by the relaxation time between two consecutive shocks, is

$$n(1/2mv^2)1/\tau$$

which leads to

$$P = 1/2\ \sigma E.E^*$$

This is the form of the density of electrical power absorbed by a load and converted into the numerous (and uncorrelated) kinetic energies of small particles, that is to say into heat.

  The electrical current has to be added to the source current, for instance in the wave equation.

4.3.2 *Dipolar absorption*

In dipolar absorption, the electrical field induces several phenomena, mainly a polarization by mechanical deformation of the molecules and a rotation to try to follow the field orientation. This movement of the dipole gives rise to a current which has to be added to the current $j\omega\varepsilon_0 E$ existing in the vacuum.

  In a non absorbing material, the molecular rotation is made inside the molecule without shocks against the surroundings. When the field intensity decreases, the molecule comes back to its original position and this movement is a current which radiates its kinetic energy. This exchange between electromagnetic energy and mechanical energy (with a delay expressed by the complex j) explains the propagation of waves.

  The charge displacement (Scheme 4.4) is assumed to be proportional to the field strength, that is εE. The corresponding current is the variation with time $j\omega\varepsilon\ E$.

electrical field E

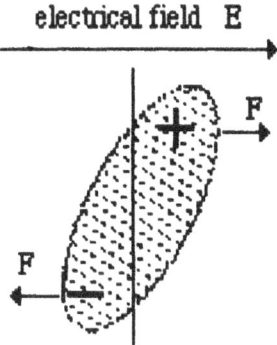

Scheme 4.4 Dipolar molecule.

The coefficient ε is the permittivity of the material. It will be denoted ε' in the case of lossy materials.

The two phenomena are far from being comparable. In the conduction process (free carriers) there is a one-way conversion of energy, from the wave to the material. As a consequence there is an accumulation of energy under the heat form. This is due to the large number of uncorrelated particles. In non lossy material, there is a conversion of electromagnetic energy to mechanical energy and vice versa. This reciprocity explains the propagation of the waves inside the material and shows that the material takes an active part in the propagation. The wavelength will be changed and the stationary waves also.

In fact, in natural media, the movement of the molecules is not only internal but it is partly oriented to the outside. The conversion process still exists but shocks on the surroundings also exist. These shocks will accumulate energy in a similar way free carriers do.

To take into account this conversion process, it is usual to transform the permittivity ε into $(\varepsilon' - j\,\varepsilon'')$. The term $\omega\varepsilon''$ is similar to σ for free carriers. The density of microwave power converted into heat power is then $\frac{1}{2}\,\sigma\varepsilon''\,E.E^*$ It may be difficult, and often useless to keep the distinction between σ and $\omega\varepsilon''$. Usually they are gathered into $\omega\varepsilon''$. They are also gathered into a new characteristic, $tn\,\delta$, defined by

$$tn\,\delta = \varepsilon''/\varepsilon'.$$

Table 4.2 gives experimental values of the permittivity of some useful materials. These electromagnetic forces, coming from outside, have to be added to the internal constitutive forces of the material. When the internal forces change, mainly with temperature, the microwave behaviour of the material, described by the permittivity, changes too. The permittivity (ε', ε'') is the electrical behaviour of natural media. It is a function of temperature (mainly when the physical state changes) and when it changes, the stationary waves change too.[2,3]

Table 4.2 Permittivity and specific heat of some useful materials

Material	$T(°C)$	ε'_r	$tn\ \delta$	$C\ (kJ/kg°C)$
Water	25	77	0.14	4.18
Ice	−10	3.2	0.001	
Oil	25	2.5	0.06	2
Milk	25	78	0.29	
Lean meat	25	40	0.3	2.9–3.5
Fatty meat	25	53.2	0.3	1.3–2.6
Cooked fish	25	46	0.26	2.8–3.6
Pyrex		4	0.001	0.84
China		5.2	0.003	0.92
Polyethylene		2.25	0.0003	
Glass, phosphate		5.17	0.004	
Paper		2–3	0.05	

Maxwell's equations need to be modified in natural media. As a result, the field distribution is deeply modified by the presence of the material. When the material is not homogeneous, the field, and as a consequence the dissipated power, may vary rapidly from one point to another producing hot and cold points.

Figure 4.1. shows the computed electric field in a food product packaged into a transparent material. The permittivity of the food is $\varepsilon_r = 5$, $tn\ \delta = 0.2$. The oven is a cube of dimension 40 cm.

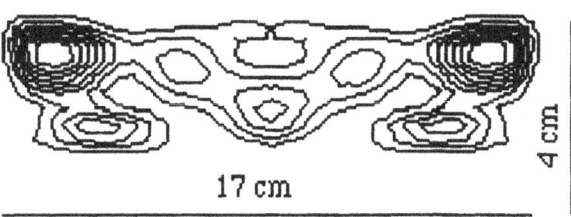

17 cm

4 cm

Figure 4.1 Line of electrical field in a food product.

The electric field configuration is a function of the oven and of the product. It is clear that the field is different from the case of the empty oven, and that the temperature produced will not be homogeneous.

The energetic equation also needs to be modified since there is no power dissipation in a vacuum. The modification is merely introduced, changing ε_0 to $\varepsilon_0\ \varepsilon_r$.

In the stored energy ε_0 becomes $\varepsilon_0\ \varepsilon_r'$ and the dissipated power P appears to be:

$$P = \tfrac{1}{2}\ \omega\ \varepsilon_0\ \varepsilon_r''\ EE^*$$

P is the volumic source term of the heat equation (see section 4.5).

4.4 Specific heat and infrared radiation

The electrical power expressed in the preceding paragraph was divided into two parts: one part was said to be exchangeable (into magnetic energy to constitute the wave propagation) and the other part dissipated into the material, which is lost from the electrical point of view. The material is submitted to a large number of disordered movements (thermal agitation) and constitutive forces (which bound the different particles and stabilize them). So it is obvious to consider, as Planck and Debye did, that, from the energetic point of view, the material is formed by a great number of mechanical oscillators (which are excited by the surrounding medium and, in the case of microwave heating, by the free charges and dipoles able to catch electrical energy).

Planck showed that the energy E_v of the harmonic oscillator at the frequency v was

$$E_v = hv/\{\exp(h_v/kT) - 1\}$$

where $h = 6.625 \ 10^{-34}$ J s is Planck's constant; $R = 8.3144$ J/°C/mole, the gas constant; $N = 6.025 \times 10^{23}$, Avogadro number, and $k = R/N = 1.380 \ 10^{-23}$ J/°C, Boltzmann's constant.

To compute the spectral density u_v of the infrared radiation, it is necessary to take into account the number of oscillators in a bandwidth dv around v. One gets:

$$u_v = du/dv = 8\pi hv^3/c^3 \ 1/\{\exp(hv/kT) - 1\}$$

The curve shows (Figure 4.2) that the radiation is located in the infrared spectrum. The maximum obeys Wien's law:

$$\lambda_m T = 2885 \ \mu m \ °T$$

It should be noted that the great number of shocks broaden the emitted radiation. The spectrum of this emission is much broader than the narrow line of the magnetron (Figure 4.2). Furthermore, the emission is at a frequency much higher than the magnetron frequency. The ratio of infrared to microwave wavelength is approximately 10 cm/10μm which is 10^4, which means that there are ten thousands shocks for a single microwave period. This gives enough time to get thermal equilibrium. This infrared radiation is emitted by the surface of any kind of material, and particularly by those which are heated in a microwave oven (which will heat the oven) and by the susceptors. To compute specific heat, Debye applies the concept of harmonic oscillators and correlates them by elastic waves. He adjusts the number of oscillators to the number of degrees of

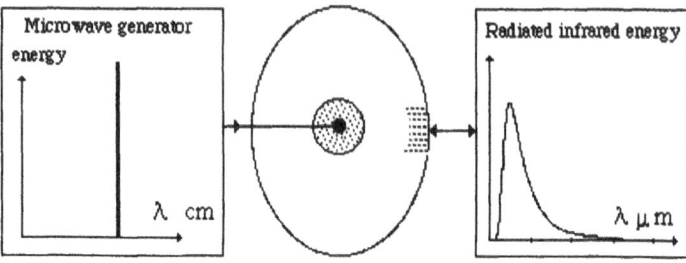

Figure 4.2 Energy transformation from microwave to infrared.

freedom which is $3N$ for a mole. This argument gives the energy level stored by thermal agitation. The specific heat C is its derivative versus the temperature T. The result is closely related to the Dulong and Petit formula at high temperature:

$$C = 3R \text{ J/mole}$$

This stored energy (which is obviously impossible to convert directly to microwave) may be modified in case of change of state or chemical reactions.

As an example, let us take the case of a microwave generator supplying 1 kW to 1 litre of water at 300 K. Each molecule catches

$$18/N = 6 \times 10^{-23} \text{ J/s} \cong 40 \text{ photon/s with } \nu = 2.45 \text{ GHz}$$
$$h\nu_{2.45 \text{ GHz}} = 1.6 \times 10^{-24} \text{ J}$$

which slightly increases the mean value of the energy:

$$kT_{300} = 4 \times 10^{-21} \text{ J}$$

4.5 Energy conservation and diffusion

The microwave energy dissipated in the material increases the temperature and diffuses. Let $\mathbf{J}$ be the thermal current of diffusion, ρ the density of the material, C the specific heat and P the volumic density of microwave absorbed energy, the energy conservation is:

$$\partial.\mathbf{J} + \rho \, C \, \partial T/\partial t = P$$

The thermal diffusion takes the form:

$$\mathbf{J} = -\kappa \, \partial \, T$$

where κ is the thermal conductivity. The diffusion equation for the temperature is

$$\Delta T - \rho \, C/\kappa \, \partial T/\partial t = -1/\kappa \, P$$

completed by the boundary conditions which express the energy exchanges through the surface.

In conventional heating, the boundary conditions express the way the material is heated (or cooled). It is usual to classify these ways into three different physical aspects which are conduction, convection, radiation (infrared). If matter is exchanged between the material and its surroundings (drying), the boundary conditions include the energy transfer linked with the matter transfer.

In conventional heating, the energy exchange of a material with its environment is restricted to its surface. The microwave internal heating should be added to the conventional one which still exists.

4.6 Microwave heating

The main features of microwave heating could be summarized as follows:

1. Microwave energy is not homogeneously absorbed in food products. But, according to the wave equation, energy decreases when propagating into the volume. This decrease it not far from exponential:

$$\exp(-x/d)$$

where x is the distance, measured from the surface of the material, where the microwave power has to estimated and d is the so-called skin depth. The skin depth d varies with the material and the frequency. Table 4.3 gives some values of d (cm), corresponding to the frequency of microwave ovens (2450 MHz).

For instance, in water ($\varepsilon_r' = 80$, tn $\delta = 0.30$), d is equal to 0.7 cm. That means that at a depth of 1.5 cm, the microwave power may be neglected.

2. Food products are not homogeneous either from the thermal point of view or from the electrical one. Some products may catch microwave

Table 4.3 Skin depth (cm) as a function of tn δ and ε_r'

tn δ/ε_r'	2	5	40	60	80
0.001	1400	880	308	250	216
0.05	28	17.6	6.1	5	4.3
0.15	9.2	5.8	2	1.7	1.4
0.30	4.6	2.9	1	0.8	0.7

energy quite well and others not. If the thermal conductivity is low, which is usually the case for foods and if the microwave power density is high, the result will be an out of equilibrium process. The temperature of hot points may reach a high level and produce particular effects (fast drying, explosions by vapour pressure, etc).

3. In most of the microwave ovens, the oven is cold and the food product is heated. As a result, the product will heat the oven and its surface temperature will decrease. In microwave heating, the internal temperature is usually higher than the surface one. The transformation produced by high temperature (like the non enzymatic, browning etc.) will not appear.

4. Microwave energy propagates towards the most convenient areas. For instance in thawing, in the case of hot and cold points, microwave energy flows into the hot points.

In conventional heating, the natural evolution is to tend towards equilibrium. This is not the case in microwave.

4.6.1 *Microwave ovens*

The two main parts of a microwave oven are the magnetron and the loaded ovens. These two parts are strongly coupled and interact with each other (Figure 4.3).

The electrical energy supplied by the DC voltage is converted into mechanical energy into the cloud of rotating electrons of the magnetron and radiated into the oven in electromagnetic energy. The oven acts as a resonator with a lot of proper modes, some of which are excited by the coupling hole. The load catches the radiated energy and converts into heat.

The main characteristics of the oven are the fields repartition which produces the temperature distribution and the matching of the load to the

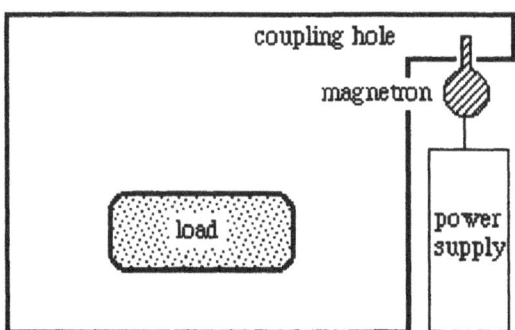

Figure 4.3 Diagram of a microwave oven.

magnetron which is measured by the efficiency. The field distribution depends on the load, the oven and the coupling hole. Stationary waves inside the oven and the load are impossible to avoid. To increase the homogeneity in mean value, manufacturers use stirrers or rotating plates. The packaging material which is located between the generator and the load could be used to get a desired temperature profile. It could convert part of the electromagnetic energy into heat or reflect it according to the need.

The coupling between the magnetron and the loaded oven is described by the Riecke's diagram (Figure 4.4). This diagram gives the power emitted by the generator and the frequency of oscillation of the magnetron (supposed to be evenly supplied) versus the impedance of the load measured on its antenna. The diagram proves that the magnetron may change its power in the ratio of 1 to 2 and its frequency in a range ±5 MHz.

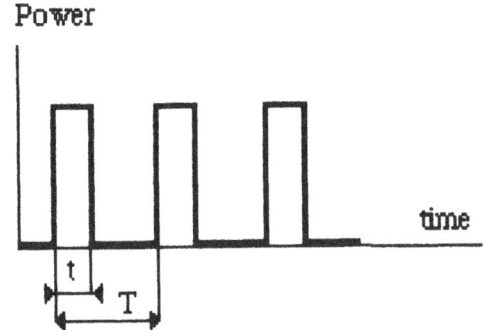

Scheme 4.5 Power delivered by the magnetron versus time.

The packaging may also play an important part in this content. It could match the food product so that the magnetron will deliver its maximum power and so that this extra power will be converted into heat in a desired area. Because the packaging material is an industrial product, with a low dispersion of its characteristics, its use might reduce the dispersion of food products and produce a more stable load.[4]

The power supply also plays an important role. For economic reasons, the DC voltage is not uniform. Scheme 4.5 shows that the mean power delivered by the generator is a function of the ratio t/T. The mean value is not determined by the generator but by the load and not on an electrical basis but on a thermal one.

A large load integrates these variations easily, but a weak one cannot. This gives a non linear result especially when the ratio is small.

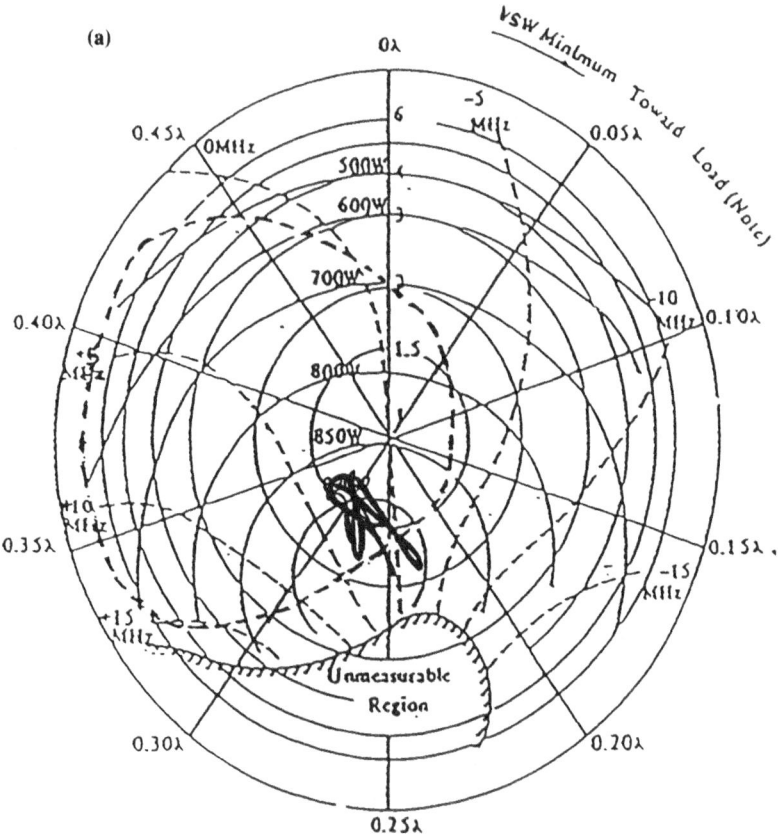

Figure 4.4 Riecke's diagram for food products with (a) plastic packaging and (b) metallic packaging in a microwave oven. Initial temperature of the food is 18°C. A light mismatching is observed with the metallic package.

4.7   Thermal and microwave ovens and the package

A thermal oven heats mainly by convection (and by infrared radiation in the case of a grill). In a convection process, the temperature of a load to be heated reaches the temperature of the oven but cannot overtake it. The heating is defined by the oven temperature and all the materials become hot whatever their nature or shape. The load heats from its surface, and inside temperature slowly increases by thermal conduction to reach an equilibrium state which is the oven temperature. This process may be counterbalanced in case of food products rich in water. The drying, which is a high energy consumer process, limits the surface temperature to some degrees below 100°C. When the drying process decreases, the surface temperature may increase at a high level and produce chemical reactions, Maillard reactions in bread and meat for instance.

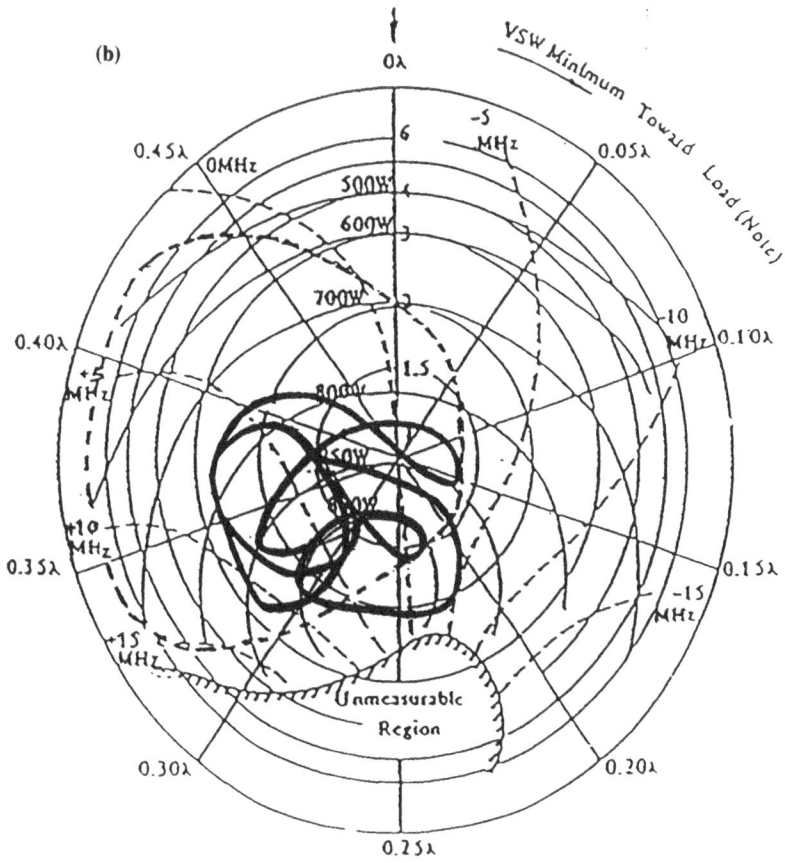

(b)

Infrared heating has properties comparable to conduction heating because the energy is caught on the surface and to microwave heating because it is a radiated energy characterized more by its power than by its temperature. The properties of a microwave oven are similar to the properties of the grill since it works with radiated energy. As a consequence, it is characterized by its power (and not by a temperature) and there is no trend towards an equilibrium, except at high temperature, when infrared emission by the surface of the product balances the microwave reception. The microwave oven is also quite different from an IR oven because the wavelength is largely greater, and the penetration is deeper. Moreover, energy does not follow the laws of geometrical optics, as it does in infrared, and there are points with high fields and points with low fields. There is another important difference which is due to the packaging and food products. The electrical properties of these products vary over a very large range. Some of them are able to catch microwave energy quite well whereas others are transparent or act as mirrors. This property induces some strange results since it may happen, for instance, that a material heats inside and not on its surface. In a microwave oven, the walls and the air are normally cold, which means that they will be heated by the oven itself. The surface temperature of this product will decrease. This result will have important consequences on the behaviour of the product, such as the acceleration of drying and the limitation of browning. With a clever use of its electrical and thermal properties, the packaging could correct these results and convert undesired heterogeneities into suitable ones. It could form a thermal oven from a microwave one or on the contrary increase these heterogeneities.

4.7.1   Packaging materials

Materials used for food packaging, in view of microwavability, should possess adapted thermal and electrical properties. Their thermal properties are quite important because they may be heated by the food product and some hot points may happen to appear and damage the package, leading to a large number of consequences, mainly diffusion. Moreover some combined ovens add grill or pulsed air which increase the surface temperature and may contribute to the damage. Suitable materials may be classified into three groups, those which are transparent to microwave, absorbers or reflectors.

4.7.1.1   *Transparent materials.*   Transparent materials like dielectric ones: paper, glass, plastics, etc., which cover the food, give rise to a weak perturbation because their thickness is small as compared to the wavelength. The electromagnetic discontinuity between the food and the surrounding air is very high because the permittivity of the food is high and not greatly modified by the package. Usually, transparent materials are also thermal insulating materials. They will have a thermal behaviour even if they lack an electrical action. For instance, a closed packaging will stop water evaporation, increase the pressure and modify the thermal gradients in the food.

4.7.1.2   *Absorbers and susceptors.*   These materials are able to catch a part of the microwave energy and transform it into heat energy. Some kinds of oxides, glass, ceramics, earthenware are able to catch microwave energy and so play an active part in the process. They may contribute to the loading of the oven and lead to a modification of the temperature profile: their ability to catch microwave energy is indicated by their coefficient $tn \ \delta$. Browning plates are made of such materials.

Susceptors are rather new components which are made of thin layers of conducting material (for instance 40 Å of aluminium) spread onto a dielectric substrate. The absorption of microwave is similar to a conduction process but, because the layer is very thin, the resistance will be high and

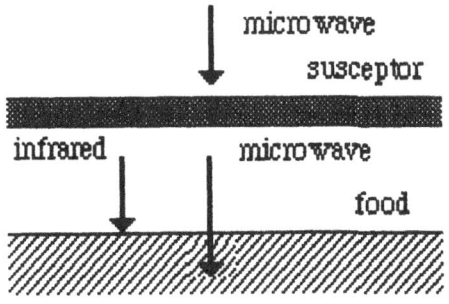

Scheme 4.6 Effects of a susceptor.

the absorbed power significant. Due to their small thickness, they are not able to keep the thermal energy which is radiated in the infrared spectrum.

Scheme 4.6 summarizes the action of susceptors. Microwave energy is partially radiated into the food to heat normally and partially transformed into infrared emission and dissipated by the surface of the product to cause browning.

Susceptors also have some disadvantages, mainly because their substrate heats and it is necessary to avoid an upper limit. It is often better to use lossy ceramics which are similar to browning plates without any problem of upper limit of temperature. The casting of ceramics is very easy and it is possible to shape them to get a desired thermal profile in the food. Results given in Figure 4.5 show that ceramics can achieve very high temperatures as compared to susceptors.[5]

Ferrites are another class of lossy materials. The way they dissipate microwave energy is different from the way explained for dielectrics. In ferrites, the magnetic part of the field is active, it rotates spins which are very small magnets. Since the magnetic field of the wave is high near a metallic surface, it is convenient to locate ferrites on a metal.

4.7.1.3 *Reflectors.* Metals are usually good reflectors which means that the main part of the energy of a wave incident on a metallic surface is reflected or scattered by it. It is necessary to understand the meaning of the main part and of the scattering. Even in a good conducting metal (silver, copper, aluminium, etc.) there is a penetration depth.

The current density I inside the metal is given by

$$I(d) = I_0 \exp(-d/\delta)$$

when the thickness of the metal is supposed large I_0 is the current density on the surface and δ is the skin depth given by:

$$\delta = (2/\omega\mu\sigma)^{1/2}.$$

(a)

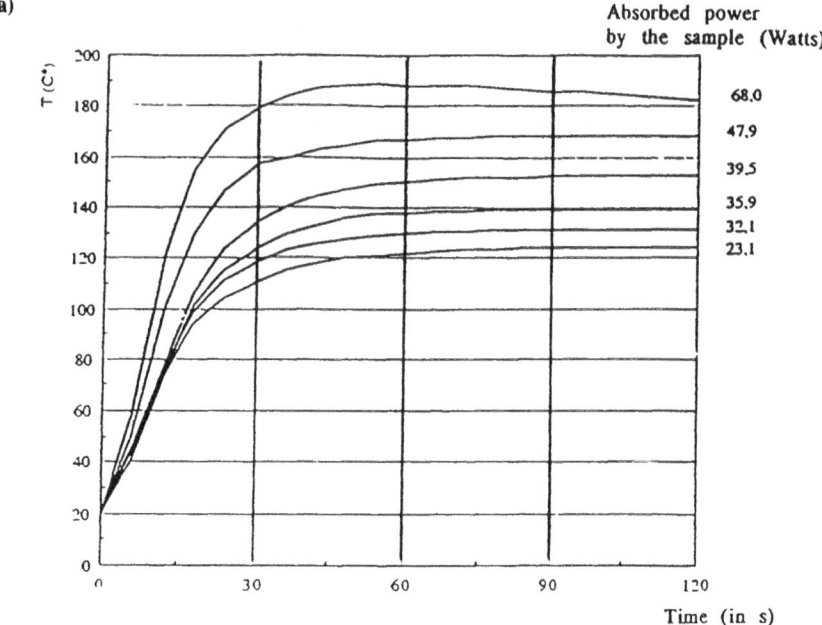

(b)

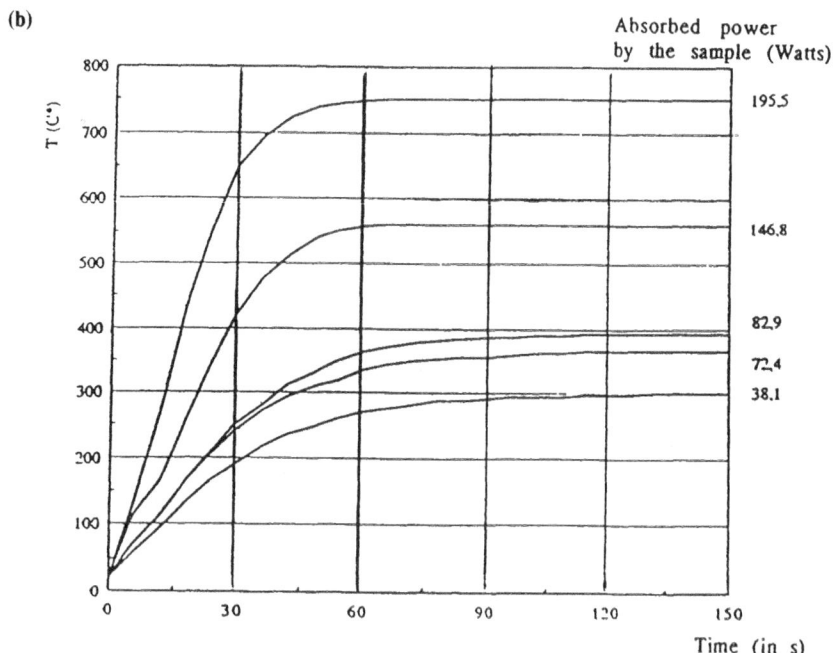

Figure 4.5 Temperature of absorber materials exposed to microwaves. Measured rise as function of the absorbed microwave power. (a) Susceptor; (b) lossy ceramic. The materials are put in a waveguide (a device used to lead or guide microwaves).

The order of magnitude of δ for good metals and at the frequency used is micrometres.

In a metallic packaging made with a thin layer of metal, these currents may have a high intensity and produce cracks and even melting. To avoid these phenomena it is better to use a not too thin layer, a tenth of 1 mm for instance.

Summarizing the properties of metallic packagings show that these materials are able to force the field distribution in a stronger way than the dielectric ones. The electric field distribution simulated by numerical methods is represented in Figure 4.6[6]. It gives a comparison of the microwave power density dissipated in the same food held in metallic and transparent packages. In a metallic box, the field distribution is mainly a function of the box itself because the boundary conditions are very strong and linked to the food. There are stationary waves in the box which are similar to the waves in the oven. The box is a kind of small microwave oven. The stirrer or rotating plate will not be very efficient. The result is different in a transparent box. The action of the wall of the oven is sensitive to the field distribution and the stirrer will be active.

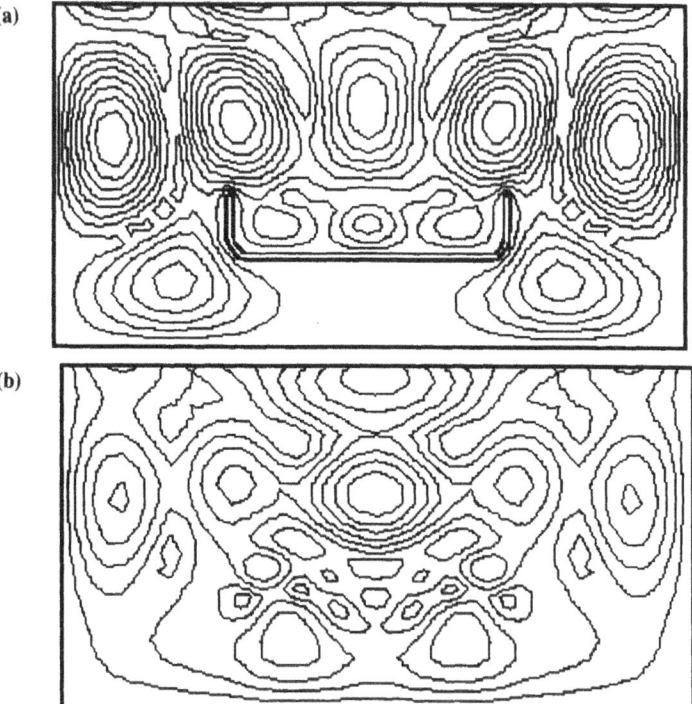

Figure 4.6 Simulation of the electric field distribution in an oven loaded with (a) a metallic package and (b) a transparent package. The food is located in the same position in each case.

A metallic foil may protect a part of a food which should be heated (for instance some sharp part which could burn, or the heating of a pizza as will be seen later). It may also let the energy penetrate only through a desired surface. For instance, if the energy penetrates only through the top of a metallic tin, it is possible to add a susceptor which simulates a classical grill to get a browning on the surface. Finally there is an interest in the physical properties of metals which have a high heat conductivity and which help to reach uniform heating.

Using metallic foil in an oven only provokes microwave perturbations but no influence upon the heating itself or the quality of the food. Usually a mismatching is observed between the oven and the magnetron and the first consequence is a decrease in the emitted power. If the oven is low loaded, the electric field intensity may be high and in that case a metallic part may facilitate the appearance of arcs. If a package has to be in metal it is better to have a large aperture when the oven is loaded to avoid high electric fields. The experimental results (see Figure 4.4) show the Riecke's diagram of the same food held in a transparent and in a metallic box. The mismatching is clearly observed but even in this case the matching is good enough to provide slightly longer heating.

4.7.2 Shape and size

To cover the influence of packaging shape and size, it is better to consider transparent and metallic materials separately.

4.7.2.1 Transparent packaging materials.
First the volume of the product to be heated should be determined. To evaluate this volume, two points are important, the penetration depth which is 1 to 2 cm and the equivalent water load which is 0.5 to 1 litre. It is then necessary to let some free space between the load and the walls of the oven. This free space will give some degrees of freedom to the stirrer/turning table in order to modify the field repartition and get a better homogeneity. Things are better when the load is much smaller than the oven. Moreover it is better if the load is not too cylindrical and not too thick. It is also necessary to avoid sharp parts or angles which can burn.

4.7.2.2 Metallic packaging materials.
It is necessary to keep in mind that power cannot propagate through metal and that it needs a rather large opening to penetrate. High cylindrical boxes with a small diameter should be avoided. Conversely, a flat and elliptical box, with an opening in the order of the wave length (12 cm) is correct (tin of 385 g). The metallic lids of transparent packagings can be used since the wave propagates through the box. They induce a stationary wave in the food between them and the lower wall of the oven and there are some hot points produced by

diffraction on the rim of the lid. These diffracted waves, which do not propagate, may be used to concentrate the energy by a process similar to lenses. Some commercial products are known, i.e. micromatch.

4.8 Monitored microwave heating

It is difficult to get a desirable heating in a microwave oven because the physical parameters of the foods are not mastered by the operator and because microwave heating is linked to rapidity while temperature uniformity is associated to long time. One way to solve this difficulty is to use active packagings. Microwave absorption of active packagings can correct direct food heating. Some examples are given to illustrate this.

4.8.1 *Pizza heater*

A pizza is composed of two main parts: tomato sauce (which catches microwave energy and heats) and the base which does not catch the energy and keeps cold. The result is an undesirable migration of the liquid into the base.

To avoid migration during heating it is advisable to use a metallic lid which prevents the waves heating the sauce and a susceptor which heats the base by thermal conduction (Figure 4.7).

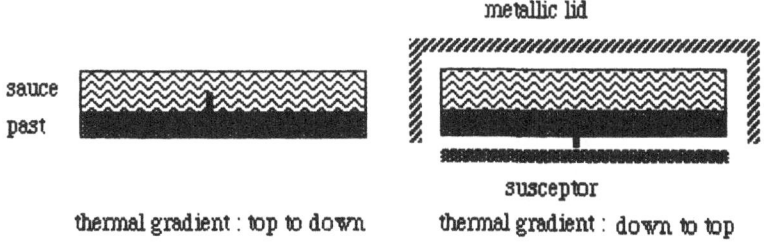

Figure 4.7 Different microwave heaters for a pizza.

4.8.2 *Use of a water bath*[7]

An extra water bath provides a supplementary load able to match the oven, to store the microwave energy and to diffuse it into the food. Convection currents into the water help reduce the hot points. Two other advantages may be gained: the water bath reduces the density power deposited into the food and limits the temperature at the boiling point (Scheme 4.7 and Figure 4.8).

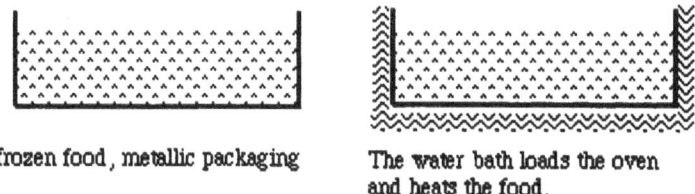

frozen food, metallic packaging

The water bath loads the oven
and heats the food.

Scheme 4.7 Effect of a water bath on the diffusion of energy and heating of a frozen food in a metal package.

(a)

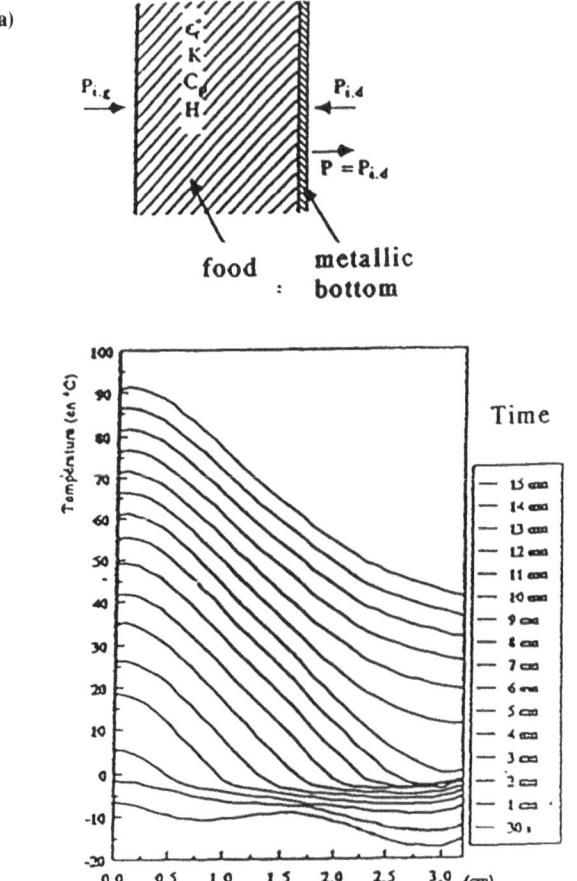

Figure 4.8 Simulated temperature in a frozen food (metallic package). (a) No water bath: thickness of food 3 cm; initial temperature −20°C; oven temperature 20°C; frequency 2.45 GHz; microwave power 5 W; treatment time 15 min.

4.8.3 Lossy ceramics

The food may be covered with a susceptor made of lossy ceramics able to convert one part of the microwave energy into infrared energy in order to achieve browning of the surface. This lid may also match the oven so that optimum power is delivered. Figure 4.9 gives some results of a mathematical modelling by the finite different method.[5]

4.9 Conclusion

Packagings and foods are parts of a whole. They have to be conceived of as complementary in order to get a desired temperature profile. It is clear that food alone cannot avoid the production of hot and cold points and

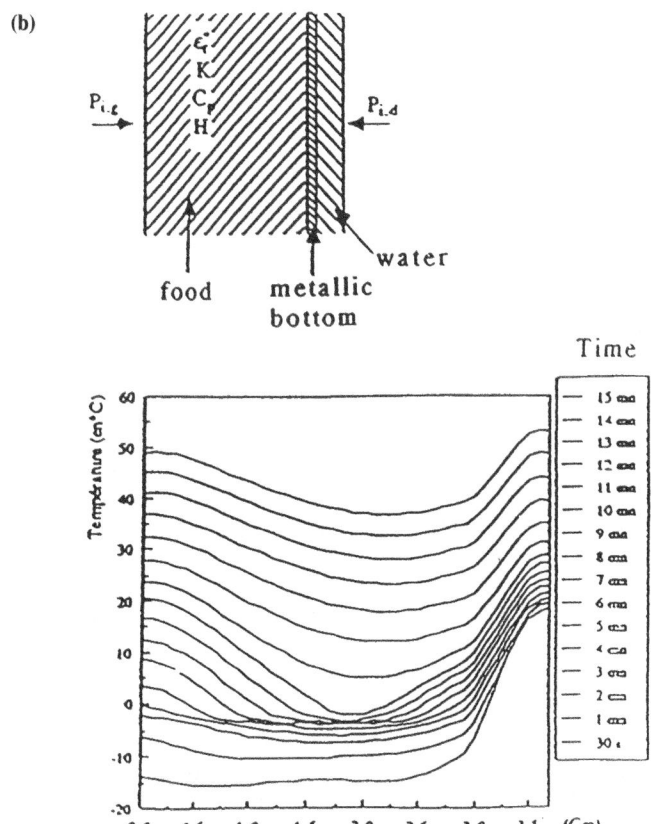

Figure 4.8 The water bath improves the uniformity of the thermal profile. (b) With water bath: initial temperature $-20°C$; oven temperature $20°C$; frequency 2.45 GHz; microwave power 2.5 W; treatment time 15 min; water depth 9.5 cm; initial temperature of water $20°C$.

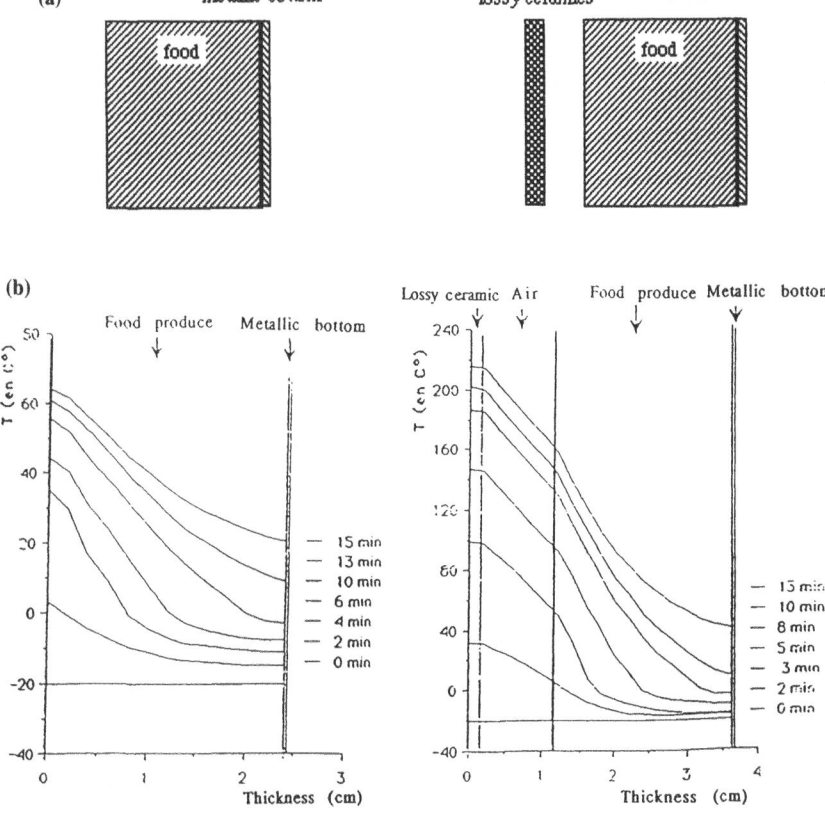

Figure 4.9 (a) Monodimensional diagram of packaged food. On the right, the lossy ceramic. (b) Simulated temperature in a frozen food with and without the lossy ceramic. The presence of the lossy ceramic produces important temperatures on the food surface and a grill effect.

that packaging with its properties of transparency, reflection, dissipation, storage, diffusion, may give a better result. The packaging material may play a role in different important ways:

- The packaging provides a volume and a shape to the load. It is better when the water equivalent load is not too far from half a litre to a litre, when the thickness is in the order of the penetration depth and when sharp areas are avoided.
- The packaging material may be either transparent, reflector or lossy. These properties enable the correction of electromagnetic and thermal properties of the food. The food is a given product, the physical properties of which are difficult to change, when, on the other hand, packagings are artificial products which may be conceived in a specific view. The packaging is a sandwich between the large diversity of the ovens and the large diversity of the food products. It is the coupling

material which can transform the microwave oven into a specific thermal one. People engaged in food packaging will find in the electromagnetic and thermal properties of the materials new research and development prospects.

4.10 List of symbols

E	Electrical field (V/m)
H	Magnetical field (A/m)
J	Current density (A/m^2)
P	Microwave power (watt)
λ	Wavelength (m)
f	Frequency (hertz)
ε_0	Permittivity of vacuum (farad/m)
$\varepsilon = \varepsilon' - j\,\varepsilon''$	Complex permittivity (farad/m)
$\varepsilon_r = \varepsilon/\varepsilon_0$	Relative permittivity
$tn\,\delta = \varepsilon''/\varepsilon'$	Loss tangent
μ_0	Permeability of vacuum (Henry/m)
σ	Electrical conductivity (ohm^{-1})
d	Skin depth (m)
$\mathbf{J}$	Thermal current of diffusion (J/m/s)
T	Temperature (°Kelvin or °C)
k	Thermal conductivity (W/m kelvin)
C	Specific constant (J/kg kelvin)
ρ	Volumetric density (kg/m^3)
h	Planck's constant = 6.625 J.s
R	Gas constant = 8.3144 J/°C/mole
N	Avogadro number = 6.025 10^{23}
K	Boltzmann's constant = 1.380 10^{-23} J/°C

References

1. Lefeuvre, S. and Audhuy, M. (1991) *Microondes et Hautes fréquences*, Cours du congrès international de nice. Chap. 1 and 2, Editions du CFE, Paris.
2. Kent, M. *Electrical and dielectrical properties of food materials*, Elsevier Applied Science, London.
3. Thuery, J. *Les microondes et leurs effets sur la matière*, Editions Lavoisier, Paris.
4. Lefeuvre, S. and Audhuy, M. (1990) *Le rôle thermique des emballages microondables* Forum IFEC: Emballage et Microonde-Paris, IFEC Promotion.
5. Bouirdene, A. (1992) *Les fours thermiques microonde*. Thesis, Toulouse.
6. Majdabadino, M. (1992) *Contribution au calcul des champs dans les fours microondes chargés*. Thesis, Toulouse.
7. Lefeuvre, S. and Bouirdene, A. (1987) *Décongélation par microonde des produits alimentaires en barquette aluminium*. Colloque SEE-Paris.

5 Effect of irradiation of polymeric packaging material on the formation of volatile compounds

Z. EL MAKHZOUMI

Abstract

Volatile compounds formed by electron beam irradiation of principal flexible food packaging materials (polyethylene, polyester terephthalate and oriented polypropylene) were examined and sampled by using a dynamic headspace technique and identified by combined gas chromatography and mass spectrometry. Hydrocarbons, ketones, and aromatic compounds were found.

A more detailed study on irradiated polypropylene indicates that the quantity of volatile compounds increased with the irradiation doses and the duration of contact with food simulants (aqueous, alcoholic or acid simulants).

The average yield of irradiation products trapped into polypropylene is estimated within a range of 1 to 50 ppm.

5.1 Introduction

The effects of electron beam irradiation on food products has been the subject of much research. This is especially concerned with the biological consequences[1] and the stability of the different food components.[2-4] In lipids, for example, and particularly unsaturated fatty acids, radiolytic decomposition is via a preferential break at the level of the carbonyl function of the double bond.[5] This decomposition induces the formation of some volatile compounds responsible for off-odours.[6]

The effect of electron beam radiation applied to plastic films depends on many parameters, among which are the nature of the packaging film (PE, PP, etc.), the number of coatings constituting the packaging, materials (mono, multilayered or composite), the formulation with respect to additives, the conditions of treatment (temperature, oxygen content), the dose used for treatment and the dose actually absorbed, as well as the contact with food. The complexity of these parameters is very likely the cause of the number and diversity of the results obtained in the past.

Nowadays, electron beam irradiation is applied to plastic materials, either to ensure a microbial disinfection for aseptic packaging, or to improve some of their mechanical properties. At weak doses (<25 kGy), a modification in the tensile strength and elongation at break is noticed.[7] It was even observed[8] that polypropylene shows some cross-linking. At medium doses (25 kGy), most polymers undergo little damage.[9] This damage can be considerably reduced by the addition of aromatic additives.[10] At strong doses (50 kGy), mechanical properties of polymers can be improved by cross-linking.[11] Electron beam irradiation has a minor impact on the permeability of polymers,[12] but, may lead to the formation of gas and volatile compounds.[12,13] Thus, knowledge of inertia between the ionised plastic packaging and the food product is essential. The determination of migration gives an indication of this inertia.

In the field of packaging in general, and that of food products in particular, the term *migration* means the transfer of substances stemming from the packaging material to the food product under the effect of physico-chemical factors of the environment.[14] At the doses generally applied (3 to 25 kGy), the migration in water increases with the dose.[15] For the heptane extractable fraction, the increase for a dose of 10 kGy was as follows: 70% for polypropylene, 35% for ethylene vinyl acetate and 20 to 30% for polyethylene. Irradiation has a minor impact on the polyester terephthalate.[16]

Electron beam irradiation of some food packaging materials produces, under certain conditions, the formation of volatile compounds.[17] These products are formed either by the degradation of the matrix or by the degradation of additives[18] (antioxidants, plasticisers, etc.). In this case, the most important necessity for food products is the absence of real or perceived contamination by migration of degradation products.

The general purpose of this work is to determine the effects of electron beam irradiation on food packaging materials (alone or in contact with food simulants), as concerns the production of volatile compounds.

5.2 Experimental

5.2.1 Materials

5.2.1.1 *Films.* The polymer films used in the tests are those often found in contact with foods:

- PET film of 25 μm thickness and without added organic antioxidants was supplied by Rhone Poulenc Films (Saint Maurice de Beynost, France).
- Low density polyethylene (LDPE) resins of 17 μm thickness was supplied by Grace S.A. (Epernon, France).
- A coextruded biaxially oriented polypropylene (OPP) of 30 μm thick-

ness with sealable outer layers was supplied by Mobil Plastics Europe (Virton, Belgium).

- Multilayer plastic films (three layers) of 70 µm thickness; only the core layer contains, in variable proportions, organic additives (antioxidants and plasticiser).

5.2.1.2 Simulants. The behaviour of the packaging film in contact with liquid simulants at 19°C for 10 days was studied after irradiation. Three food simulants were used: double distilled water, 15% (v/v) ethanol and 3% (w/v) acetic acid.

5.2.2 *Methods*

5.2.2.1 Irradiation. Irradiation was carried out at CARIC (Irradiation Center, Orsay, France), using a linear electron beam accelerator (CGR meV Equipement) with an irradiation energy of 10 meV, a power of 10 kW and a frequency of 250 Hz.

The films were irradiated in air at room temperature. The treatment was a single phase irradiation. Polyvinyl chloride films were used for absolute calibration of the electron source.

5.2.2.2 Analysis. Irradiated film (40 cm^2) were cut into small pieces and placed for desorption in the oven of a combined DCI-GC (Delsi Instrument, Argenteuil, France). Volatile compounds were trapped on Tenax GC (Delsi Instrument) by using a dynamic headspace system (Delsi Instrument) and helium at a flow rate of 25 mL/min. The Tenax GC trap was cooled to −40°C. The trapped volatiles were desorbed at 260°C and swept into a gas chromatograph (DELSI-NERMAG Model DI-700) equipped with a capillary column (RSL-150, 30 m, 0.32 mm i.d., Alltech). An initial temperature of 40°C for 30 sec was used, followed by heating at a rate of 10°C/min to 220°C. Mass spectra were recorded in the electron impact mode at an ionisation voltage of 70 eV. The volatile compounds were identified from their retention times in GC analysis, using the NIH/EPA and mass spectra of authentic compounds.

5.3 Results and discussion

Electron beam irradiation of PET, PE and OPP induces the formation of volatile compounds. Two types were identified: products formed by radiation reaction (hydrocarbons, aromatics) and compounds formed by oxidation reaction (ketones, aldehydes, etc.). Quantitative analysis and the kinetics of formation of the products which could alter the organoleptic quality of food were studied on OPP. Two important aspects of ionisation were studied: the effects of the film composition on the production of aromatic compounds on the one hand, and the behaviour of the packaging film in contact with food simulants on the other. The liquid simulants were not studied. The products resulting from the degradation of irradiated polymers, the oxidation products and the kinetics of degradation are described as follows:

5.3.1 Degradation products

5.3.1.1 *PET.*   Irradiation has no marked effect on a PET film at 5 kGy. Only one compound (acetone) appears in a relatively low amount (see Figure 5.1 and Table 5.1). This moderate reactivity results from the absence of organic additives especially antioxidants, and the presence of stabilising phenyl groups in the matrix.[10,16] If the quantitative analysis

Table 5.1 Volatile compounds released from polyester films ionised at 5 kGy

Peak number*	Compounds
1	Acetaldehyde
2	Acetone
3	2-Butanone
4	Acetic acid

*Peak number correlates with the number in Figure 5.1.

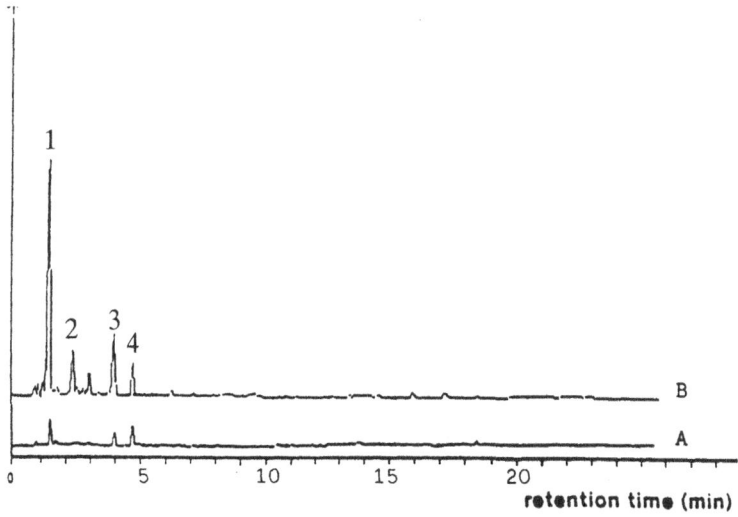

Figure 5.1 Reconstructed total ion current chromatogram of volatiles from polyester. A: unirradiated; B: irradiated at 5 kGy. Compound identification shown in Table 5.1.

shows that the acetaldehyde is present in a small amount which does not alter the flavour of foods, then the PET film can be a good candidate for irradiation at low doses (5 kGy), of pre-packed food products.

5.3.1.2 *PE.* Electron beam irradiation at 5 kGy of polyethylene induces the appearance of many volatiles, of which twenty-two are identified (see Figure 5.2 and Table 5.2). Of these products 42% are linear hydrocarbons. These results probably originate from direct radiation effects such as the breakage of C–C bonds under radiation reactions accompanied by the formation of alkyl radicals followed by hydrogenation.[13]

5.3.1.3 *OPP.* Twenty four compounds were identified after irradiation of OPP (see Figure 5.3 and Table 5.3). Hydrocarbons and aromatic compounds can result from the effect of radiation. The aromatic compounds identified in polypropylene were: 1,3-di-*tert*-butylbenzene and 1,3-di-*tert*-butyl-2-hydroxy-benzene. The formation of these compounds is tightly linked to the presence of different additives (especially the antioxidants and plasticisers) in polymers.

The position of the *tert*-butyl groups (*meta* position) shows that these aromatics can be formed only from Irgafos-168 [Tri(1,3-di-*tert*-butyl-2-hydroxy-benzene) phosphite]. Indeed, only this molecule can give the aromatic nucleus with *tert*-butyl groups in meta position. In fact, it is

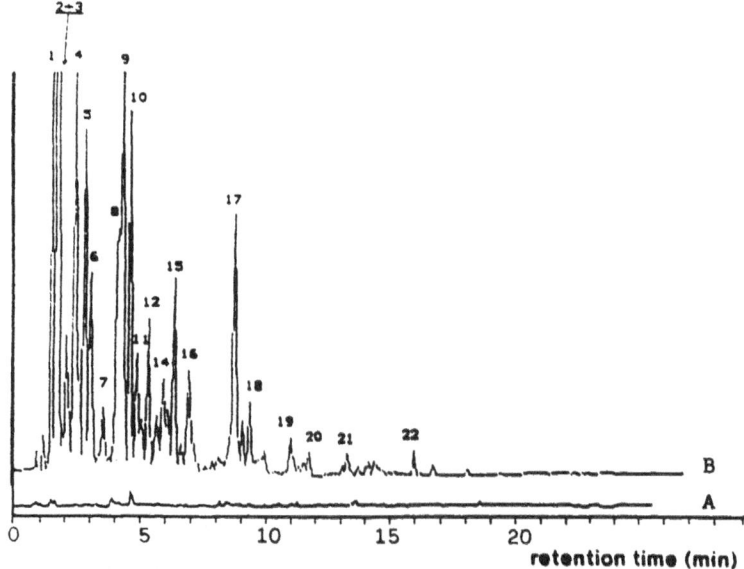

Figure 5.2 Reconstructed total ion current chromatogram of volatiles from polyethylene. A: unirradiated; B: irradiated at 5 kGy. Compound identification shown in Table 5.2.

Table 5.2 Volatile compounds released from polyethylene films ionised at 5 kGy

Peak number*	Compounds
1	Acetaldehyde
2	2-Butane
3	Propanol
4	Acetone
5	2-Methylbutane
6	*Tert*-Butanol
7	Formic acid propyl ester
8	2-Butanone
9	2-Methylpentane
10	Hexane
11	Acetic acid
12	2-Methyl-1,3-dioxolane
13	Formic acid butyl ester
14	3,3-Dimethylpentane
15	3-Methylhexane
16	Heptane
17	3,4-Dimethylhexane
18	Octane
19	3-Heptanone
20	Nonane
21	6-Methyl,5-heptene,2-on3
22	Nonanal

*Peak number correlates with the number in Figure
5.2.

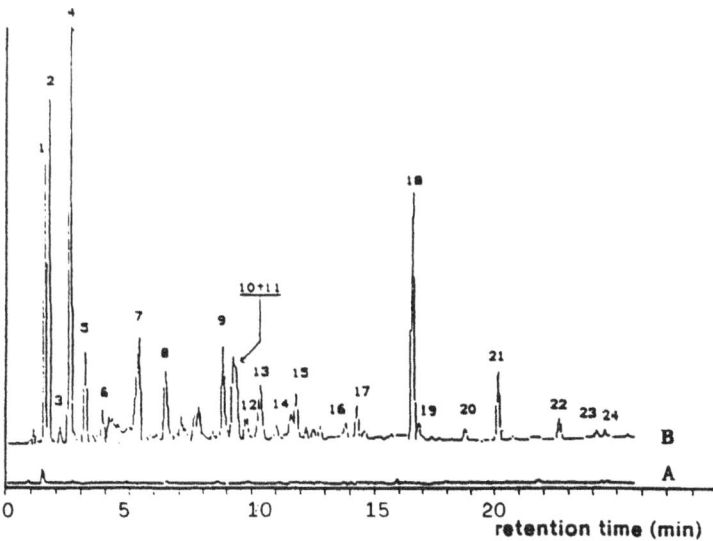

Figure 5.3 Reconstructed total ion current chromatogram of volatiles from polypropylene.
A: unirradiated; B: irradiated at 5 kGy. Compound identification shown in Table 5.3.

Table 5.3 Volatile compounds released from polypropylene films ionised at 5 kGy

Peak number*	Compounds
1	Acetaldehyde
2	2-Butane
3	Ethanol
4	Acetone
5	*Tert*-butanol
6	2-Methyl,2-propene
7	Acetic acid
8	4-Methyl,2-pentanone
9	2,4-Pentanedione
10	Hexanal
11	2,2-Dimethyl,propanoic acid
12	Octane
13	4-Hydroxy,4-methyl,2-pentanone
14	Cyclohexanol
15	5-Methylhexanal
16	6-Methyl,5-heptene,2-one
17	Octanal
18	Nonanal
19	Undecane
20	1-Dodecene
21	1,3-Di-*tert*-butylbenzene
22	Dodecanal
23	2,6-Di-*tert*-butyl-1,4-benzoquinone
24	1,3-Di-*tert*-butyl-2-hydroxybenzene

*Peak number correlates with the number in Figure 5.3.

difficult to accept that Irganox-1010 [tetrakis 3-(3',5'-di-*tert*-butyl-4-hydroxy phenyl)-propanoate] which possesses *tert*-butyl substituents at positions 2 and 6, gives 1,3-di-*tert*-butyl-2-hydroxy-benzene by migration of the *tert*-butyl from position 6 (*ortho*) to position 4 (*para*).

The formation of non aromatic compounds probably follows a radical oxidative process involving peroxyl radicals. These radicals attack the polymer and form hydroperoxides which will develop (under additive effects) to give either aldehydes or ketones or carboxylic acids.[19]

The identification of 2,6-di-*tert*-butyl-1,4-benzoquinone in the polymer one day after irradiation shows the start of an oxidation reaction. The initiator of this reaction is very likely Irganox-1010. Indeed, an investigation of the photochemical behaviour of Irganox-1010 shows that it behaves like a hydrogen donor towards free radicals.[20]

5.3.2 Kinetics of degradation

Taking into account the number and the nature of volatile compounds released by irradiated copolymer polypropylene films, it is important to follow their evolution especially for the compounds which could alter the

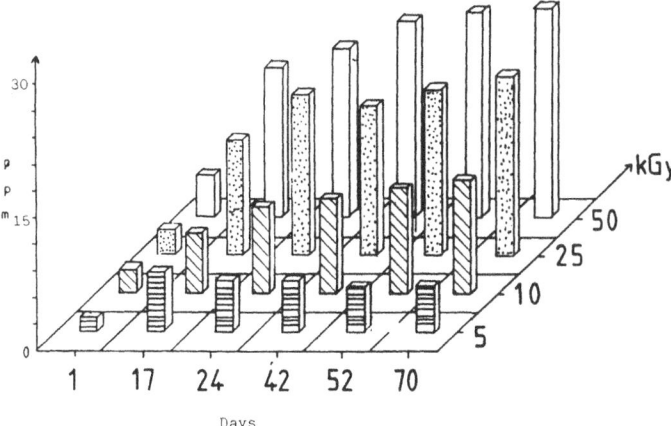

Figure 5.4 Kinetics of benzoquinone formation.

flavour and taste of foods or (and) induce possible toxicological problems. This is achieved in order to verify that the degradation process is continuing during the storage period.

The quantitative analysis of eight studied compounds during three months shows that in most cases, the quantity of volatile products increases with the irradiation dose (5, 10, 25 and 50 kGy) on the one hand, and with the time during the fifteen first days on the other. This increase is followed either by a stabilisation of benzoquinone (Figure 5.4) or a decrease in acetone concentration or that of 2,4-pentanedione. The particular case of 1,3-di-*tert*-butylbenzene may be noted (Figure 5.5). This product remains

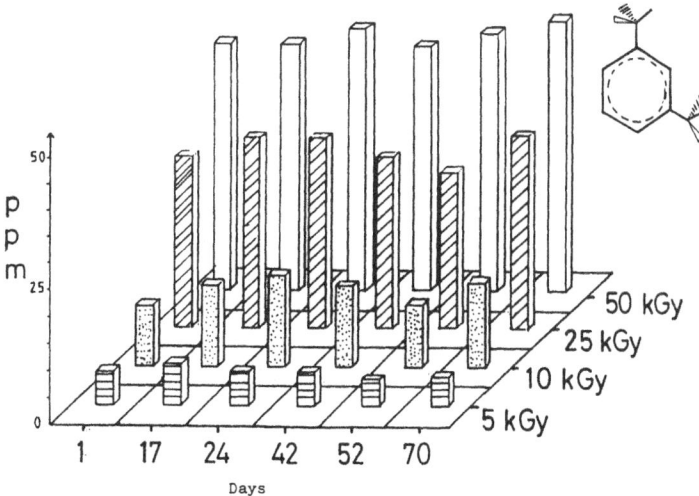

Figure 5.5 Kinetics of 1,3-di-tert-butyl-benzene formation.

almost constant three months after treatment whatever the irradiation dose used (50 ppm at 50 kGy).

A prospective application of the results of this kinetics study might be important for the users (or manufacturers) of polypropylene. Indeed, since some compounds remain trapped in the polymer, we can conceive of using them as irradiation detectors. This supposes that there is no modification of the film formulation, especially for additives.

5.3.3 Effects of the composition in additives

Table 5.4 shows that irradiation provokes the release of some undesirable aromatic compounds. Indeed simple rings like benzene and phenol were identified. This is dependent on the plasticiser as no such rings are formed when the film does not contain plasticisers [film I]. Indeed, irradiation induces the breaking of either an oxygen–carbon bond, or oxygen–phosphorus bond which causes the formation of very stable radicals leading to the formation of benzene and phenol. This stability is ensured

Table 5.4 Influence of film composition on the release of aromatic compounds

Films	Benzene	Phenol	DTBB*	DTBP**
A	×	×	–	–
B	–	–	–	–
C	×	×	–	–
D	–	×	–	–
E	–	–	–	–
F	–	×	–	–
G	–	×	–	–
H	×	×	–	×
I	–	–	×	×
J	–	×	–	–
K	–	×	–	×

×: presence; –: absence; *1,3-di-*tert*-butylbenzene; **2,4-di-*tert*-butyl-phenol.

by the electronic delocalisation of the aromatic rings.[19] The identification of 3-methylheptane and 2-ethylhexanol proves that the site most exposed to irradiation is the oxygen atom.

Table 5.4 shows that the modification of some parameters, especially the quantity of the antioxidants and of the plasticisers, permits the acquisition of packaging materials which tolerate ionisation without formation of undesirable compounds (B and E films).

5.3.4 Effects of irradiation on the packaging material in contact with food simulants

This study is informative on the resistance of the packaging films in direct contact with food simulants (aqueous, alcoholic or acid), to an irradiation treatment. The results show that the same compounds were identified as for the irradiation of the film alone. However, some quantitative differences were observed (Table 5.5). The contribution of an aqueous solution is generally the increase in the quantities of the products formed. This increase varies according to the nature of the product. Indeed, although the liquid simulant seems to have a very limited effect on 1,3-di-*tert*-butyl-benzene and 1,3-di-*tert*-butyl-2-hydroxy-benzene, this is not the case for the ketonic compounds which increase significantly. The acetone quantity is doubled when the film is in contact with an acid liquid simulant. This is probably due to the aqueous solution favouring the formation of radiolysis compounds. For food packaging, the contact surface is an important parameter. Attention is paid to the determination of the quantity corres-

Table 5.5 Quantity (ppm) of volatile compounds in OPP after irradiation by a dose of 50 kGy with and without contact with food simulant (averages of three samples)

Compounds	No simulant	Water	Ethanol	Acetic acid
Acetone	12.94	18.33	14.73	21.12
2-Pentanone	3.73	4.63	4.28	3.74
2,4-Pentanedione	11.84	16.72	13.24	14.74
1-Dodecene	7.16	5.78	6.03	3.68
1,3-DTBB[a]	41.70	45.40	45.30	46.30
2,6-DTB-1,4BQ[b]	7.45	8.95	6.13	7.98
1,3-DTB-2-HB[c]	71.40	66.30	73.40	68.50

[a]1,3-di-*tert*-butylbenzene
[b]2,6-di-*tert*-butyl-1,4-benzoquinone
[c]1,3-di-*tert*-butyl-2-hydroxy-benzene

Table 5.6 Quantity of volatile compounds in OPP after irradiation by a dose of 50 kGy, in the case of all volatiles migrating (averages of three samples)

Compounds	Q (ppm)	Q ($\mu g/dm^2$)
Acetone	12.94	3.49
2-Pentanone	3.73	1.01
2,4-Pentanedione	11.84	3.20
Dodecane	7.16	1.93
1,3-DTBB*	41.70	11.27
2,6-DTB-1-4-BQ*	7.45	2.01
1,3-DTB-2-HB*	71.41	19.29

*See Table 5.5

ponding to each identified compound for a defined contact surface (dm^2).

In the case of OPP, and in the most unfavourable conditions, where all volatile compounds migrate towards food product, the results are listed in Table 5.6. To know if these quantities are enough to make the flavour and taste of packaged food products deteriorate, some organoleptic analysis should be made. Indeed, the olfactory threshold of sensitivity to all these products is specific to the compound detected and the food product.

5.4 Conclusion

Volatile compounds are formed in PET, LDPE and OPP after irradiation. Twenty-two compounds were identified for LPDE, 40 for OPP and only acetone was identified for PET which can be a good candidate for irradiation of prepacked food products. The kinetics of degradation shows that some compounds remain trapped in the polymer and can be used as irradiation detectors. The presence of aromatic compounds promotes an investigation of their behaviour. Indeed, these compounds are able to migrate into a packed food product and affect its quality. To assess quality change, sensory analysis on possible migrants from packaging material is necessary to gather information about deterioration effects.

Acknowledgements

This work was carried out in the Laboratoire d'Agro-alimentaire et Conditionnement at the University of Reims Champagne-Ardenne and in the laboratories of ADRIAC (Reims). The author is indebted to Professor Pascat, General manager of ADRIAC and to Dr Bureau.

References

1. Saint-Lebe, L., Raffi, J. and Henon, Y. (1982) Rapport C.E.A. - R - 5162. Unpublished.
2. Hayashi, T. (1986) *Jap. Agric. Res. Quarterly.*, **19**, 295–301.
3. Simmons, J. (1987) *Radiat. Res. Rev.*, **111**, 374–377.
4. Leduc, A. (1986) Utilisation des rayonnements ionisants pour la conservation et la stérilisation des denrées alimentaires. Thesis Université of Paris XI.
5. Gruik, K. and Kiss, I. (1987) *Acta Alimentaria.*, **16**, 111–127.
6. Merrit, C. (1972) *Radiat. Res. Rev.*, **3**, 353–368.
7. Varsanyi, I. (1975) *Acta Alimentaria.*, **4**, 251–269.
8. Guimon, C. (1979) Polyoléfines greffées d'acide acrylique, *Colloque de la SPE et GFP*, Paris-France, pp. C-1–C13.
9. Skiens, W.E. (1980) *Radiat. Phys. Chem.*, **15**, 47–57.
10. Elias, P. (1979) *Chem. and Ind.*, **19**, 336–341.
11. Jacobs, P. (1981) *Polym. Plast. Technol. Eng.*, **17**, 69–82.
12. Rojas de Gante, C. and Pascat, B. (1990) *Pack. Technol. and Sci.*, **3**, 97–115.

13. Azuma, K., Hirata, T., Tsunoda, H., Ishitani, T. and Tanaka, Y. (1983) *Agric. Biol. Chem.*, **47**, 855–860.
14. Lox, F. and Pascat, B. (1989) in *L'Emballage des Denrées Alimentaires de Grande Consommation* Multon, J.L. and Bureau, G. eds, Tech et Doc, Lavoisier, Paris, pp. 57–75.
15. Lox, F. (1986) *Proceedings 5th IAPRI Conf.* Bristol, UK, pp. 1–8.
16. Killoran, J. (1972) *Radiat. Res. Rev.*, **3**, 369–388.
17. Azuma, K., Tsunoda, H., Hirata, T., Ishitani, T. and Tanaka, Y. (1984) *Agric. Biol. Chem.*, **48**, 2009–2015
18. Allen, D., Crowson, A. and Leathard, D. (1990) *Chem. and Ind.*, **30**, 16–17.
19. Jaworska, E., Kaluska, I., Strzelczak-Burlinska, G. and Michalik, J. (1991) *Radiat. Phys. Chem.* **37**, 285–290.
20. Kowal, J. (1984) *Poly. Deg. and Stab.* **7**, 175–188.

6 Package coating with hydrosorbent products and the shelf-life of cheeses*

M. MATHLOUTHI, J.P. de LEIRIS and A.M. SEUVRE

Abstract

The quality of ripening of a soft cheese depends on different factors amid which are the upholding of water activity close to saturation without condensation and the transfer of oxygen and CO_2 across the packaging material at a rate compatible with the respiration of the surface flora of the cheese. The conventional package for soft cheeses is a complex associating cellulose film and waxed paper. This hydrophilic wrapping material allows a shelf-life of only 6 weeks. Improvement of the conventional package was made possible by coating the cellulose film with an amount of 4 g/m^2 of modified starch included in the nitrocellulose varnish. The modified package was patented and gave the same performances as the cellulose film plus waxed paper.

To monitor the transfer of water vapour, oxygen and carbon dioxide, coated hydrophobic polymers seem advantageous. The presence in the coating of an optimized amount of high potency hydrosorbent product helps in both the upholding of a high level water activity (0.98) and the slowing down of water diffusion across the package.

The thickness of the hydrophobic synthetic polymer is reduced to minimize the waste weight and control the permeabilities to CO_2 and O_2. To improve the efficiency and optimize the organoleptic quality of the product a microperforation of the film is sometimes required. The mastering of the functionality of the packaging material is inspired from the active packaging approach.

6.1 Introduction

Since the publication of the book *Food packaging and Preservation, Theory and Practice*,[1] where we stressed the importance of water interactions in the packaged food for increasing its shelf-life, a new approach

*Paper given at the IFTEC Symposium Food Packaging Interactions and Packaging Disposability, The Hague, 15–18 November 1992.

called[2] active packaging has been made. We already noticed[1] that both the packaged food and the packaging material are sensitive to water vapour sorption. The modelling of water vapour transfer across the film was found[3] sufficient to account for the shelf-life of the packaged food because the transfer across the polymer seems to follow a slower kinetics than the equilibrium of relative humidity inside the package. Thus, when the water activity of the food product is close to 1 and the rate of water vapour transfer is low, there is a need to prevent the condensation of water at the surface of the product. Such a condensation may cause a necrosis of the flora or a microbial spoilage especially for high moisture foods.

Moisture in excess could be adsorbed by a desiccant or a modified packaging material. Thus, the relative humidity in the headspace is lowered to a level which protects the product from quality losses. Desiccants are frequently included in moisture-sensitive solid pharmaceutical products[4] to scavenge moisture and prevent them from moisture damage. The removal of headspace water was recently found[5] to be one of the most exciting new developments in flexible packaging technologies. Different techniques of moisture scavenging were proposed. One of these consists in two layers of regenerated cellulose films between which glycerol is sandwiched.[5] Another approach consists in using inpackage sachets to produce fixed relative humidities[6] with salts like potassium chloride (84–85%), sodium chloride (74–76%) or humectants like xylitol (78–79%) or sorbitol (72–74%). The Japanese Company Showa Denks was reported[2] to commercialise a 'Pichit film' composed of two sheets of polyvinyl alcohol (PVA) sealed end to end and filled with propyleneglycol which plays the role of water trap. The PVA film is permeable to water vapour and impermeable to propylene glycol. An important ΔA_W (0.9 for the product – 0 for the glycol) is the origin of surface dehydration and the 3 to 4 day increase in shelf-life of refrigerated fresh fish. Moreover, the desiccant film may be washed, dried and reused up to 10 times.[2] The same company released[7] moisture-controlling sheets called 'Red Keeper'. These sheets are made of a water absorbent high-polymer sandwiched between non-woven fabric and film for both food packaging and paper. The non-woven fabric absorbs meat juice without drying up the surface. It seems to be a solution to the problems posed by conventional meat packaging.[7] The increase in the rate absorbency of the liquids exuded by such products as meat, poultry and other humid products in individual packages is a subject of continuous interest. An evidence of that is the number of patents on the subject amid which is a recent US patent[8] on an absorbent pad with incorporated additives which provide superabsorbency, bactericidal characteristics or deodorization.

The adsorption of water in multilayer films proceeds from the same approach as active[2] or 'smart'[5] packaging. Moreover, it should be noticed that most of other active packaging technologies (oxygen scavengers, ethanol or CO_2 release) are more or less dependent on moisture. The control of water transfer may also be achieved by use of edible films.[9] It is especially the case for bilayer films consisting of stearic-palmitic acid on the one hand and hydroxypropyl methylcellulose on the other.[10] This film was used for the retardation of water activity equalization when placed between the constituents of a food with markedly different water activities such as tomato paste and crackers.[10] Composite edible films composed of beeswax and methylcellulose,[11] wheat gluten[12] or chocolate[13] prove to have good water vapour barrier properties.

6.2 Water activity and shelf-life

Whatever the barrier, the transfer of water vapour depends on the gradient of A_w on both sides of the barrier. Different models were proposed[14] to account for the moisture transport prediction in order to maintain a certain A_w compatible with the shelf-life duration of a packaged food. Solving of the proposed models requires the knowledge of water vapour sorption isotherm for the studied food. A simple approach of shelf-life prediction was based[15] on the fact that the flow of water vapour (dM) through a plastic package may be expressed by Fick's first law:

$$dM = QA \, (p_e - p_i)dt \qquad (6.1)$$

where Q is the rate of water vapour transfer across the packaging material, p_e and p_i being the partial pressures of water outside and inside the package, A the inner surface of the package and dt the time.

If instead of the mass of water vapour dM the ratio dW of moisture to dry matter $G(dW = dM/G)$ is used in equation (6.1), we obtain:

$$dW = \frac{QA}{G} \, (p_e - p_i)dt \qquad (6.2)$$

At constant temperature p_e and p_i may be replaced by A_{we} and A_{wi}. The water content dW is connected to A_{wi} or p_i through the relation $p_i = f(W)$ which is the equation of water vapour sorption isotherm. Replacing p_i by $f(W)$ and assuming that $p_e = P_a$ (atmospheric pressure), the integration of equation (6.2) leads to an expression of shelf-life:

$$t = \frac{G}{QA} \int_{W_i}^{W_{cr}} \frac{dW}{P_a - f(W)} \qquad (6.3)$$

The prediction of the shelf-life of the products requires that W_i and W_{cr}, the initial and critical moisture levels, are known as well as $f(W)$, the sorption isotherm. Several equations[16-21] were proposed to account for the adsorption of water by foods. They vary from the simple linear model[19] to the well-known BET[16] model which is based on the kinetics theory of gases and usually applied to determine the monolayer value for foods and biological products. However, the most valid representation of water vapour sorption isotherm is the fitting with experimental results as was recommended by Heiss.[21]

6.3   Water vapour sorption of cheeses

The sole knowledge of water content and water activity of the food is not sufficient to predict its shelf-life. It is also necessary to know its composition and the biochemical reactions developed during storage. Although all food products are sensitive to water transfer, some are more sensitive and should be treated as living systems. This is the case for fresh fruits and vegetables[22] and also fermented cheeses because of the requirements of respiration of the flora at the surface of the cheese.

The A_w in cheese is mainly controlled by low molecular weight constituents.[23] The concentration of NaCl and the degree of proteolysis are particularly important.[24] Another factor which was found[25] to play a major role in the ripening of Emmental cheese is the structure of water as influenced by the temperature of ripening and the breakdown of casein into small peptides and amino acids during hot cellar storage. In fact, ripening is not finished in the cellar. It continues in the packaging material during distribution and storage in the consumer's refrigerator.

The shelf-life and quality of a ripened soft cheese mainly depend on the quality of the packaging material used. A hydrophobic material impermeable to water vapour is favourable to an enhanced proteolysis which very rapidly leads to a flowing soft cheese unsuitable for commercialization and consumption. A hydrophilic material permeable to water vapour provokes an important weight loss and a drying detrimental to its acceptance by the consumer. The permeability to other gases, especially O_2 and CO_2 should be considered because of the respiration of the *Penicillium* flora at the surface of the cheese. All these considerations make the packaged soft cheese a complex system which develops a living flora at the surface, a controlled transformation (proteolysis, lipolysis) of the dry substance and an exchange of gas and vapour across the package (Figure 6.1).

The transfer of water vapour during the ripening of the cheese is of crucial importance as regards the weight loss, the quality of cheese and its shelf-life. All methods permitting monitoring of water vapour transfer through the wrapping material allow the control of conditions of ripening and the quality of the product. Amid these methods, the coating of films with hydrosorbent polymers which act as a reservoir and respond to the need of an A_w close to saturation without condensation seems to be one

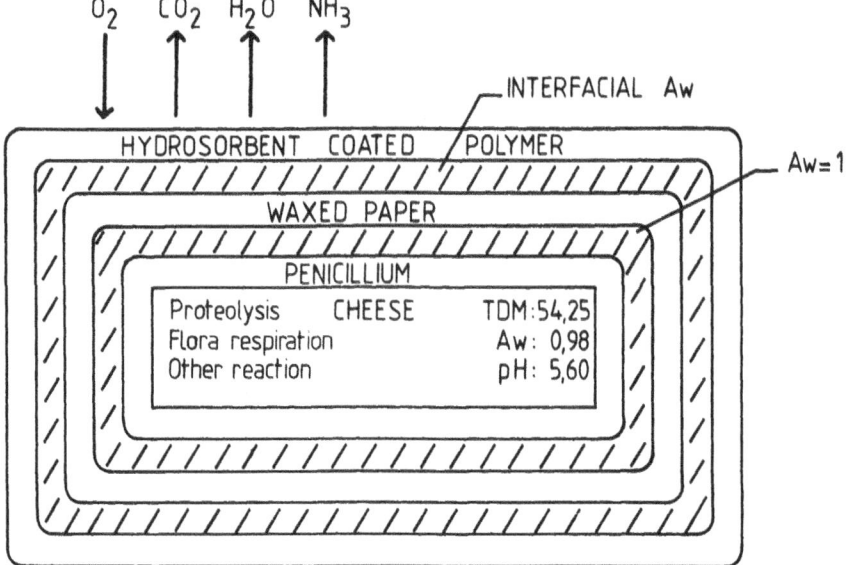

Figure 6.1 Schematic representation of gas exchange of a soft cheese with mould flora on the rind during ripening in a package.

of the most adapted solutions for soft cheeses and wrapping with gas-tight opaque films proves to be suitable for ripening of hard cheeses.

6.4 Coating of soft cheese packaging materials

Soft cheeses with flora on the rind are generally ripened in a few weeks (4–5). The cheese is packaged 10 days after the preparation of the curd. Its evolution in the packaging material consists essentially in the breakdown of proteins under the action of proteolytic enzymes from the fungal starter (*Penicillium caseicolum*). One of the most critical parameters during the ripening of a packaged soft cheese is water activity. The A_w of a Camembert cheese is about 0.98. It has to be maintained[24] as high as 0.98–0.99 in the rind. This is particularly difficult to achieve as cheese is generally transported, distributed and stored in ambiences with an average value of 60% RH without any particular control of relative humidity, although temperature is generally maintained around 6 to 7°C. The observation of the water vapour sorption isotherm of a Camembert cheese at 6°C shows the very high sensitivity (change in water content) of the cheese when A_w reaches the region of 0.97 to 0.99 (Figure 6.2).

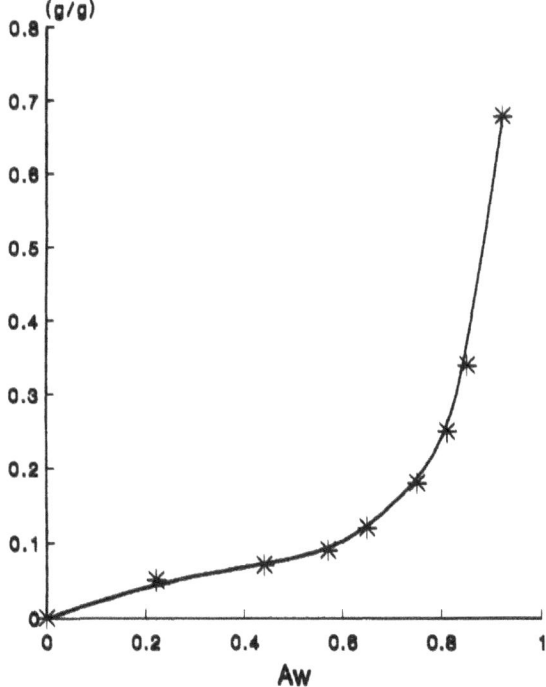

Figure 6.2 Water vapour sorption isotherm (6°C) of a Camembert cheese (g/g: g water/g dry matter of cheese).

6.4.1   *Conventional packaging materials*

The conventional wrapping of soft cheeses is mainly achieved with two simple materials: Kraft paper board (32 to 40 g/m²) waxed on both sides with 6 to 15 g/m² of paraffin and type MS or LMS regenerated cellulose film. These basic materials are in complexes like the following:

– two waxed papers associated with paste stripes
– cellulose films (MS, LMS, WS or WSZB) assembled to waxed paper by paste stripes.
– oriented polypropylene (OPP) 15 to 20 μm thickness, also pasted to waxed paper. OPP is generally microperforated to increase its water vapour permeability.

Other polymers were used such as HDPE (12–20 ξm), polyester terephthalate (PET) transparent or metallized, with a thickness varying from 6 to 12 ξm, pasted to waxed paper and microperforated and aluminium foil (7–15 ξm) pasted to waxed paper. However, the classical packaging material for Camembert remains the complex composed of a cellulose film (22 μm thickness) pasted to waxed paper. Analysis of the

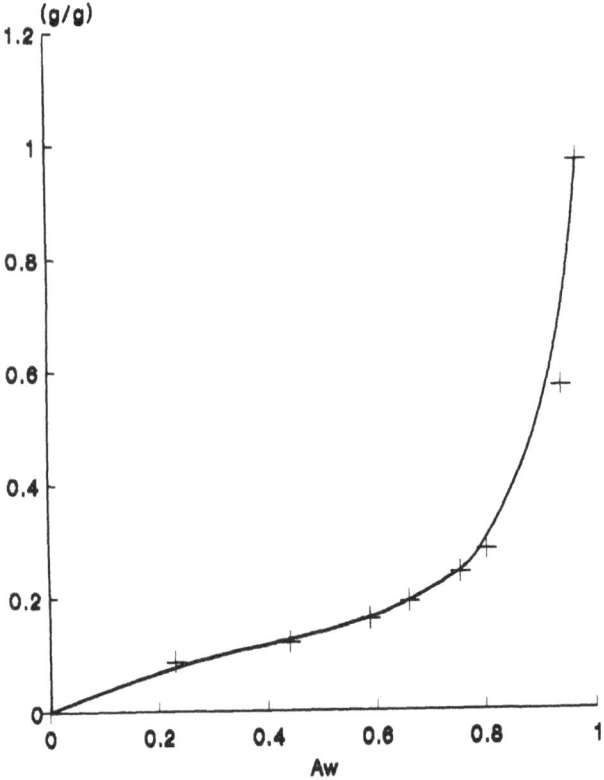

Figure 6.3 Water vapour sorption isotherm (22°C) of a cellulose film (g/g: g water/g dry matter).

behaviour of this material towards water vapour shows that the interference of waxed paper in water transfer and retention is negligible whereas the cellulose film, like cheese itself, shows a high sensitivity to A_w change in the region of 0.96–0.99 A_w (Figure 6.3).

At the surface of the rind of a soft cheese like Camembert, the respiration of the mould *Penicillium caseicolum* needs about 7 mg of O_2 per hour for a cheese of 100 g weight and 130 cm^2 surface stored at 15°C and 95% RH.[26] If the cheese is stored at 4°C, the requirement of oxygen for the respiration of the mould is estimated to be about 2300 cm^3/m^2/day. This estimation for a non packaged cheese could hardly be used to determine the permeability characteristics of the packaging material. Indeed the conditions are different after the packaging of the cheese, the intensity of the respiration changes during the ripening and the permeability of a flat film determined in the standard conditions is far from being representative of the same film folded and placed in a humid ambience around the cheese. The cellulose films generally used for the wrapping of soft cheeses swell

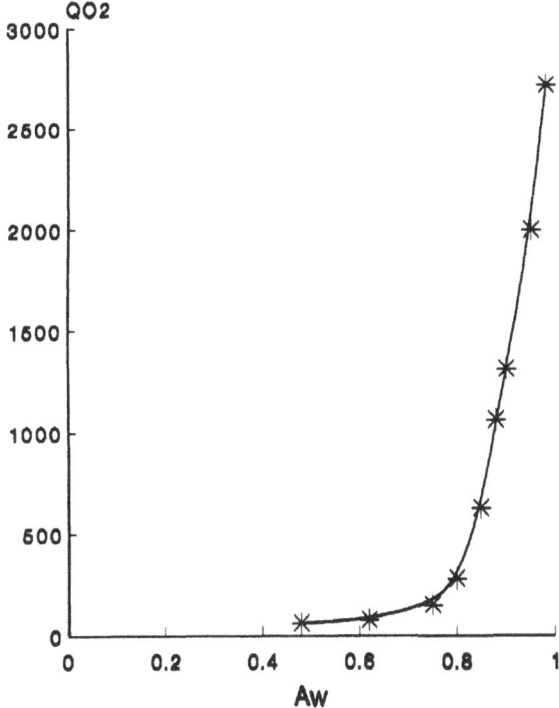

Figure 6.4 Permeability to oxygen (QO_2) in $cm^3/m^2/day$ for a cellulose film (330 LMS) 22 μm thick as a function of A_w.

when water activity is increased. Their permeability to oxygen (Figure 6.4) is increased and follows the same behaviour as water vapour sorption (see Figure 6.3). At an A_w of 0.98 in equilibrium with the cheese, the permeability to O_2 at 23°C for a complex composed of a 22 ξm thick LMS cellulose film pasted with waxed paper is 3050 $cm^3/m^2/day$. This is largely satisfactory for the respiration requirements of the flora, even though a lowering of the permeability may be provoked by a decrease in temperature to 6–7°C, the value generally used for storage. This complex proved to be particularly suitable for Camembert. However, the water vapour sorption isotherm of LMS cellulose film (330 LMS) shows a hysteresis phenomenon[27] and a variation of its permeability to water vapour depending on the direction of diffusion of moisture (from outside the film to inside or from the inner to the outer side) and this may cause in certain conditions of storage (low RH) a desiccation of the cheese. As the respiratory process is vital for a good development of flora and a good ripening of the cheese, it is of particular importance that both water vapour and oxygen transfer be optimized.

6.4.2 *Improvement of the conventional package by coating*

As may be observed in Figure 6.4, oxygen transfer across cellulose film is highly dependent on A_w. The water sorption behaviour of the conventional package also shows some variability (swelling, hysteresis, change in WVTR). An improvement of the cellulose film, which is at the origin of the variability, may be achieved by coating of the inner face of this film with a hydrophilic product acting as a reservoir of water and aiming to minimize the hysteresis behaviour of the water vapour transfer through the package. Moreover, the coating helps in preventing the condensation of water at the surface of the cheese which may be prejudicial for the quality (mould flora necrosis). The first attempts at improvement of the packaging of cheeses by coating the cellulose film go as far back as 1976, the date of study of this problem by Rhône-Poulenc Films.[28] The results were patented and the licence transferred to Charles Morin S.A. (Sarrebourg, France), a company which specialized in the conversion and printing of flexible films. The registered mark of the coated cellulose film reported in this work is Parpack-cello®. It consists of the coating of cellulose films with nitrocellulose containing particles of modified starch or caseinates used for their water holding capacity.

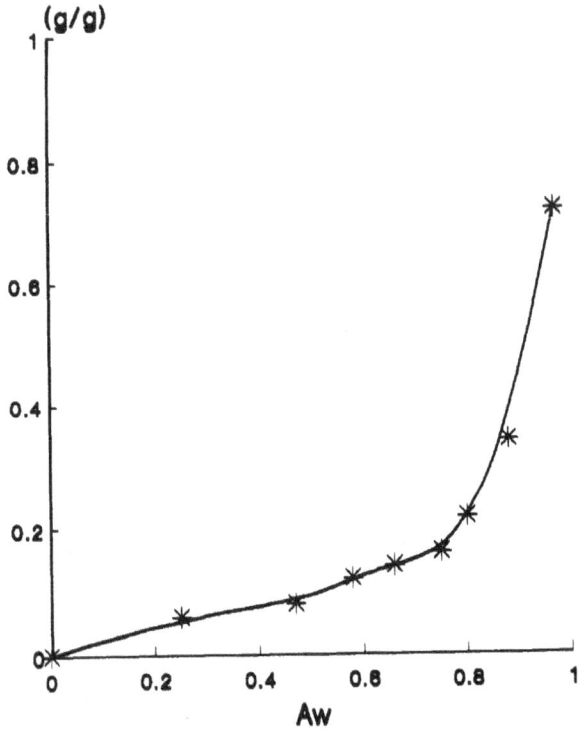

Figure 6.5 Water vapour sorption isotherm (23°C) of Parpack-Cello® (g/g: g water/g dry matter).

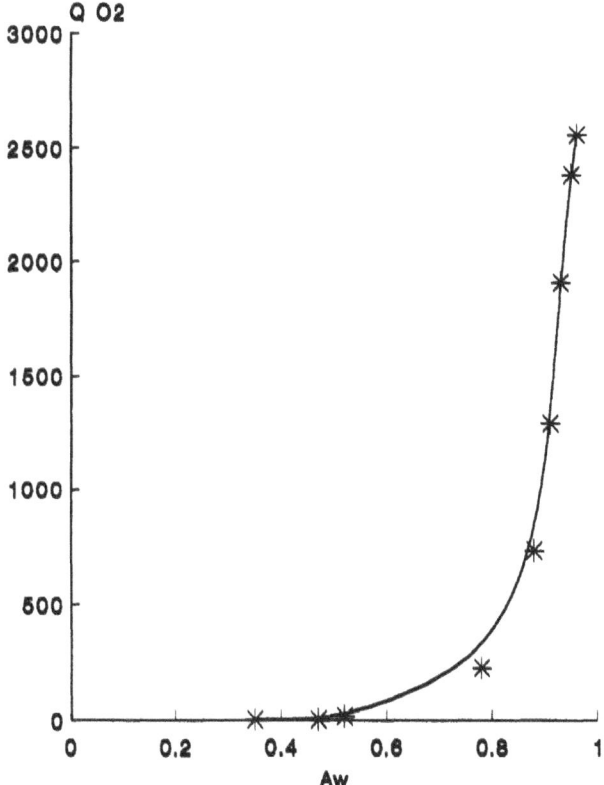

Figure 6.6 Permeability to oxygen (QO_2) in cm^3/m^2/day for Parpack-Cello® as a function of A_w.

The water vapour sorption isotherm (23°C) of Parpack-cello® shown in Figure 6.5 was found to be comparable to that of the non coated cellulose film. The permeability to oxygen of Parpack-cello® also increases rapidly with the increase in A_w (Figure 6.6) as was observed for cellulose. Indeed the quantity of the hydrophilic product used in the coating ($\cong$4 g/m^2) does not affect the equilibrium relative humidity. It only slightly affects the kinetics of water vapour transfer. The hydrophilic product plays its role of reservoir. The transfer of water vapour is slightly delayed, then it increases with the increase in water activity (Figure 6.7) like the non coated cellulose.

6.4.3 *Packaging of soft cheeses: comparison of conventional and coated packages*

To ensure a shelf-life of about 6 weeks for the soft cheese Camembert and an acceptable quality for the consumer, the proteolysis must not start rapidly and the transfer of oxygen required for the respiration of the flora

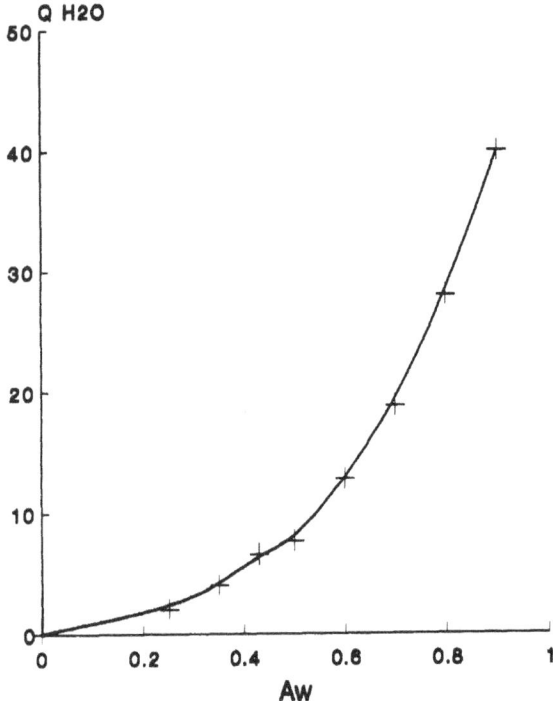

Figure 6.7 WVTR (QH_2O) in g/m^2/day for Parpack-Cello$^®$ as a function of A_w.

must be well adequate for the degree of ripening of the cheese. This is achieved when the moisture of the package and its permeability to O_2 as well as to water vapour gradually increase with the degree of ripening to respond to an increased activity of the flora (*Penicillium caseicolum*) which needs more oxygen and releases more CO_2 and water vapour. The coated package Parpack-cello$^®$ was found to show this behaviour (see Table 6.1) during a 36 days trial of ripening of a Camembert packed in it. It was especially noticed that the WVTR of Parpack-cello$^®$ is comparable to that of the conventional package composed of a cellulose film (330 LMS) pasted with waxed paper (Table 6.2). Comparison of the WVTR of these two films with one or two layers of waxed paper shows the essential role of the cellulose film in controlling the permeability to water vapour (Table 6.2).

A trial of ripening of Camembert cheese in an industrial cheese cellar was performed. The conditions of storage were 6°C and about 85% RH. Comparison was made between conventional and Parpack-cello$^®$ packages as well as a double layer of waxed paper. One of the criteria traditionally used to check the evolution of ripening is the weight loss. The results obtained with Parpack-cello$^®$ and the classical package (cellulose film + waxed paper) are shown in Figure 6.8. A weight loss of about 6% was

Table 6.1 Effect of ripening duration on moisture and $Q\ O_2$ of Parpack-Cello®

Time (days)	Moisture (%)	$Q\ O_2$ $(cm^3/m^2/24\ h)$
5	20	740
15	33	1607
36	45	2864

Table 6.2 Water vapour permeability across the experimental packaging materials used for Camembert ripening

Material	$Q\ H_2O$ $(g/m^2/24\ h)$
Paper (2 layers)	127
Paraffin coated paper	180
Parpack-cello[R]	16
Cellulose film 330/LMS + paraffin coated paper	15

found for both packages after 22 days of storage. This value is generally accepted by the cheesemakers as a criteria for a good ripening. Indeed, it is a rule of thumb that a weight loss higher than 7% provokes a drying of soft cheese and it may be prejudicial to the surface flora (risk of condensation) if the loss in weight is low. Sensory evaluation by a trained panel in the cheese factory of the Camembert cheeses ripened in different packages was made. The results are shown in Table 6.3. All packages were found satisfactory as concerns the degree of ripening after 42 days. The texture which is closely linked to the degree of ripening and which may be a

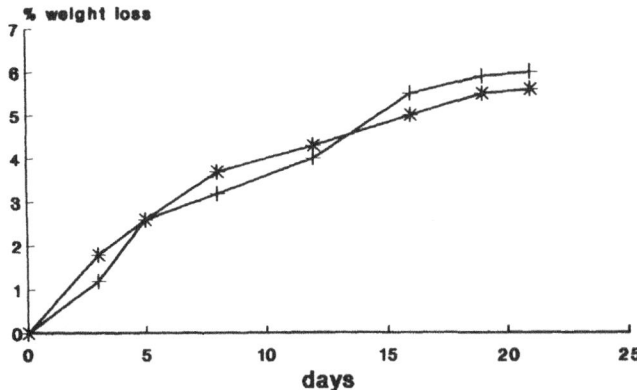

Figure 6.8 Percentage weight loss of a Camembert cheese as a function of ripening duration (+ = (cellulose + waxed paper); * = Parpack-cello®).

Table 6.3 Sensory evaluation of Camembert packaged in Parpack-cello® and control packages after 42 days of ripening

Sample	Observations			
	% Ripening	Texture	Taste	Odour
Control 1 (Cellulose LMS + paper)	100	Soft	Good	Pleasant
Control 2 (Cellulose film 320 WSZB)	100	Highly flowing	Fairly good	Slight ammonia
Control 3 (Double paper)	100	Firm	Fairly good	Pleasant
Sample 1 (Normandie) (Parpack-cello®)	100	Soft	Good	Slight NH_3
Sample 2 (Lorraine) (Parpack-cello®)	100	Slightly flowing	Fairly good	Slight NH_3

decisive factor in acceptance by the consumer was firm for the double waxed paper package, the new complex (Parpack-cello®) being acceptable. The taste and odour were acceptable for all cheeses with a slight odour of ammonia for samples with an excessive proteolysis.

The coating of a regenerated cellulose film to improve its water holding capacity did not change its characteristics tremendously. Indeed, the water retention of the modified starch used for the coating is only moderate (about 2 g/g dry matter) and it undergoes an important reduction after mixing with the nitrocellulosic varnish. Moreover adding a hydrophilic product to another hydrophilic polymer which is highly permeable to water vapour is certainly less efficient than the association of a hydrophobic polymer with low WVTR to a hydrosorbent product which plays a role of reservoir for water vapour.

6.4.4   *Coating of PET with a hydrosorbent polymer*

The use of Parpack-cello® allowed Camembert to reach the same shelf-life (6 weeks) as the conventional package. There is a need for French soft cheeses like Camembert, Brie or Coulommiers, if it is desired to export them abroad, to have a shelf-life extended 4 or 6 more weeks. For this, the new package should prevent desiccation and water condensation, ensure a control of water vapour transfer and allow a permeability to O_2 and a release of CO_2 compatible with the respiration of the hydrophilic *Penicillium* flora. This may be achieved with a hydrophobic film like polyester terephthalate with a low thickness providing that a highly hydrosorbent product be associated to it by coating. Moreover, the use of a thin (3 to 6 μm) film of PET is compatible with the package waste weight reduction concern for the protection of the environment. To minimize the interactions of the hydrosorbent material with the cheese, it is necessary for only a few grams per square meter of package to be used. This is achieved with sodium polyacrylates which are known to hold more than 500 times their volume of water. But, what is true for liquid water and applicable to baby's napkins for example might not be as important when water vapour is adsorbed. Observation of the water vapour sorption isotherm at 6°C, temperature of ripening of the cheese, shows that a maximum of 3.5 g/g is reached for A_w = 0.987 (Figure 6.9). The water sorption isotherm was also determined for PET and its retention of water found negligible even at high A_w values. The hydrosorbent polymer is first washed in a solution of NaCL (0.9% (w/w)) to minimize the migration of the sodium acrylate monomer. After drying and grinding, coatings with different percentages of hydrosorbent polymer were prepared in a nitrocellulosic varnish, their sorption behaviour evaluated from their isotherm

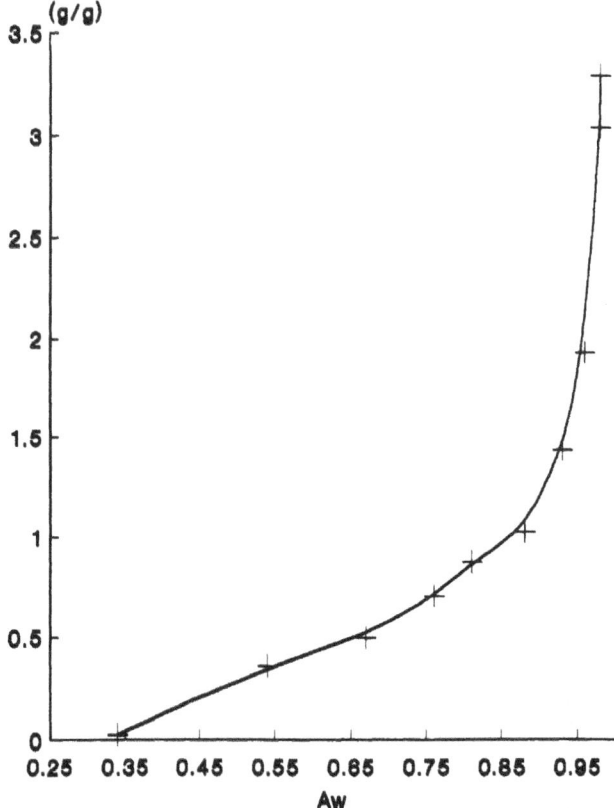

Figure 6.9 Water vapour sorption isotherm (6°C) of hydrosorbent polymer (g/g: g water/g dry matter).

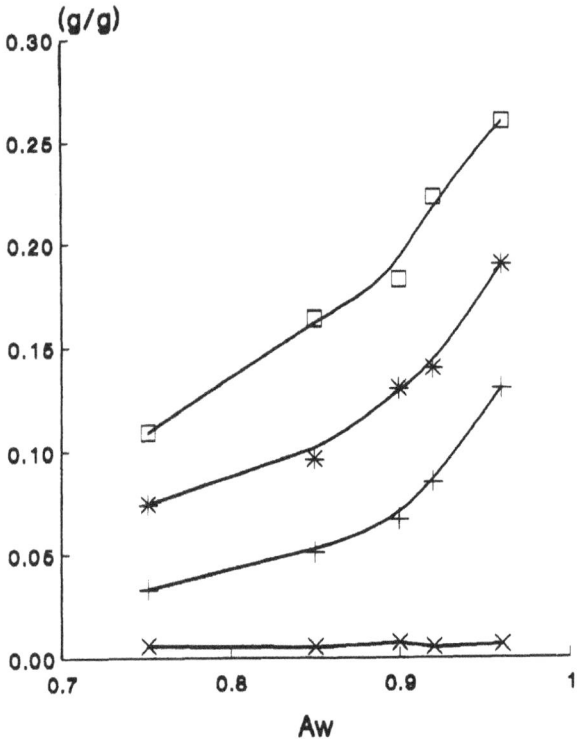

Figure 6.10 Water vapour sorption isotherm (6°C) of varnish (×), 10% (+), 20% (*) and 30% (□) of hydrosorbent polymer in the varnish used to coat the PET film (g/g: g water/g dry matter).

(6°C) curve (Figure 6.10). As expected, the higher the percentage of Na–polyacrylate in the varnish, the higher the water adsorption. The varnish has no affinity for water. An optimal coating was found for a hydrosorbent content of 4 g/m^2 of package. The water vapour sorption isotherm of the coated PET (6 ξm thickness) at 6°C, the temperature of cheese ripening, is shown in Figure 6.11. It exhibits a general aspect comparable to that of the sole hydrosorbent (see Figure 6.9). However, because of the entrapping of the hydrosorbent polymer in varnish, the maximum water content for the coated PET at A_w = 0.98 is only the tenth (0.35 g/g) of what was found for the pure Na-polyacrylate.

To the coated PET (6 μm thickness) film a waxed paper was associated to separate the cheese from the coating on the one hand and prevent the diffusion of lactic acid from the cheese on the other. The water vapour sorption isotherm (6°C) of the packaging complex (coated PET + waxed paper) was determined in the region of 0.75–0.98 A_w (Figure 6.12). The maximum water content is almost the same for the sole coated PET. Although this value of moisture content (0.35 g/g) for A_w = 0.98 is lower

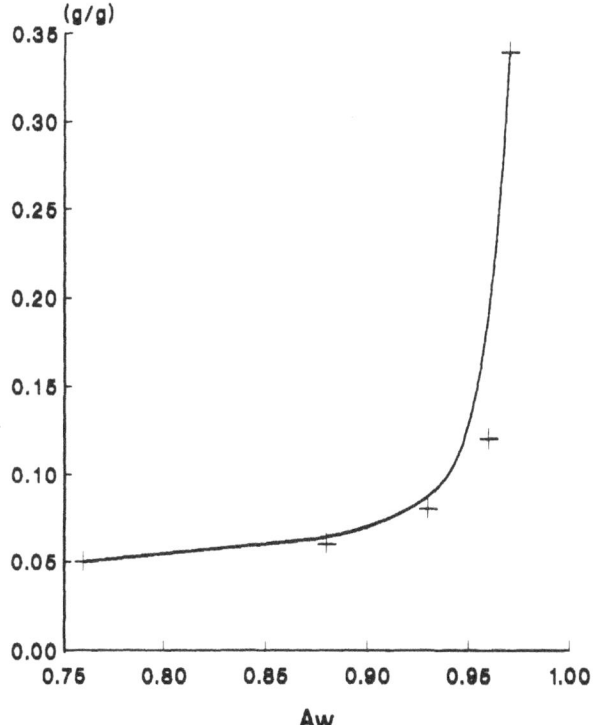

Figure 6.11 Water vapour sorption isotherm (6°C) of the coated (4 g/m² hydrosorbent polymer) PET film (g/g: g water/g dry matter).

than that of the conventional package (Figure 6.3), the advantage of the coated PET is that the hydrophobic polymer slows down water vapour transfer while the hydrosorbent material rapidly entraps water vapour in the case of increased water activity and prevents condensation. The permeabilities to O_2 and CO_2 were determined for the packaging complex at 6°C (see Table 6.4). These results are not comparable to those obtained with the regenerated cellulose film at 23°C, but experience showed that these conditions of gas transfer are high enough to ensure respiration of the mould flora at a relatively reduced rate and prolong the shelf-life of the cheese.

Table 6.4 Permeability to O_2 and CO_2 of the new packaging complex (coated PET + Waxed paper) (6°C)

	$Q\ O_2$ (cm³/m²/24 h)	$Q\ O_2$ (cm³/m²/24 h)
Sample 1	260	1310
Sample 2	280	1140

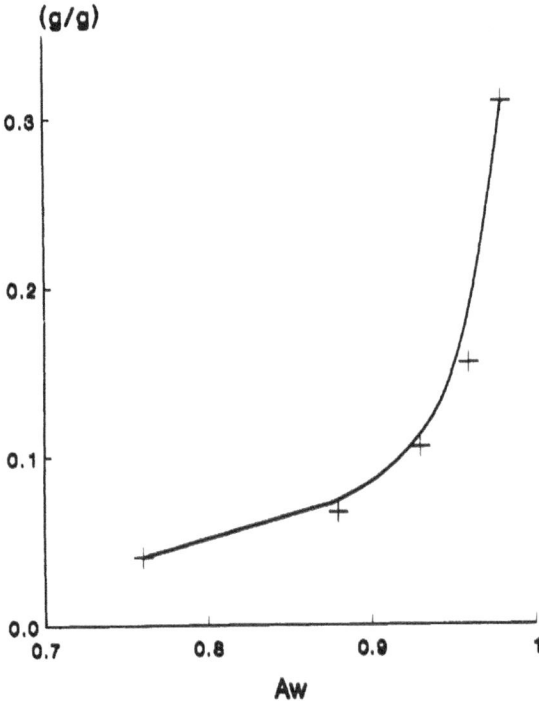

Figure 6.12 Water vapour sorption isotherm (6°C) of the PET complex (coating + waxed paper) (g/g: g water/g dry matter).

6.4.5   *Increase in shelf-life of soft cheeses packed in coated PET complex*

The permeabilities of coated PET complex originate only from the barrier characteristics of PET of 6 μm thickness. The waxed paper and the coating do not interfere in gas transfer. The hydrosorbent coating only affects the kinetics of water vapour transfer. The optimization of soft cheese flora respiration is not obtained by the coating of all types of synthetic polymers. Polypropylene cast could be used for its very high permeability[29] to O_2 and CO_2 but its low WVTR (8.2 $gm^{-2}day^{-1}$, ΔRH = 90%, 38°C) is rather a limiting factor. Oriented PVC is insufficiently permeable to gases and water vapour. Polyester with 6 μm thickness seems satisfactory.[29] However, the optimization of gas transfer may require that the PET film be microperforated. This was achieved for the study of the shelf-life of soft cheeses packed in coated PET complex. Three packaging materials were compared (the conventional cellulose film pasted with waxed paper, the coated PET complex with and without microperforation).

Prior to experimentation in cheesemongery, a trial was made with a model soft cheese, consisting of a cellulose pulp dipped in glycerol with

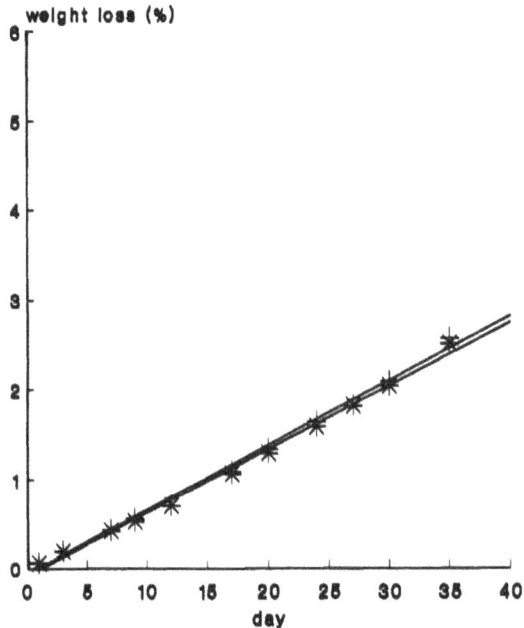

Figure 6.13 Percentage weight loss of the model soft cheese as a function of storage time
(+ = PET; * = coated PET).

A_w = 0.98 and pressed to have the shape and dimensions of a Camembert
cheese. The weight loss of the model soft cheese was controlled for a
period of 36 days of storage in the refrigerator (6°C 70–85% RH) after
wrapping in plain PET (6 μm) or coated PET pasted with waxed paper.
The results are shown in Figure 6.13. The same weight loss of about 3%
is obtained for both materials which confirms that the polyester film
controls water vapour transfer. However, weighing of the packages after
the 36 days of storage shows that PET gains only 30 mg/m^2 whereas 540
mg/m^2 were gained by the coated PET complex, which confirms the role
of moisture reservoir of the coating.

When the size of cheese is increased, the surface of the packaging
material is also augmented and the water retention by the coating becomes
appreciable. This is the case for the soft cheese Brie which weighs 3.200
kg, about 30 times that of Camembert, the surface being 10 times higher.
An attempt was made to prolong the shelf-life of Brie from 45 to about
80 days. Comparison was made between the conventional package and the
coated PET complex with and without microperforation. The traditional
criterion of ripening control, i.e. weight loss, was measured. The results
(Figure 6.14) show that the loss is stabilized at about 6% for the conventional package, 3% for the microperforated PET complex and 2% for the
non perforated complex after 60 days. The weight loss is mainly due to

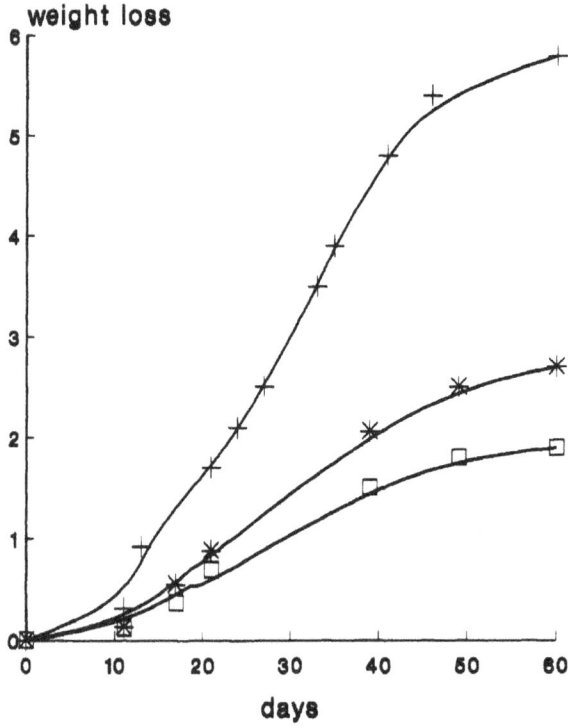

Figure 6.14 Percentage weight loss of Brie cheese as a function of ripening duration (+ = control; * = perforated PET complex; □ = coated PET complex).

Table 6.5 Sensory evaluation of Brie cheese after 45 and 75 days of ripening

Sample (package)	Observations				
	Flora	Texture	Taste	Fag-end	Rank
After 45 days					
Control	White	Slightly firm	Good	Firm	1
Coated PET	Grey slow growth	Soft	Slightly sour	Soft	3
Perforated complex	White	Slightly firm	Good	Soft	2
After 75 days					
Control	White-yellow	Firm	Acceptable	Hard	2
Coated PET	Slightly yellowish	Flowing	Ammonia	Soft	3
Perforated complex	Yellowish growth in the holes	Soft	Acceptable to good	Hard	1

water vapour transfer. The difference between the three packages is essentially due to the low WVTR across PET and to the increased water retention by the hydrosorbent polymer. The microperforation is rapidly filled with the mould flora in the first days of ripening. Sensory evaluation of the ripened Brie cheese in the three packages was carried out 45 and 75 days after the wrapping. The results are reported in Table 6.5. The criteria for the organoleptic value of the cheese were chosen by the panel of the cheese factory. The conventional package was preferred after 45 days of ripening, but the increased shelf-life (75 days) was only found good for the Brie cheese wrapped in a microperforated coated PET complex.

A comparable experiment was carried out with a third soft cheese called Coulommiers with an average weight of 350 g. The results are shown in Figure 6.15 for weight loss and Table 6.6 for sensory evaluation. Apart from the coated PET, which exhibits a low weight loss comparable to what was found with Brie, both the microperforated complex and the conventional package show a regular increase in function of storage time. The weight loss after 60 days is almost double the generally accepted 7 to

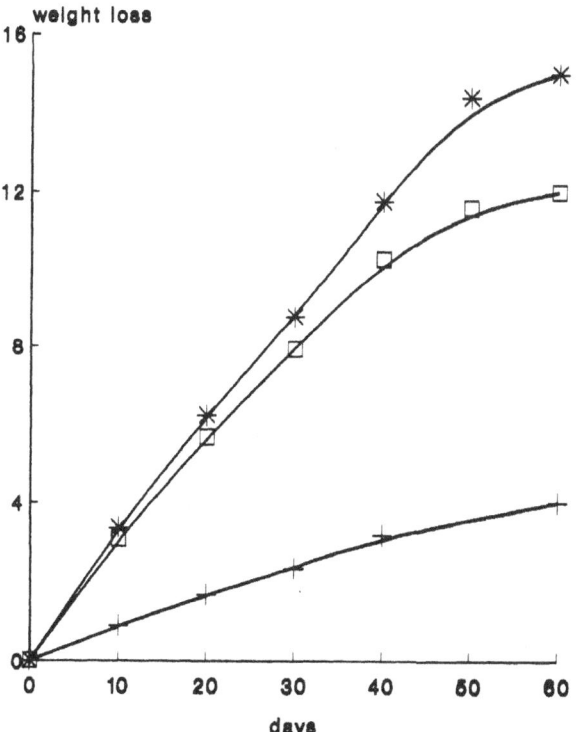

Figure 6.15 Percentage weight loss of Coulommiers cheese as a function of ripening duration (* = perforated; □ = control; + = coated PET complex).

Table 6.6 Sensory evaluation of Coulommiers cheese after 30 and 60 days of ripening

Sample (package)	Observations			
	Flora	Texture	Taste	Rank
30 days				
Control	White	Slightly hard	Slight bitterness	1
Coated PET	Yellowish	Soft	Slight sourness NH_3	3
Coated PET perforated	White growth in the perforation	Hard	Dry	2
60 days				
Control	White–grey	Soft to pasty	Pasty sour	2
Coated PET	Yellowish	Flowing	Sour	3
Coated PET perforated	White–dry	Middle soft rind hard	Dry	1

8% found after the regular shelf-life of about 5 weeks. However, the microperforated coated PET allows the production of a Coulommiers cheese ranked 1 (see Table 6.6) with a flora which remains white, a soft texture in the middle and an acceptable taste. Although the weight loss is relatively high, the ripening in the perforated PET complex proves to be interesting for increasing the shelf-life of Coulommiers especially if it is desired to export it.

6.5 Conclusion and prospects

Extension of the shelf-life of soft cheeses was made possible by the choice of a packaging complex compatible with the respiration of the surface flora (*Penicillium caseicolum*) and the preservation of an organoleptic quality acceptable by the consumer. In order to maintain a high water activity (0.98) at the surface of the cheese together with good rates of O_2 and CO_2 transfer, the coating of a thin film of PET with a hydrosorbent layer was found particularly efficient. The association of hydrophilic and hydrophobic materials together with the control of permeability to gases through the thickness and/or the perforation of the polymeric film was found well adapted to solve the problems of protection and control of Brie and Coulommiers cheeses during ripening. This solution should open the frontiers for exportation if no other customs barrier is set up.

Although the coated PET complex has already been patented by Brodart S.A.[30] (converter and printer of flexible films) the profession of cheesemakers in France will certainly take some time before applying it. Some more studies are needed to optimize the choice of the hydrosorbent product, and some precautions should be taken in order to conform to the regulations. However, the approach followed for the improvement of cellulose film or for the development of the PET complex should interest the cheesemakers. This interest was not met in the case of another innovative technology of cheese ripening we developed some years ago.[31] Certainly it was demonstrated that the ripening of French Emmental cheese in a gastight polyester-aluminium-polyethylene laminate was interesting from the organoleptic and economical point of view. However, this technology was not adopted in France. It was easier to find cheese dairies in Austria to apply it.

Use of metallized gastight complexes, simple hydrophilic films which are highly permeable to gases and water vapour or the association of a microperforated hydrophobic film with a hydrophilic coating, produce packages as different as the cheeses to which they are applied. What is needed in food packaging is that the food product be well understood if it is desired to choose a packaging material adapted to its protection and to the increase of its shelf-life. This is particularly true for such living products as soft cheeses with a mould flora at the surface.

If the existing materials cannot solve the problem posed by the duration of shelf-life needed or the requirements of oxygen, CO_2, aroma or water availability, the answer may be the functionalization of the wrapping material. This functionalization could be inspired from the well-known 'drug release' technology adopted in pharmaceutical industries. It can easily be imagined that biopolymers like polysaccharides could be used through ionic or covalent bonds as carriers of antioxidants, antimicrobials, or CO_2 or any other ingredient which may be released during the storage of the product. Functionalized biopolymers are compatible with the food product and could be sandwiched between two films of synthetic polymers or coated on the surface of a packaging material. One of the functions to assign to the package which should play a major role in preservation and extension of shelf-life is the scavenging and/or release of water vapour at the surface of the product. Some progress is still to be made in this area.

Acknowledgements

This study was supported by Brodart S.A. (Arcis-sur-Aube) France. Dr. Y. Albayati (Brodart S.A.) is thanked for his participation.

References

1. Mathlouthi, M. (1986) *Food Packaging and Preservation, Theory and Practice*, Elsevier Applied Science Publishers, London.

2. Labuza, T.P. and Breene. W.M. (1989) *J. Food Process Preserv.*, **13**, 1–69.
3. Peppas, N.A. and Khanna, R. (1980) *Polymer Eng. Sci.*, **20**, 1147–1156.
4. Kontny, M.J., Koppenol, S. and Grahame, T. (1992) *Intern. J. Pharm.*, **84**, 261–271.
5. Robertson, G.L. (1991) Smart Packaging films: New opportunities for food processors, in *Proceedings of the 18th Asian Pack Congress*, Singapore pp. D16–D21.
6. Shirazi, A. and Cameron. A.C. (1992) *Hort. Science*, **27**, 336–339.
7. Anon (1989) *Packaging Trends in Japan*, **89**(11), 4.
8. Kannankeril, C.P. and Cruikskank, B.A. (1991) US Patent no 5, 176, 930.
9. Gontard, N. and Guilbert. S. (1992) Chapter 9, this book.
10. Kamper, S.L. and Fennema, O. (1985) *J. Food Sci.*, **50**, 382–384.
11. Greener, I.K. and Fennema, O. (1989) *J. Food Sci.*, **6**, 1393–99 and 1400–406.
12. Gontard, N., Guilbert. S. and Cuq, J.L. (1992) *J. Food Sci.*, **57**, 190–195.
13. Biquet, B. and Labuza, T.P. (1988) *J. Food Sci.*, **53**, 989–998.
14. Taoukis, P.S., Elmeskine, A. and Labuza, T.P. (1988) Moisture transfer and shelf-life of packaged foods, in *Food Packaging Interactions*, Hotchkiss, J.H. (ed.), ACS Symposium Series 365, pp. 243–261.
15. Rudolph. F.B. (1987) *Lebensm. wiss-u.-Technol.*, **20**, 19–21.
16. Brunauer, S., Emmett, P.H. and Teller, E. (1938) *J. Am. Chem. Soc.*, **60**, 309–319.
17. Oswin, C.R. (1946) *J. Chem. Ind.*, **65**, 419–426.
18. Halsey, G. (1948) *J. Chem. Phys.*, **16**, 931–937.
19. Labuza, T.P., Mizarhi, S. and Karel, M. (1972) *Trans. ASAE*, **15**; 150–55.
20. Tubert, A.H. and Iglesias, H.A. (1986) *Lebensm. Wiss.-u-Technol.*, **19**, 365–68.
21. Heiss, R. (1968) Haltbarkeitund Sorptionerhalten wasserwarmer Lebensmittel, Springer-Verlag, Berlin-Heidelberg.
22. Chapter 8, this book.
23. Rüegg, M. and Blanc, B. (1977) *Milchwissenschaft*, **32**, 192–201.
24. Rüegg, M. and Blanc, B. (1981) Influence of water activity on the manufacture and aging of cheese, in L.B. Rockland and G.F. Stewart (eds), *Water Activity: Influences on Food Quality*. Academic Press, New York, pp. 791–811.
25. Mathlouthi, M., Conry, M., Jaillant, G. Maitenaz, P.C. (1980) *Lebensm, Wiss.-u.-Technol.*, **13**, 264–268.
26. Kiermeier, F. and Wolfeseder, H. (1972) *Zeitsch. Lebensm. Unters. und Forsch.*, **150**, 75–83.
27. de Leirus, J.P. (1986) *Water activity and permeability*, in Food Packaging and Preservation, Theory and Practice, M. Mathlouthi (ed.), Elsevier Applied Science Publishers, London, pp. 213–233.
28. de Leiris, J.P. (1992) Les emballages actifs, in *Colloque Conditionnement alimentaire, Innovation, Environnement.* I.S.E.C.A. Pouzauges, pp. 1–30.
29. Mathlouthi, M. and de Leiris, J.P. (1993) Packaging of solid foods, in *Encyclopaedia of Food Science, Food Technology and Nutrition*, Macrae R., Robinson R.K. and Sadler M.J. (eds), Academic Press, London.
30. Brodart, M. (1990) French Patent no 9016208.
31. Seuvre, A.M. and Mathlouthi, M. (1982) *Lebensm. Wiss.-u.-Technol.*, **15**, 258–262.

7 Trehalose – a multifunctional additive for food preservation

C.A.L.S. COLAÇO and B. ROSER

Abstract

Trehalose, a non-toxic disaccharide of glucose, is responsible for the remarkable survival of a group of inconspicuous and little known desert organisms, called cryptobionts, which can desiccate totally during drought and yet recover completely when rehydrated. Our recent work has shown that this protection against desiccation damage can be reproduced *in vitro* when biomolecules are dried in the presence of this simple sugar. Trehalose drying technology has been utilised in a number of health care, pharmaceutical and biotechnological applications, where it has enabled the development of novel, robust and versatile formats.

When added to foods prior to air drying, trehalose completely protects the foods from denaturation and also prevents loss of aromatic volatiles that give fresh foods their characteristic aromas and flavours. Furthermore, the addition of trehalose, a very stable non-reducing sugar, results in the inhibition of intrinsic Maillard reactions during the processing and storage of dried food products. Thus, besides the benefits of stability and reduction in bulk, common to other dried food products, trehalose-dried foods also contain less toxic by-products and have a higher nutritional content than conventionally processed foods.

7.1 Introduction

The industrialisation of modern societies has inevitably resulted in their becoming increasingly urbanised. This shift from rural to urban communities has required the development of extensive food distribution networks and of preservation methods to enable perishable food products to be stored and transported.

Refrigeration is the most popular method of food preservation for storage, and refrigerated container lorries the main method of food distribution. However, both refrigeration and motorised transport exact a heavy environmental toll with excessive energy consumption and the emission of environmentally harmful greenhouse gases. In today's environmental and energy conscious climate, these prices are being recognised as unaccept-ably high and alternatives are thus urgently required. Moreover, in some countries more than 30% of perishable food crops are lost to spoilage during transport or storage, making the need for alternative, more efficient methods for the preservation of foodstuffs even more necessary.

The drying of foodstuffs presents a particularly attractive alternative, as not only is stability assured by the absence of water, but the tremendous reduction in weight and volume achieved by water removal also greatly reduces the transport requirements.[1] This is evidenced by the rapid growth of freeze-drying as a method of food processing in the wealthier industrial-ised countries. However, this technology still requires refrigeration for both the production of the freeze-dried foodstuffs and their storage, making it an inadequate solution to the problem of the preservation of foodstuffs in energy and environmental terms.

The most energy and cost efficient method of drying foods is air-drying (Table 7.1). As a method of food preservation, air-drying has the addi-tional advantage of producing foods that are stable at ambient tempera-tures, thus eliminating the need for refrigeration for both the production and storage of dried foodstuffs. However, although air-drying of foodstuffs as a method of preservation is an ancient technology, it is still severely limited by the quality of the dried foodstuffs produced. Besides the frequent denaturation of the dried food product, air-drying often results in a significant loss in nutritive value as well.[1,2] The latter is largely a result of the loss of protein quality due to the Maillard reaction between the naturally occurring sugars and proteins in the dried foods.[3a,b] This is a particularly acute problem during drying as the cascade of spontaneous deleterious chemical reactions is considerably accelerated at low water activities.[4,5]

Thus the non-enzymatic browning resulting from the Maillard reaction that commonly typifies the spoilage of foodstuffs on storage, is often a problematic feature of food processing involving drying.[1,6] The denatura-tion of food as a consequence of drying is, however, not always a problem

Table 7.1 Costs of drying methods

Drying method	Fixed costs ($.kgh)	Manufacturing costs (¢/kg)
Air drying	610	17
Fluid bed	1000	17
Drum	1040	23.4
Spray	1360	19.4
Vacuum	5860	49
Freeze drying	11228	95.2

*Department of Food Science and Technology, University of Read-ing, UK. Approximate figures 1983 adjusted to 1990 prices.

as it may sometimes confer desirable new properties on the dried food products. Thus, raisins, prunes, dried apricots and dried fish are all unique food products in their own right and are not thought of as inferior substitutes for the fresh product. Nevertheless, nobody is convinced by claims that dried milk, egg or meat have the same quality as the fresh produce. Any new drying process that could combine the virtues of air-drying such as cheapness, simplicity and speed with preservation of the nutritive and 'fresh' characteristics of food would be of great value. Such a process would be particularly attractive as an efficient and cost-effective alternative to current methods of preservation as it would completely eliminate the current need for refrigeration.

7.2 Trehalose and cryptobiosis

In the arid regions of the world, certain inconspicuous and little known plants and animals have solved the problem of drying out without damage during droughts in the most remarkable way. This is strikingly demonstrated by their ability to promptly resume normal life and metabolism upon simple addition of water. These so called 'cryptobiotic' or 'anhydrobiotic' organisms are found in many phyla in both the plant and animal kingdoms. The better known examples include baker's yeast, *Saccharomyces cerevisiae*, soil nematodes such as *Ditylenchus dipsaci*, tardigrades like *Adoribiotus coronifer*, the brine shrimp, *Artemia salina*, and resurrection plants such as *Selaginella lepidophylla* (Figure 7.1) and *Myrothamnus flabellifolius*. It is important to emphasise that these cryptobiotic organisms do not freeze-dry but air-dry at ambient temperatures. Thus the resurrection plants *Selaginella lepidophylla* and *Myrothamnus flabellifolius* routinely dry out during summer droughts in the deserts of southern USA and South Africa respectively and under these conditions drying occurs at temperatures up to 50°C. In the dry state, these organisms appear biologically dead with no detectable metabolic activity and are resistant to withstand extreme conditions of temperature, pressure and ionising radiation.[7,8] Dry cryptobionts have been successfully rehydrated after storage for 120 years in a museum.[8,9] Moreover, these organisms can be repeatedly dried and rehydrated, without cumulative ill effects, suggesting that the preservation of molecular structures and function during the dry phase must be close to 100%. The protective mechanisms which have evolved in cryptobionts therefore clearly protect a wide variety of structural and metabolic molecules and this remarkable life cycle illustrates the principle that air drying of biomolecules can be entirely harmless under the right conditions.

The metabolism and structure of these organisms is similar to that of the vast majority of other plant species which die after the loss of >20–30%

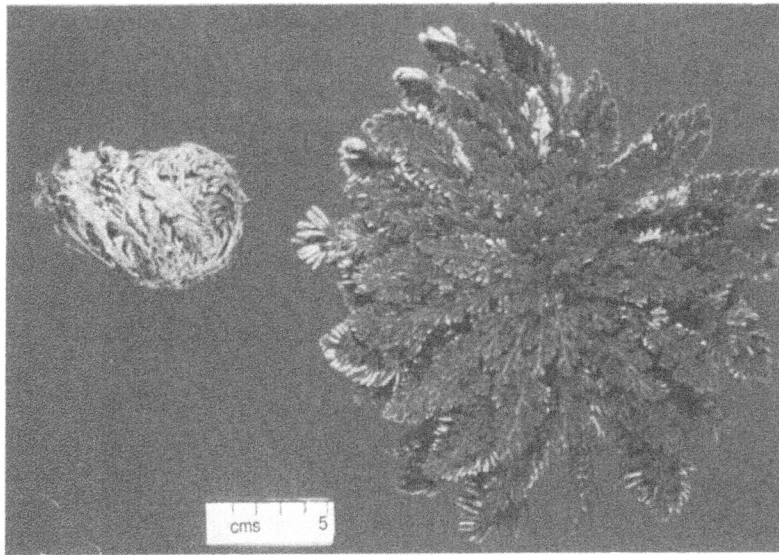

Figure 7.1 Cryptobiotic organism: The 'resurrection plant' *Selaginella lepidophylla* can exist for many years in the desiccated form (left). It becomes a vigorous, actively growing plant within a few hours of watering (right). This cycle of drying and rehydration can occur repeatedly without damage.

of body water. However, a unique feature common to many cryptobionts is the presence of large quantities of a simple but rare sugar, trehalose, in the tissues of dry cryptobionts.[10] Trehalose (α-D glucopyranosyl $(1 \rightarrow 1)$ α-D glucopyranoside) is a simple disaccharide of glucose (Figure 7.2). The linkage of the two glucopyranose rings via the 1-carbons makes the sugar completely non-reducing and the resulting α1–α1 glycosidic bond is exceptionally stable. The parallel evolution of trehalose accumulation as

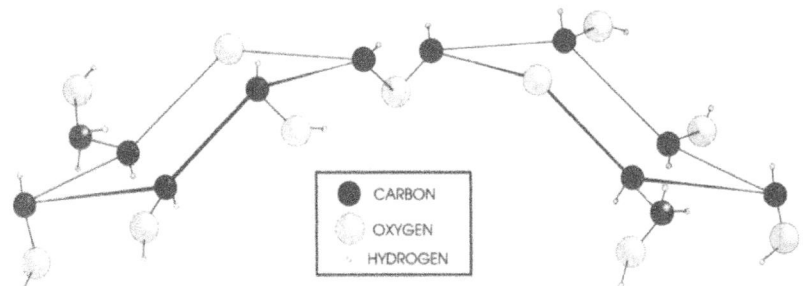

CARBON
OXYGEN
HYDROGEN

Figure 7.2 Trehalose, α-D-glucopyranosyl α-D-glucopyranoside. Trehalose derives its name from Trehala manna, the cocoon of a parasitic beetle of the genus *Larinus*. This contains large amounts of trehalose, is found on aromatic thorn bushes in the Middle East and is thought to be the 'Manna from Heaven' referred to in the Old Testament. Certainly this unusual combination of trehalose and protein would have provided excellent nourishment for the Israelites in the desert!

a mechanism of avoiding desiccation damage in such a wide diversity of organisms in the plant and animal kingdoms suggested to us that the sugar itself might possess the protective properties that confer on it the ability to substitute effectively for structural water. Trehalose could thus provide the basis for drying virtually any biological molecule at ambient temperatures without damage and the idea clearly merited further study.

At Quadrant, experiments over the past few years have been directed at the question of whether trehalose was not only necessary, but also sufficient, to confer on biomolecules the desiccation tolerance observed during cryptobiosis. We addressed this question by attempting to reproduce in the laboratory the complete functional preservation of a very wide range of biological molecules. These represented most classes of biomolecules required for metabolism, all of which are clearly protected in dry cryptobionts. The results were unequivocal. We have yet to find a biological molecule, molecular complex or mixture of biomolecules that cannot successfully be dried, without damage, from trehalose-containing buffer and then rehydrated with recovery of close to 100% of biological activity, even after storage at extremely high temperatures in the dry form.[11-15]

A striking demonstration of the ability of trehalose to protect biomolecules against damage due to water loss is the drying of the labile enzymes used in molecular biology.[12] These are currently stored in freezers at −20°C and transported on ice in insulated polystyrene foam containers.

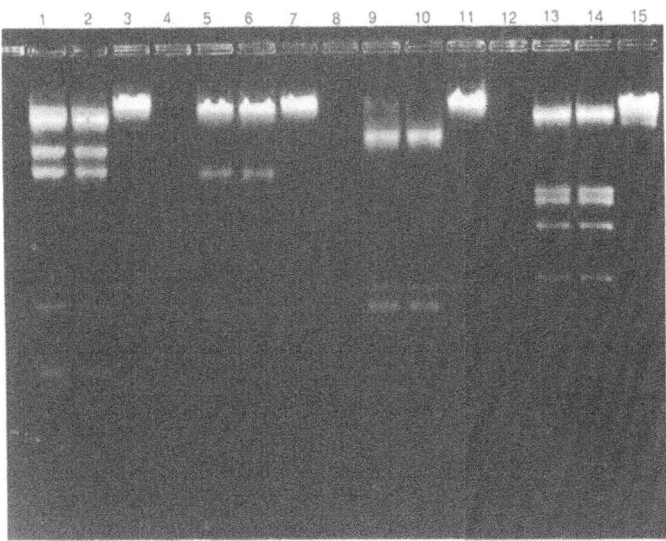

Figure 7.3 Stabilisation of restriction enzymes by trehalose. *Eco*R I (tracks 1–3), *Bgl* II (tracks 5–7), *Pst* I (tracks 9–11) and *Hin*d III (tracks 13–15) were dried either with trehalose (tracks 2,6,10) or without trehalose (tracks 3,7,11) and then compared with five units of non-dried enzyme from the −20°C freezer (tracks 1,5,9).

All of the DNA modifying and restriction enzymes we tested could be dried at room temperature from buffer solutions containing trehalose without loss of activity (Figure 7.3). In contrast, all activity was lost if trehalose was omitted from the solutions being dried and none of them could be stabilised significantly by freeze drying even in the presence of trehalose.

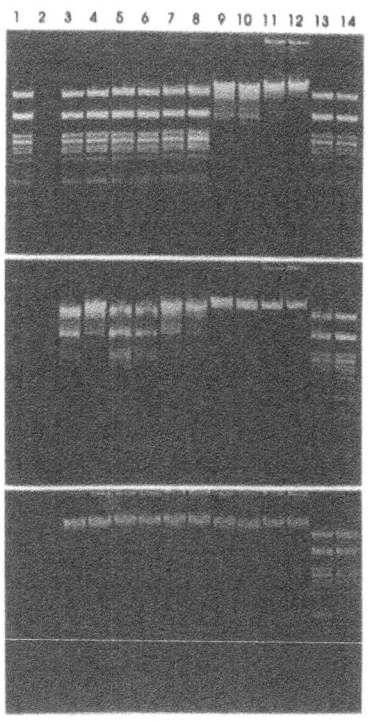

Figure 7.4 Comparison of the high temperature stabilities of enzymes dried using different additives as stabilisers. Five units of fresh *Pst* I (track 1) as a positive control was compared in cutting 1 μg lambda DNA with 5 unit and 2.5 unit doses of enzyme dried in different stabilisers and stored at high temperatures for one month. The disaccharide alcohols were identified as good stabilisers from a screen of known excipients. Maltose and sucrose were tested because of their widespread use as food additives and bulking agents. Tracks 3,4 glucopyranosyl mannitol (GPM); tracks 5,6 glucopyranosyl sorbitol (GPS); tracks 7,8 Palatinit (the alcohol of isomaltulose); tracks 9,10 maltose; tracks 11,12 sucrose and tracks 13,14 trehalose. Upper panel. When stored at 37°C, the samples dried in sugar alcohols and trehalose cut DNA with fidelity. Enzyme dried in either maltose or sucrose failed to digest the DNA. Middle panel. When stored at 55°C, trehalose-dried enzyme cut perfectly (tracks 13,14). The enzyme dried in the sugar alcohols now gave only partial digests (tracks 3–8). Lower panel. When stored at 70°C, all failed totally except for enzyme dried in trehalose (tracks 13,14) which still cut with fidelity. Other enzymes so far successfully dried and stored at high temperatures include numerous restriction enzymes, T4 DNA ligase, Taq DNA polymerase, Sequenase and the Klenow fragment of *E. coli* DNA polymerase I. Photograph copyright © Quadrant Holdings Cambridge Ltd, 1992.

Remarkably, accelerated-ageing studies on trehalose-stabilised dry restriction enzymes show that these labile proteins still cut DNA accurately even after storage for 35 days at 70°C in ovens that are too hot to touch (Figure 7.4). Furthermore, the dried samples can be stored at 55°C for at least 18 months or cycled repeatedly between high and low temperatures with no loss of activity. This extraordinary stability is unique to trehalose; no other sugars or other widely recommended excipients confer similar

Table 7.2 Stability of Pst I and *Hind* III dried at room temperature in various stabilising agents and stored at higher temperatures. The restriction enzymes Pst I and *Hind* III were dried in a universal cutting buffer containing the sugar or sugar alcohol at 0.3 M. After storage for the indicated time and temperature the activity recovered was titrated for its ability to cut 1 μg of bacteriophage lambda DNA at 37°C for 1 h.

Stabiliser	Storage temperature (°C)	Time (days)	Activity
Trehalose	37	35	+++
	55	35	+++
	70	35	+++
Sucrose	37	35	− .
Lactose	37	35	−
Maltose	37	35	−
Palatinit	37	35	++
	55	35	−
GPS	37	35	++
	55	35	−
GPM	37	35	++
	55	35	−
Inulin (c)	37	7	−
Inulin (d)	37	7	−
PVP	37	7	−
Dextran	37	7	+
Ficoll	37	7	−

All stabilising agents were tested at 37°, 55° and 70°C.
+++ = full activity
++ = partial digest (< 25% activity remaining)
+   = some cutting (< 10% activity remaining)
−   = no detectable activity

protection (Figure 7.4 and Table 7.2). Reducing sugars, such as lactose and maltose, failed under the mildest conditions tested, 35 days at 37°C, as did the non-reducing disaccharide sucrose. The chemically more stable non-reducing sugar alcohols showed better stabilities, but all failed within 35 days at 55°C. Other widely recommended stabilising agents such as inulin, ficoll, PVP and dextran were all ineffective (Table 7.2), as were all the monosaccharides, monosaccharide alcohols and other sugars tested.

7.3 Trehalose-dried foods

The ability of trehalose to protect the very labile enzymes used in molecular biology, even during extended storage under extreme conditions, suggested that this sugar might also have a useful preservative effect on drying unstable mixtures of biological molecules such as are found in natural foods. Furthermore, being a major component of commonly eaten foods such as mushrooms or yeast products, humans and other omnivores have evolved specific intestinal enzymes which digest trehalose to two glucose molecules. This means that, as a dietary sugar, trehalose is completely non-toxic and furthermore, not being particularly sweet, it does not significantly change in flavour of foods to which it is added. We have mixed trehalose with liquid or liquidised foods and find that they were well preserved on drying. For example, drying of blended fresh eggs at 30–50°C with trehalose dissolved in the liquid, produces an odourless yellowish orange powder which can be stored at room temperature and rehydrated easily to give a product virtually indistinguishable from fresh blended egg.

The texture, colour, taste and cooking properties of a wide variety of foods are similarly well preserved when they are dried with trehalose. Fresh fruit purées containing trehalose could easily be dried to yield shelf-stable dry powders. The dry powders themselves had very little smell and certainly did not smell like the fresh products. However, when water was added, the viscosity and texture of the fresh purées returned and after a short time, the products began to emit the unique aroma of the fresh fruit. When dried without trehalose, the products were more difficult to reconstitute and had lost much of their fresh flavour and aroma. Since the clean, perfumed smell of fresh produce is an important sensory property that identifies them as 'fresh' rather than preserved, this process provides a means of supplying convenient, stable, instant versions of perishable foods. It results in naturally fresh dried foods or beverages which otherwise lose these properties when dried using standard methods.

The application of trehalose technology is not just limited to the processing of liquid or liquidised foods. Fresh fruit slices and fresh herbs were also well preserved when dried after brief immersion in trehalose solutions (Figure 7.5). The colours of the fresh foodstuff were again preserved in the dried food product, even on extended storage and when rehydrated the characteristic flavour and aroma of the fresh product were regained. The control samples of dried herbs, dried without trehalose, became brown and woody and lost all flavour and aroma on storage. The control samples of the banana slices, dried without trehalose, gave a brown crispy product (Figure 7.5) which did not rehydrate, whereas the trehalose-dried samples easily rehydrated to yield soft fruit slices with the colour and texture of fresh banana slices. Furthermore, although the dried slices did not brown during storage, the rehydrated banana slices began to brown

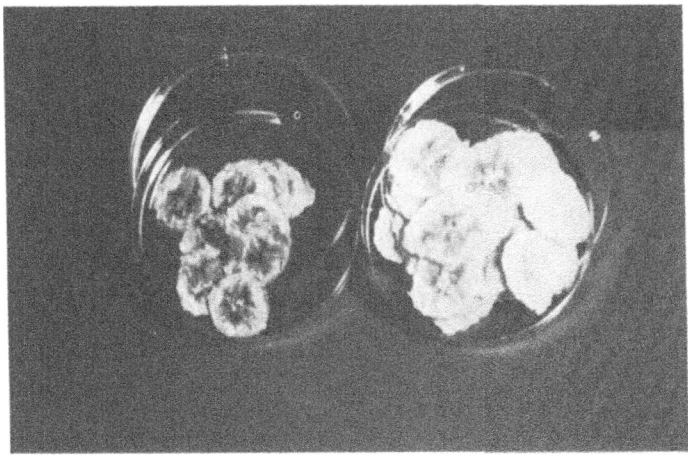

Figure 7.5 Fresh fruit slices dried using trehalose technology. Banana slices soaked in a trehalose solution before drying (right) retained the colour and texture of fresh banana unlike the control samples (left).

on exposure to air as a result of preoxidase activity as does the fresh product. The preservation of structure and function of food molecules when dried in trehalose appears to be as complete as the preservation of other labile biomolecules used in scientific or clinical procedures, such as enzymes or antibodies.[11,12]

7.3.1 *Trehalose preserves food volatiles*

The finding that trehalose added to purées of fresh fruits, before drying, retained and preserved the volatile aromatic molecules responsible for their characteristic aromas and flavours in the dried product was an unexpected result. When dry, the fruit powders emit essentially no aroma. However, when the powders are rehydrated with water, not just the colour, viscosity and texture of the fresh purées are recovered, but the aroma and flavour of the fresh fruit as well (Figure 7.6). The control powders, dried without trehalose, were more difficult to reconstitute and emitted the denatured flavour notes of cooked fruit when rehydrated. Sensory evaluation panels testing trehalose-dried fruit purées remarked that the products tasted 'fresh' and lacked the 'cooked' flavour and aroma of the controls that had been dried under identical conditions but without the addition of trehalose prior to the drying process.

To confirm these anecdotal observations, the volatile aromatics released by rehydrated dried mango purée were collected, analysed by gas chromatography and the peaks of interest identified by mass spectroscopy. In addition to the increase in number and intensity of native volatile aroma-

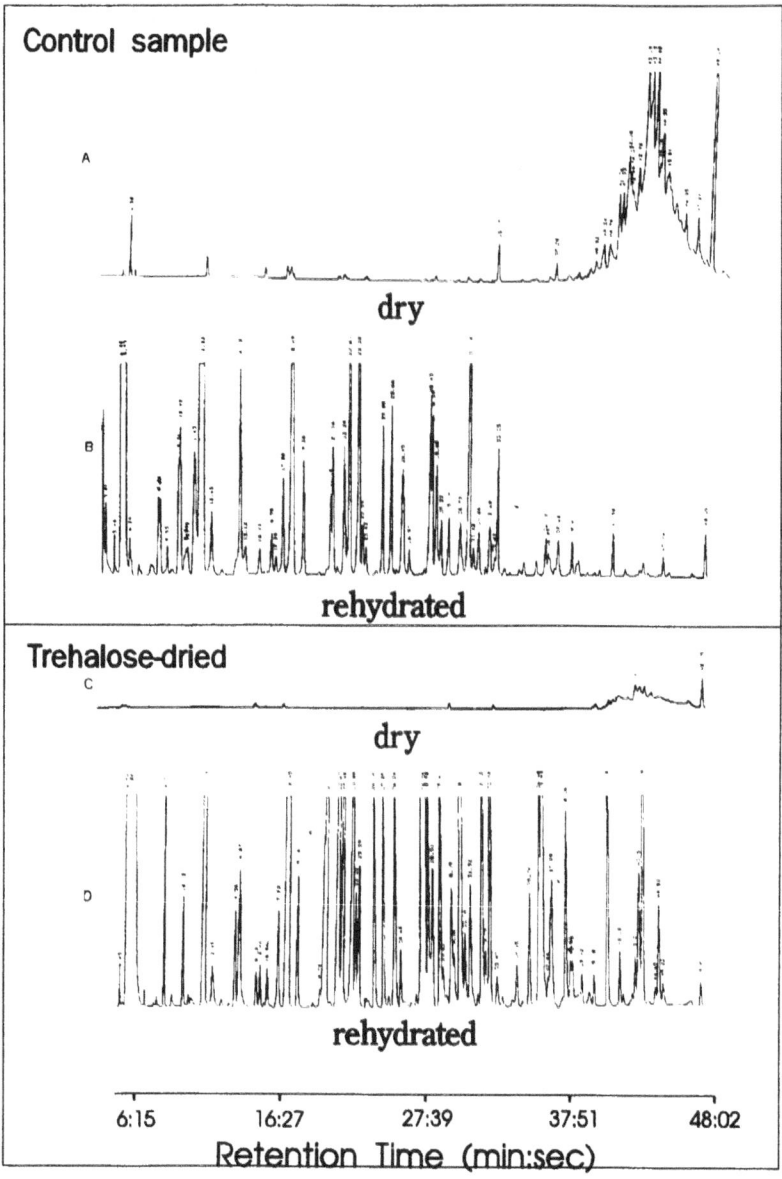

Figure 7.6 Gas chromatogram of head space analysis of banana volatiles. Fresh, ripe banana of the variety 'Cavendish' was puréed and divided into two batches. One was blended with 10% w/w trehalose powder which dissolved in the purée. Both batches were then vacuum-dried at 37–40°C. After storage for several months in polyethylene bags at room temperature, samples were subjected to head space analysis and gas chromatography by an independent laboratory (Institute of Food Research, Reading, England). (A) Purée dried without treha-lose emits volatiles in a cluster of peaks with a mobility of 40–47 min demonstrating a slow loss of flavour volatiles during storage of the control dried banana samples. (B) The loss of flavour volatiles by the control dry banana samples is further demonstrated by the lack of peaks with mobilities greater than 35 min in the head space analysis chromatography profile

tics (Figure 7.7), there was also a significant decrease in the amounts of the degradation compounds furfural and α-humulene produced in the trehalose-dried samples (Figure 7.7 arrowed). These compounds are known to be characteristic degradation products formed as a result of the Maillard reaction between the naturally occurring sugars and proteins in most foodstuffs. Trehalose thus appears to inhibit this reaction during the drying process. As the 'cooked' flavours and aromas are mostly due to these and other Maillard reaction products, such as furans, furanones, etc., this inhibition of the Maillard reaction by trehalose probably accounts for the anecdotal observations of the sensory evaluation panels. Though trehalose, being a very stable, non-reducing sugar, would itself not undergo the Maillard reaction, how it inhibits the Maillard reactions between the natural sugars and proteins of foodstuffs being dried is unclear and an area of particular interest and active research at Osmotica Foods.

The ability of trehalose to retain and preserve the volatile aromatic molecules responsible for the characteristic aromas and flavours of fresh herbs and fruits during drying enables the development of revolutionary new food processes to maximise this effect and produce superior dried powders for use as flavouring additives. To maximise the effects of trehalose, we have developed a novel drying process and instrumentation called the Captive Atmosphere Partial Pressure Drier.[14] The basic principle of the process is the use of the selective removal of water vapour to lower the partial vapour pressure in the drying chamber and affect the drying of the food product. This selective removal of water vapour does not affect the presence of other volatiles thus enhancing their entrapment and retention in the dried foodstuff.

7.3.2 Nutritional characteristics

Trehalose occurs naturally in many widely consumed foods. It is present in mushrooms,[16] yeast products such as bread,[17] beer,[18] wine[19] and vinegar,[20] in honey[21] and some seeds.[22] Certain mushrooms which show cryptobiotic properties (Figure 7.8) are indefinitely stable on storage at room temperature when dry and promptly rehydrate to give a fresh product, contain up to 20% of trehalose.[23] Undoubtedly the most familiar example of a commonly eaten cryptobiont is Baker's yeast. Its high trehalose content[10] allows it to survive indefinitely on the kitchen shelf as

---

of the rehydrated control samples. (C) The banana samples dried with trehalose emits very few volatiles showing that loss of flavour on storage is minimal. (D) When rehydrated, purée dried with trehalose emits volatiles over the whole range of mobilities from 6 to 45 min and both the number and size of the peaks is much greater than in the control samples (B). The trehalose-dried purées retain the majority of the volatile species which give the fresh fruit its unique character.

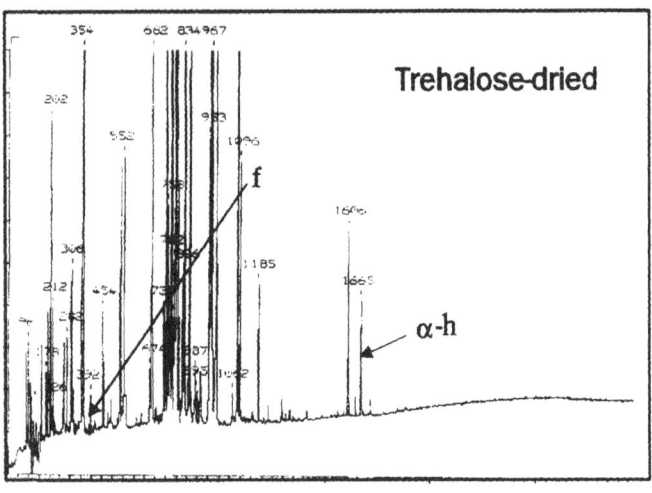

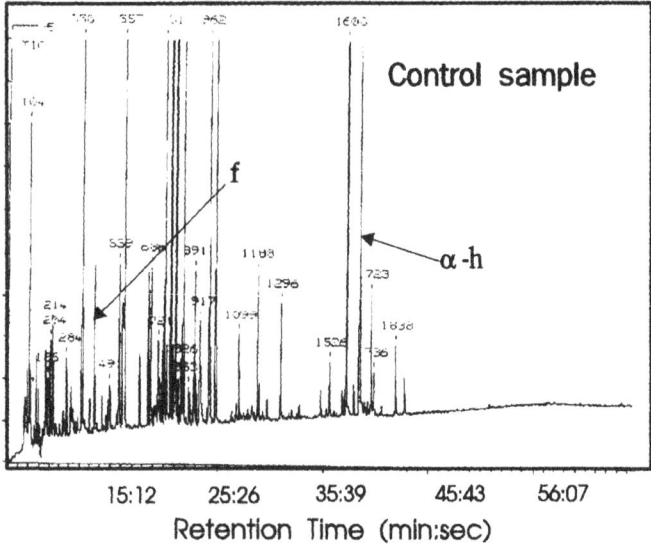

Figure 7.7 Gas chromatogram of head space analysis of mango volatiles. Fresh, ripe mango of the variety 'Malcurado' was puréed and divided into two batches. One was blended with 10% w/w trehalose and the batches were then vacuum-dried at 37–40°C. After storage for several months in polyethylene bags at room temperature, samples were subjected to head space analysis and gas chromatography by an independent laboratory (Institute of Food Research, Reading, England). As well as the markedly greater retention of volatiles in the mango purée dried in the presence of trehalose, there was a significant decrease in the size of the peaks for the Maillard reaction products α-humulene and furfural (arrowed).

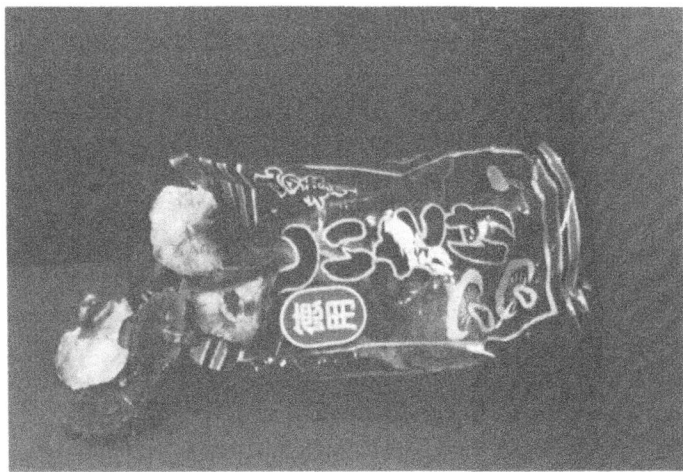

Figure 7.8 Shiitake mushrooms. These mushrooms produce and store large amounts of trehalose which enables them to dry naturally and be kept on the shelf and yet rapidly rehydrate to give a fresh product for cooking.

a dry product and yet it begins the metabolic production of CO_2 in dough making, within minutes of rehydration.

Omnivores and herbivores, including humans, have a specific enzyme, trehalase, in the intestinal brush border which splits trehalose into two glucose molecules.[24] Dietary trehalose is thus rapidly hydrolysed to the normal body constituent glucose and absorbed. Many other disaccharides cause vomiting, flatulence and diarrhoea if eaten in significant amounts, as they are not digestible. Trehalose thus does not have these undesirable dietary effects and genetic deficiency of intestinal trehalase associated with trehalose intolerance is rare, being considerably less frequently observed than lactose intolerance.[25] Indeed the presence of the enzyme trehalase in the proximal tubules of the kidney and the plasma[24] means that this sugar can even be injected into the bloodstream without harm. Since the ability of trehalose to preserve biological molecules during drying extends to small molecules such as vitamins,[26] trehalose-dried foods may have a higher nutritional value than foods dried without trehalose. Trehalose is equivalent to an equal amount of glucose in calorific value. Trehalose has already received government approval as a novel food substance in the UK and does not require approval in Holland and Denmark. Similar approval should be a formality in other countries in view of its bland nature and widespread occurrence in substantial quantities in common foods.

7.4 Mechanism of action

The detailed molecular mechanism of the protective action of trehalose is not yet clear. However, this is an area of active research and certain principles have emerged which provide some insights into the mechanism of action of trehalose. Three theories have been put forward to explain the mechanism of action of trehalose, none of which appear to be exclusively correct on the basis of current experimental evidence.

7.4.1 Water replacement hypothesis

The ability of sugar molecules to bind protectively to the surface of molecular structures has been ascribed to their ability also to form hydrogen bonds; the so called 'water replacement hypothesis'.[7,27] The 3-dimensional shape of virtually all biological macromolecules in solution depends on stabilisation of the structure by a shell of water molecules hydrogen-bonded to their surfaces. For example, proteins are reported to have between 0.25 and 0.75 grams of water hydrogen-bonded to each gram of protein in solution.[28] Some evidence that sugar molecules can indeed replace structural surface-bound water has been obtained by using ^{1}H NMR analysis solutions of ovalbumin containing sucrose.[29]

Unlike most other disaccharides, trehalose has no direct internal hydrogen bonds. All four internal bonds are with the two water molecules which form part of the native dihydrate structure. This arrangement gives the molecule unusual flexibility about the disaccharide bond, which may allow trehalose to conform more closely to the irregular surface of macromolecules than other, more rigid, disaccharides in which the rings are directly hydrogen bonded to each other. Consistent with this possibility, data on the perturbation of phospholipid bilayers has previously been interpreted as direct evidence for specific binding of trehalose to phospholipid head groups.[7] However, these results have recently been ascribed to lipid or detergent contaminants in the sugar preparations used.[27] We have been unable to show any direct interactions between trehalose and model proteins either in equilibrium dialysis or dielectric relaxation studies and our results thus do not support the water replacement hypothesis. However, these studies are, by definition, carried out using trehalose solutions and may thus not reflect the situation at the extremely high concentrations of trehalose attained in the final stages of drying.

7.4.2 Glass transformation hypothesis

The well-known ability of sugars to solidify from solution as a glass rather than by crystallisation has also been suggested as an important element in the mechanism by which trehalose confers desiccation tolerance on cryptobionts.[30,31] Extrapolating from the glasses formed during super-cooling of liquids, a number of workers have argued that the formation of glass is the sole requirement for preservation of biomolecules.[32-34] A further property of the glass transition is to prevent the loss of small hydrophobic volatile esters during drying and storage thus ensuring their release only on rewetting and dissolution of the glassy matrix.

Unlike many other sugars that also undergo glass transformation treha-lose produces glasses which are not hygroscopic. In contrast, sucrose glasses are very hygroscopic and, even at modest and commonly experi-enced ambient humidities, they readily attract water and become tacky. This absorbed water dissolves the surface layers of sucrose and, since the crystalline state is more stable, the sucrose glass gradually recrystallises. This process is well known in the food industry as 'weeping' and is responsible for the loss of clarity and quality of clear boiled sweets and other products based on sugar glass. Unlike the stable clear glass produced by trehalose which remains in the glassy state indefinitely, glasses of other sugars that recrystallise become porous and therefore permeable to volatile molecules which are slowly lost from the dried products.

The glass-transition theory has received support from food processing studies and in particular with respect to the retention of volatiles in the preparation of biscuits.[34] Most of these studies, however, do not contain actual measured experimental data but are theoretical considerations based on extrapolations from measurements on glass transition tempera-tures made in a variety of different systems.[32,34] Furthermore, the validity of these values for glass transition temperatures, measured by differential scanning calorimetry, the recommended experimental method,[32] has recently been questioned.[35] However, notwithstanding the latter reserva-tions, we have measured the glass transition temperatures of the actual trehalose-dried samples that we have assayed in our stability studies. The values we obtain are consistently lower than room temperature. However, when biological activity is assayed, the same samples are stable for at least 16 months and not just at room temperature but even at temperatures of +70°C! It is clear from these studies that glass transformation is certainly not the complete explanation for the effects of trehalose. However, the limited mobility of molecules in glasses, the proposed basis of their protective effect, may also be a feature of 'rubbery' states of trehalose as experimental measurements have shown that mobility in trehalose solutions begins to be restricted at temperatures well above the glass transition temperatures.[31]

7.4.3 Chemical stability hypothesis

Trehalose is one of the most stable known sugars. Because the disaccharide bond is formed between the 1–1 carbon atoms of the glucopyranose rings it is a completely non-reducing sugar and the glycosidic bond has a low free energy of activation (K_0 of 119 sec^{-1}), making the disaccharide structure very stable to hydrolysis. Apart from sucrose, completely non-reducing disaccharides are uncommon. However, in contrast to trehalose, the glycosidic bond in sucrose has a high free energy of activation (K_0 4.44 × 10^4 sec^{-1}), which renders the disaccharide readily labile to hydrolysis in the presence of mild acid to yield glucose and fructose, both strong reducing sugars. As one of the consequences of the progression of the Maillard reaction in dried foods is a fall in pH, this susceptibility to acid hydrolysis is a particularly acute problem with foodstuffs dried in sucrose. Except under extreme conditions, trehalose does not hydrolyse to yield glucose and thus cannot undergo the Maillard reaction with proteins or peptides in the dried foodstuffs.

The results of our studies on the head space analysis of fruit purees dried in trehalose suggest that the presence of this sugar during drying actually inhibits the Maillard reactions between the naturally occurring sugars and proteins in the foodstuffs being dried (Figure 7.6). This unexpected property of trehalose is a particularly significant factor with respect to food processing involving dehydration, as a number of the spontaneous chemical reactions in the Maillard cascade are greatly accelerated at low water activities.[36] As these reactions result not only in the production of high levels of undesirable compounds such as furans, furfurans, imidazoles, N-nitroso derivatives and melanoids but also significant losses in nutritive value,[2,6] the inhibition of the Maillard reaction would thus be a significant advantage of the use of trehalose in the drying of foodstuffs.

7.5 Conclusion

The remarkable ability of trehalose to completely protect cryptobiotic plants and animals from desiccation damage can be applied on an industrial scale to the drying of foods. Trehalose itself is bland, non-toxic and commercially attractive as 'Nature's way of drying'. While at present the trehalose technology has mainly been applied to the drying of foods which are liquid or puréed, our preliminary results indicate that the process can be successfully extended to sliced foods with only minor modification of technique. An additional benefit of the use of trehalose in food processing is the inhibition of the Maillard reaction between the naturally occurring proteins and sugars in the foodstuffs being dried. As the Maillard reaction results in both the production of potentially toxic by-products, as well as the loss of the reacting proteins and sugars, trehalose dried foods should not only contain less toxic by-products of processing but also have a higher nutritional content. Finally, the ability to dry complex foods such as whole fruit, vegetables and even cultivated seafood is within reach by using modern genetic engineering techniques to confer cryptobiotic properties on these organisms.

On a more light hearted note, given its undoubted potential as an additive for food preservation, it is perhaps fitting that the name trehalose is derived from *Trehala manna*, the cocoon of the beetle genus *Larinus*. This pupa of the parasitic beetle is found on aromatic shrubs in the Middle East and is thought to be the biblical manna that nutritionally sustained the Israelites in the desert during their journey to the promised land.

References

1. MacCarthy, D. (1986) *Concentration and Drying of Foods.*, Elsevier Applied Science, London.
2. Somogyi, J.C. and Miller, H.R. (1989) Nutritional impact of food processing. *Bibl. Nutr. Diet.*, No. 43. Karger, Basel.
3. (a) Erbersdobler, H.F. (1986) Loss of nutritive value on drying. In *Concentration and Drying of Foods*, (ed. MacCarthy, D.), Elsevier Applied Science, London.
3. (b) Erbersdobler, H.F. (1989) Protein reactions during food processing and storage – their relevance to human nutrition, in *Nutritional Impact of Food Processing*, (eds Somogyi, J.C. and Miller, H.R.), *Bibl. Nutr. Diet.*, No. 43. Karger, Basel.
4. Labuza, T.P. and Saltmarsh, M. (1981) The non-enzymatic browning reaction as affected by water in foods, in *Water Activity: Influence on Food Quality* (eds Rockland, L.B. and Stewart, G.F.), Academic Press, New York.
5. Nursten, H.E. (1986) Maillard browning reactions in dried foods, in *Concentration and Drying of Foods*. (ed. MacCarthy, D.), Elsevier Applied Science, London.
6. O'Brien, J. and Morrissey, P.A. (1989) Nutritional and toxicological aspects of the Maillard browning reaction in foods. *Crit. Rev. Food Sci. Nutr.*, **28**, 211–248.
7. Crowe, J.H., Crowe, L.M. and Mouradian, R. (1983) Stabilization of biological membranes at low water activities. *Cryobiology*, **20**, 346–356.
8. Young, S. (1985) The dry life. *New Scientist* 31 Oct, 40–44.
9. Weisburd, S. (1988) Death-defying dehydration. *Science News*, **133**.
10. Gadd, G.M., Chalmers, K. and Reed, R.H. (1988) The role of trehalose in dehydration resistance of *Saccharomyces cerevesiae*. *FEMS Microbiol Letts.*, **48**, 249–254.
11. Blakeley, D., Tolliday, B., Colaco, C.A.L.S. and Roser, B. (1990) Dry, instant blood typing plate for bedside use. *The Lancet*, **336**, 854–855.
12. Colaco, C., Sen, S., Thangavelu, M., Pinder, S. and Roser, B. (1992) Extraordinary stability of enzymes dried in trehalose: Simplified molecular biology. *Biotech.*, **10**, 1007–1012.
13. Roser, B. (1987) Protection of Proteins and the like. International Patent application no PCT/GB86/00396 (and corresponding patents and applications worldwide).
14. Roser, B. (1989a) Drying water-containing foodstuffs. International Patent application no PCT/GB88/00511 (and corresponding patents and applications worldwide).
15. Roser, B. (1989b) Preservation of viruses. International Patent application no PCT/GB89/00047 (and corresponding patents and applications worldwide).
16. Birch, G.G. (1970) Mushroom sugar in food technology. *Process Biochemistry*, **5**, 9.
17. Oda, Y., Uno, K. and Ohta, S. (1987) Selection of yeasts for breadmaking by the frozen dough method. *Appl. & Environ. Microbiol.*, **58**, 921–943.
18. Patel, G.B. and Ingledew, W.M. (1974) Internal carbohydrates of *Saccharomyces carlsbergensis* during commercial larger brewing. *J. Inst. Brewing*, **79**, 392–396.
19. Bertrand, A., Dubernet, M.O. and Ribereau Gayon, P. (1976) Trehalose, main disaccharide present in wines. *Comptes Rend. Hebdom. des Seances de l'Acad. des Sci.*, **280**, 1907–1910.
20. Olano, A. and Gomez Cordoves, M.C. (1985) Non volatile constituents in vinegars: Inositol, trehalose, total polyphenols and catechins. *Anal. Bromatol.*, **35**, 171–176.

21. Bogdanov, S. and Baumann, E. (1989) Determination of the sugar composition of honeys by HPLC. *Mitteil. Gebiete Lebensmittel. und Hygeine.*, **79**, 198–206.
22. Mikolajczak, K.L., Smith, C.R. Jr. and Wolff, I.A. (1970) Phenolic and sugar components of Armavirec variety sunflower *Helianthus anuus* seed meal. *J. Agricult. & Food Chem.*, **18**, 27–32.
23. Yoshida, H., Sugahara, T. and Hayashi, J. (1988) Free sugars, free sugar alcohols and organic acids of wild mushrooms. *J. Jap. Soc. Food Sci. & Technol.*, **33**, 426–433.
24. Riby, J., Sunshine, S. and Kretchmer, N. (1990) Renal trehalase: Function and development. *Comp. Biochem. Physiol.*, **95A**, 95–99.
25. Madzarovova-Nohejlova, J. (1973) Trehalase deficiency in a family *Gastroenterol.*, **65**, 130–133.
26. Kampinga, J. personal communication.
27. Clegg, J.S. (1986) The physical properties and metabolic status of *Artemia* cysts at low water contents: The water replacement hypothesis, in *Membranes, Metabolism and Dry Organisms*, (ed. A. Carl Leopold), Cornell University Press, pp. 169–187.
28. Saenger, W. (1987) Structure and dynamics of water surrounding biomolecules. *Ann. Rev. Biophys. Biophys. Chem.*, **16**, 93–114.
29. Hanafusa, N. (1985) The interaction of hydration water and protein with cryoprotectant, pp. 1–6 in *Fundamentals and Application of Freeze Drying to Biological Materials, Drugs and Foodstuffs, Int. Inst. Refrig. Commiss. Mtg. Tokyo*, **1.3**.
30. Burke, M.J. (1986) The glassy state and survival of anhydrous biological systems, in *Membranes, Metabolism and Dry Organisms*, (ed. A. Carl Leopold), Cornell University Press, pp. 358–363.
31. Green, J.L. and Angell, C.A. (1989) Phase relations and vitrification in saccharide-water solutions and the trehalose anomaly. *J. Phys. Chem.*, **93**, 2880-2882.
32. Franks, F. (1990) Freeze Drying: From empiricism to predictability. *Cryoletters*, **11**, 93–110.
33. Hirsch, A.G., Williams, R.J. and Merryman, H.T. (1985) A novel method of natural cryoprotection. Intracellular Glass Formation in deeply frozen *Populus*. *Plant Physiol.*, **79**, 41–56.
34. Levine, H. and Slade, L. (1988) Water as a plasticiser: physico-chemical aspects of low-moisture polymeric systems. *Water Sci. Rev.*, **3**, 79–185.
35. Ablett, S., Izzard, M.J. and Lillford, P.J. (1992) Differential scanning calorimetric study of sucrose and glycerol solutions. *J. Chem. Soc. Faraday Trans.*, **88**, 789–794.
36. Rockland, L.B. and Stewart, G.F. (1981) Water activity: influences on food quality. Academic Press, New York.

8 Packaging of fruits and vegetables: recent results

F. RIQUELME, M.T. PRETEL, G. MARTÍNEZ,
M. SERRANO, A. AMORÓS and F. ROMOJARO

Abstract

In this chapter we detail the composition of the equilibrium atmosphere which is produced when fruits and vegetables are packed in plastic films of a given permeability. The types of films used in packaging and the influence of MAP on respiratory activity, ethylene synthesis, chemical composition and the organoleptic characteristics of several species of fruits and vegetables are discussed. The most recent application of this technology is described, highlighting the most important effects on the physiology, biochemistry and quality of the packaged products and the possibility of prolonging their shelf-life. Factors which determine microbiological safety and the organisms most commonly found in the packaged products are described. The influence of certain external factors such as temperature, relative humidity and light on produce preservation in MAP is discussed.

8.1 Introduction

There has been a steady decrease in the amount of fresh fruit and vegetables consumed in the EC. Many explanations have been offered including the changes in social and eating habits and a drop in the organoleptic qualities of fruit and vegetables due to the new methods of marketing and distribution.

For the produce to reach the shops in optimum conditions, it should be harvested at the correct moment and kept in such a way that its evolution to senescence is slowed down. Exposure to low temperatures immediately after harvest is useful but is not sufficient for maintaining quality during all the marketing steps. For this reason, other techniques associated with cold storage have been studied. One of these is the use of plastic wrappings which act in two ways: to prevent the fruit from drying out by creating an atmosphere with a high relative humidity and to modify the headspace atmosphere yielding high levels of CO_2 and low levels of O_2. The action of low temperatures to slow down the evolution of fruit towards senescence can be aided by the use of 'modified atmosphere packaging' (MAP).[1]

The correct use of MAP complements and improves traditional post-harvest preservation methods by protecting produce from harmful factors which lower quality and thus facilitates their marketing.

8.2 Definition of MAP

MAP consists in preserving fruits, or vegetables either whole or sliced, under plastic films of known permeability. It is based on modifying the initial gaseous conditions of the packed fruit's surroundings as a consequence of both its own metabolic activity and the semipermeable barrier of the packaging.

The metabolic activity of the produce is maintained after harvesting, and gaseous interchanges with the surrounding atmosphere continue. As a consequence of respiratory activity, the composition of the atmosphere under the film is modified: the concentration of O_2 is decreased whereas that of CO_2 and water vapour increase. A gradient is created between the fruit and the exterior, which facilitates the transfer of gases between the interior and exterior of the packaging.

A dynamic equilibrium is established between the gases produced by the fruit and those of the microenvironment surrounding it. In this equilibrium, O_2 consumption and CO_2 emission are equal to the release of these gases through the packaging at a determined temperature.[2] These fruit–microenvironment and microenvironment–atmosphere interchanges occur simultaneously.[3] Thus, to obtain a stable atmosphere for plastic-packaged fresh vegetables, the following equilibria must be established:

– The intensity of oxygen absorption by the product should be equal to the total outflow of O_2 through the plastic.
– The intensity of CO_2 emission by the product should be equal to the total outflow of CO_2 through the plastic.
– The respiratory intensity of the product together with the permeability of the film, temperature and relative humidity are the factors which determine the O_2 and CO_2 equilibrium conditions inside the packaging.

8.3 Types of plastic films used for MAP

These last few years have seen an increase in the use of flexible plastic films due to the development of new materials and new systems of application. In order to reach equilibrium in the atmosphere, optimal preservation conditions have been sought by using the most suitable degree of film permeability and systems to eliminate or reduce some of the components of the atmosphere (scrubbers). The development of these new forms of fruit and vegetable packaging materials has led to the production of films with new characteristics based on the modification of different factors: the material itself, combinations to form laminates of different thicknesses, microporosity and microperforations by means of which the permeability, transmission of water vapour, antimisting properties, strength and sealing facility, etc. can be modified.

Numerous advances were made in the design and manufacture of polymeric films with a wide range of gas permeability. This has been paralleled by the development of absorbers and adsorbers of O_2, CO_2 and C_2H_4 and other ways of maintaining the desired atmosphere within the packaging.[2] These advances have aroused the interest of industry in the creation and maintenance of atmospheres by means of polymeric films. The number of types of films nowadays available for food packaging is very high although relatively few are used in the preservation and marketing of fresh vegetable products. Among the most commonly used, we may mention PVC, polyethylene, polypropylene and polystyrene. These polymers present a wide range of permeability which cover most necessities from barrier-type films with a very reduced permeability to highly permeable films. These varying permeabilities are obtained by specially treating the film or its constituents.

Another aspect which is at present receiving serious interest is the use of films that can modify their own permeability in response to changes in the external atmosphere, or to modifications in the composition of the internal atmosphere. If these 'intelligent' films can really adapt their characteristics, then the influence of such changes as those which occur in the long marketing process (changes in temperature, relative humidity, light, etc.) can be reduced to a minimum. However, the development of a new type of film should be based on the preservation conditions and the type of atmosphere produced by the fruit or vegetable due to their physiological activities. A suitable film with the correct rate of permeability must be chosen for each product so that the desired equilibrium in the modified atmosphere is reached.

Even the use of a suitable film for a particular product cannot always prevent the damage caused by an excessively modified atmosphere. Indeed, the packaging material might be insufficiently permeable when changes occur in storage conditions during the long marketing process. These changes, which affect temperature and relative humidity, alter the respiratory intensity of the packaged product and the permeability of the film. However, with the most widely used packaging materials, the changes in the physiological activity of the vegetables are always more intense than what can be predicted in the film. Thus, PE shows a higher response to temperature change than polystyrene.[4] It was found[5] that the permeability of PVC to CO_2 is 1.5 to 2 times greater at 10°C than at 0°C. The ratio is about 2 to 3.5 at 22°C. This change in permeability directly affects the concentration of gases within the packaging material and, consequently, the shelf-life of the product.

Considering the transfer of gases through the packaging materials used for the MAP, attention should be paid to the modification originating from the intrinsic permeability of these materials. This depends on their permeability characteristics, their thickness, and to a certain extent to the external and internal conditions to which the packaging is exposed. The possible existence of tears, cracks, perforations and a deficiency in the seams of the packaging must also be taken into account. These may be due to accidental damage or to a mechanical defect in the pinching mechanisms for creating the seams or to insufficient heat being applied. However, these defects in the film could improve the modified atmosphere by increasing gas transfer, providing that the permeability has been duly evaluated and is suitable for the produce in question. This possibility has been considered since the first studies of controlled atmospheres. The silicon windows of the Marcellin system actually provide a way of increasing the interchange between the external and internal atmosphere in a controlled manner. The application of this idea to MAP led to the development of new materials such as microperforated films or active membrane label systems.

Weight loss in the packaged product should also be taken into account. With low permeability films for water vapour, final weight loss is in the order of 1–2% while 'cellophane' type films, with a high water diffusivity and water vapour permeability, can produce losses of 8–9%. Another important aspect is the efficacy of the sealing mechanism. Although the equilibrated atmosphere is directly linked to storage conditions and to product–package interactions, it is also important that the plastics used are sufficiently homogeneous and the sealing mechanisms sufficiently secure to avoid product loss due to bad packaging and handling.

The risk of deterioration in the quality of the packaged product and the associated high costs mean that there must be strict quality control in the packaging process and the final product. The fundamental parameters which should be controlled are the composition of the internal atmosphere (O_2, CO_2 and C_2H_4) and the tightness and resistance of the seals. Control by a statistical sampling of these parameters will make it possible to know the effectiveness of the process.

8.4   Effect of MA on the respiratory activity of fresh fruits and vegetables

It is widely accepted that one of the main effects of MA on fruit and vegetable metabolism is a decrease in their respiratory intensity corresponding to a decrease in the consumption of substrates, CO_2 production, O_2 consumption and release of heat.[6] These changes result in a slowed metabolism, which allows the product to be kept for longer periods.[2] The

decrease in respiratory rate basically depends on the species being preserved and the gaseous composition, inside the packaging at equilibrium.
For example, 'garriguette' strawberries show a decrease in respiratory rate
when the concentration of CO_2 increases to 20% and O_2 decreases to 3%
at equilibrium, which allows the respiratory rate to be reduced to a value
of 42% of a non packed control. Similarly, 'California' pepper kept for 7
and 14 days at 20°C under films of differing non-selective permeability
shows a fall in respiratory rate in the least permeable films or, when stored
at different temperatures, at the lowest temperature.

   This inhibitory effect due to the high CO_2 concentration may not be
irreversible if the physiological activity of the fruit has not been altered.
For example, the respiratory activity of three varieties of apricot recovers
once they are removed from the bags and placed in a normal atmosphere.[7]
Similar results have been obtained with 'Bartlett' pear exposed to concentrations of 0.5, 0.25 and 0.3% of O_2 for 10 days. However these need
3 days to reach the same values of a fruit exposed to the air, probably
because of a residual effect of treatment.[8] The decrease in respiratory
activity shown by fruits and vegetables exposed to MA is proportional
to O_2 concentration. However, this concentration should not fall below
1 to 3% otherwise an anaerobic metabolism could begin with the
decarboxylation of the prunic acid to acetaldehyde, CO_2 and ethanol.[2]
This critical O_2 level depends on the species, variety, temperature and
preservation time. Oxygen is a substrate not only for cytochrome oxidase,
an enzyme involved in the respiration process, but also for several other
oxidases such as ascorbico-oxidase, phenolase, peroxidase, AIA-oxidase,
lipoxigenase, amino-oxidase and cytochrome P450. Thus, low concentrations of O_2 can modify the metabolisms in which these enzymes act.[9]

   The fact that aerobic metabolism might start with a drop in product
quality led to the investigation of the minimum O_2 concentration responsible for the triggering of the transition between aerobic and anaerobic
respiration. This in turn led to the definition of the 'extinction point' (EP),
a concentration of O_2 at which ethanol production ceases.[10] Since alcohol
is a metabolite which also occurs in fruit under aerobic conditions, an
alternative index, 'anaerobic compensation point' (ACP), has also been
proposed. This defines the concentrations of O_2 at which CO_2 is minimum.
For 'Bartlett' pear and a culture of 'Passe Crassanne' pear cells submitted
to different concentrations of O_2 for 2 to 4 hours, these values were 1.6
to 1.7% and 1.1 to 1.3% respectively.[11]

   In broccoli an O_2 concentration of 2.5% not only causes the respiration
rate to fall by 50% with respect to the maximum value reached in the open
air but also virtually eliminates the 'climacteric'.[12] Similar reductions have
been observed in some varieties of nectarines stored for 3 days in 0.5%
O_2 at different temperatures.[13] Significant reductions in CO_2 production
have been found in 'Bartlett' pear exposed to atmospheres of 0.5 and

0.25% O_2 at 0°C. However, when the concentration of O_2 is reduced to 0.03% there is an increase in respiration probably due to the onset of anaerobic respiration.[8] A fall in the respiration rate, in the internal concentration of CO_2 and an increase in the resistance to CO_2 diffusion occurs when peach is exposed to 0.25% and 0.02% O_2,[14] while orange at 0°C shows no decrease in CO_2 production and generally presents a greater internal concentration of CO_2 and resistance to CO_2 diffusion.[15]

The factors which determine the tolerance of a fruit to low O_2 atmospheres were studied recently. In 'Granny Smith' and 'Yellow Newton' apples and 'Angelero' plum the respiration rate decreases, while that of '20th Century' apple increases. The resistance to CO_2 diffusion increases in all cases as does the concentration of internal CO_2, with the exception of the plum 'Angelero', which shows no difference from fruit exposed to air.[16]

CO_2 levels ranging over about 20% or more can also affect respiration. Depending on the plant material used and the O_2 concentration, it may induce anaerobic respiration. CO_2 concentrations below 20% can decrease the respiration rate of apple,[17] pear[18] and strawberry.[19] The inhibitory effect of CO_2 on respiration is not clear since high concentrations of CO_2 (60% CO_2 + 20% O_2 + 20% N) produce different responses depending on the horticultural crop and its stage of development. In some products there is a pronounced fall in respiration (measured by O_2 uptake) and in others such as lettuce, spinach, cucumber and aubergine, there is practically no change.[20,21]

8.5   Effect of MA on ethylene biosynthesis

O_2 is necessary for ACC (1-amino cyclopropane-1-carboxylic acid) to be converted into ethylene[22] and for metabolisation of the ethylene produced in the union with the receptors.[23,24] Low O_2 concentrations can inhibit the action of EFE (ethylene-forming enzyme) and increase the levels of ACC, lessen the phenomenon of autocatalytic ethylene production and limit the increase in ACC synthase activity, which is stimulated by the maturation hormone.[25] The decrease in ethylene production and the limitation of its effect, especially to delay maturation and senescence, was applied to the preservation of fruits and vegetables by use of CA/MA with low levels of O_2 to the following products: apple,[2,26–28] nectarine,[29] pear,[8] plum,[16] peach[14] and broccoli.[30]

As a general rule, ethylene production falls, along with conjugated ACC, ACC-synthase activity and EFE, while levels of ACC increase. ACC and ABA levels fall when Chinese cabbage is exposed to atmospheres with 1% CO_2 for 12 weeks at 0°C, although this concentration is not sufficiently low to reduce the conversion of ACC into ethylene.[31] The way in which CO_2 acts on ethylene production is complex and the interaction between both gases has not been clearly established. CO_2 is considered to be a competitive inhibitor of ethylene action.[32] In plant tissues, CO_2 speeds up the metabolization of ethylene into ethylene oxide and the competitive inhibition would function like a system completely oxidizing the ethylene into CO_2. On the other hand the oxidation of ethylene to CO_2 can be inhibited by a 'feedback inhibition mechanism'.[23] The possible interactions of CO_2 in the ethylene metabolic pathway are numerous. It was shown that low levels of CO_2 regulate its synthesis and high levels inhibit its action, although the actual mechanisms responsible for these effects are not understood.[33] The inhibitory effect of CO_2 on ethylene production was observed in several climacteric fruits such as apple,[17,28,34-36] avocado,[36,37] pear[8] and figs.[38] However, in other cases, high concentrations of CO_2 favour ethylene production. This is the case with broccoli,[12] banana, aubergine, cucumber and lettuce.[21] CO_2 can inhibit ethylene formation by limiting the formation of ACC[39] and inhibiting the conversion of ACC into ethylene.[17] In apple, the induction of ACC-synthase and the EFE activity are inhibited when the fruit is exposed to high concentrations of CO_2.[28] The effects of a low concentration of O_2 on ACC-synthase inhibition and ethylene production are more pronounced when CO_2 is added to the mixture of gases.[27] In plums, apple and avocado, EFE induction was inhibited, although its activity increased.[36]

Since CO_2 acts on the metabolic pathway of ethylene, the action of this gas on the enzymatic systems involved (ACC-synthase and EFE) and on intermediate metabolites (ACC and MACC) has been studied in apple,[17,27,28,36,40] tomato[41] and broccoli.[12] In plugs of pre-climacteric apricot exposed to different mixtures of CO_2 and O_2, EFE is partially inhibited when the concentration of O_2 falls, and reaches the maximum inhibition with a (20% CO_2–1% O_2) mixture while the same activity is maintained as in the control in atmospheres of CO_2 enriched air. Levels of total ACC, both free and conjugated, decrease in the mixtures containing high concentrations of CO_2 and are restored as O_2 levels fall.[7]

8.6 Influence of MA on chemical composition

Fruit maturation and senescence is characterized by important physiological and biochemical changes in the tissues, which modify the fruit's sensory characteristics. The modification of the compounds responsible for the colour, taste and odour determine the final fruit quality. Low O_2 concentrations and/or high CO_2 concentrations can alter the metabolism of some constituents and/or the rate at which they are broken down or formed. Anaerobic respiration or the physiological damage caused by CO_2 alter the composition of the product and considerably alter its quality.

These undesirable effects mean that O_2 and/or CO_2 concentrations should be within certain limits. Thus, it is necessary to take into consideration other factors which affect the preservation conditions, among which the most important are the variety, state of maturity, quality and temperature. Recent reviews of CA/MA for different fruits and vegetables detail the minimum concentrations of O_2 and maximum of CO_2, the O_2–CO_2 mixtures, optimal temperatures and the positive or negative effects on the produce's physiology and biochemistry and the physiological disorders which may arise during preservation.[2,42–45]

It was shown that low O_2 concentrations and/or high CO_2 concentrations delay maturation and strongly affect texture and colour. In general, softening slows down and the fruit has a consistent texture at the end of the experiment. This effect is more pronounced as CO_2 concentration in the mixture increases than when O_2 concentration decreases. The rate at which chlorophyll (green) diminishes and carotenoids (reds and yellows) and anthocians (blues) increase is also delayed as a result of the modified atmospheres. Although high CO_2 concentrations can reduce the formation of phenolic compounds, phenoloxidase activity and phenol oxidation,[46] it has been shown that when the O_2 and CO_2 concentrations are not within the tolerance limits, browning phenomena usually occur. Also, high levels of CO_2 prevent normal colouring in fruit when they are transferred to the open air.[6]

Modifications in chemical composition depend on the concentrations of O_2 and/or CO_2 and on the plant material. While fruit maintains a high level of titratable acidity, some vegetables (broccoli, cauliflower and lettuce) show the opposite effect.[47] In general, when a fruit is exposed to mixtures of O_2 and CO_2 over long periods, there is a decrease in the formation of volatile compounds, which affects the development of a good aroma. However, the transfer to the open air cancels this effect, assuming that the pre-climacteric stage has been passed.[48] It appears that the lack of good aroma and even the presence of undesirable ones which strongly affect the sensory quality of the product are due to the triggering of anaerobic respiration when the O_2 and CO_2 concentrations are too low or too high, respectively. Recent studies [27] confirm these changes which affect quality and examine the modification of other constituents of great interest as regards the application of a modified atmosphere. For apples, only a delay in softening and colour development is seen in atmospheres with 1% O_2. A concentration of 3% O_2 does not affect the evolution of texture and only slightly affects colour, additional CO_2 (3–6%) being necessary to obtain similar results.[27] Exposing of 'Bartlett' pears to 1%, 0.5% or 0.25% O_2 at 0.5 or 10°C for 10 days does not affect the soluble solid content, acidity and pH; only at 10°C is there a certain delay in softening and a yellowing of the colour. 20%, 50% and 80% concentrations of CO_2 give similar results, although pH is slightly increased (decreased acidity). Ethanol and acetaldehyde levels rise slightly with 0.25% O_2 and considerably with 50% and 80% CO_2.[8]

Low O_2 atmospheres (0.25% or 0.02%) at two different temperatures (0.5 and 10°C) have no effect on the soluble solid content of pears, apples and plums, pH or external aspect whereas titratable acidity remains constant and softening and skin colour changes are delayed. An alcoholic off-flavour, related to the alcoholic content of the fruit, is observed.[16] Low O_2 concentrations may have an insecticidal action. If fruit quality is not affected, this may be used for short periods to avoid the use of pesticides or cold treatment for several days. However, the O_2 concentrations needed to control insects or nectarines affect the fruit quality. Indeed, ethanol, acetaldehyde and internal browning increase, whereas texture and titratable acidity improve and there is no variation in pH or the soluble solid content.[29] On the other hand, the exposure of peach to 0.25 and 0.02% O_2 at 0° and 5°C for 40 days delays or prevents internal browning although there is an accumulation of ethanol and acetaldehyde. No changes in skin colour, texture or the soluble solid content are observed.[14]

Fruit softening during maturation occurs as a consequence of cell wall decomposition, which is associated with the activity of several hydrolytic enzymes. In avocado, a concentration of 2.5% O_2 prevents an increase in polygalacturonase, cellulase and acid phosphatase activity and delays the formation of a certain number of polypeptides which appear during the normal maturation of the fruit.[49] In a non-climacteric fruit such as strawberry, treatment with very low O_2 and/or high CO_2 concentrations prevents softening, decay and colour loss[19,50] but does not affect titratable acidity, pH, the soluble solid content or ascorbic acid.[19] When O_2 concentration is around 0.25% or below and CO_2 is 20% or above there is an accumulation of ethanol, acetaldehyde and ethyl acetate, which has an adverse effect on quality.[19,50] While 'Valencia' orange can tolerate O_2 concentrations of 0.02% at 5 or 10°C for days without the appearance of external or internal defects, a 60% CO_2 concentration has a strongly adverse effect characterized by browning of the skin. Neither treatment affects the colour of the juice, the soluble solid content, pH, titratable acidity or the ascorbic acid content although ethanol and acetaldehyde levels rise.[15]

Atmospheres containing 15 or 20% CO_2 prolong the post-harvest life of figs since losses through decay are diminished and external appearance improved. Furthermore, despite the increase in the ethanol and acetaldehyde contents, the relatively high concentration of soluble solids means that there are no detectable off-flavours at 0°C.[42] Beneficial effects of an increased CO_2 level have also been observed in okra,[51] sugar peas[52] and broccoli[30] although a concentration exceeding 10% CO_2 can induce undesirable odours and physiological injury. Similar CO_2 concentrations shorten the life of Iceberg lettuce as a result of browning when they are transferred to the open air. There is also an increase in phenylalanine ammonia-lyase, ionically bound indole acetic acid (IAA) oxidase, while soluble IAA oxidase decreases.[53]

8.7   Recent applications of MAP

MAP was recently applied to different types of fruits and vegetables in order to study the effects of different plastic polymers on the physiology, biochemistry and quality of the packaged product and the possibility of prolonging their post-harvest life. Since these studies are reviewed in the bibliography,[2,42,43,54] we shall limit ourselves to mentioning only the most recent applications of this technology to individual commodities. 'Conference' pear in preclimacteric stage packed in low density polyethylene for 3 days at 20°C reaches an equilibrium atmosphere of 4–9% CO_2 and 5% O_2 which completely inhibits chlorophyll degradation and slows softening. However, this delay in maturation prevents the normal development of the sensory characteristics of this variety.[55] Similar results are observed in 'Doyenne du Comice' pear and the risk of skin injury as the pears ripen in the MA pack limits the use of this technique.[56] Tomatoes harvested at breaker stage and packed in polymeric film for 23 days at 15°C with an active MA of 3.5–4% O_2 and CO_2 showed no change in quality when ripened in ambient conditions. The internal atmosphere was slightly modified after 24 h of packaging (2.5% O_2 and 8% CO_2) but reverted to the initial equilibrium conditions. MAP slowed maturation and delayed changes in acidity, soluble solids, texture, colour and polygalacturonase.[57]

The storage of almonds for periods exceeding 9 months necessitates the use of polymeric films of low O_2 permeability together with refrigeration if physical, chemical and sensory changes are to be avoided.[58] The use of a low density polyethylene (LDPE) film of 75 μm thickness and a permeability of 1300 ml O_2 ml/m^2.day.atm and 0.27 g H_2O ml/m^2.day.atm prolongs the life of muskmelon and preserves their quality. For prolonged transport, water vapour and ethylene should be reduced by the use of NaCl and $KMnO_4$, respectively, once the fruits are packed in waxed cardboard shipping boxes.[59] The maximum shelf-life of Triumph persimmons in MA using 0.08 mm low density polyethylene film is 8 weeks under vacuum and 20 weeks under nitrogen. After these durations there is an accumulation of ethanol and acetaldehyde which causes tissue browning.[60]

Using individual seal-packaging of high density polyethylene (HDPE) 'Eureka' yields a slower maturation of lemons and a better quality than that of unsealed fruits after 6 months storage at 13°C. The fact that ethanol content does not increase in the sealed fruit package shows that HDPE permits adequate gas interchange.[61]

CO_2 levels in film bags containing 'Starkinson' apples rise to 73% after 23 days storage at 2°C. These conditions result in browning problems and the destabilization of the several anthocians present in the peel.[62]

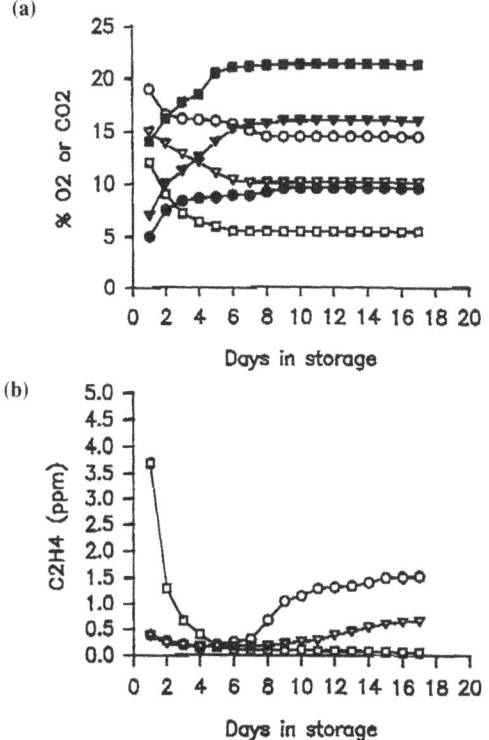

Figure 8.1 Changes in (a) O_2 (open symbols) and CO_2 (filled symbols); and (b) C_2H_4 with time within sealed packages of apricots (var. Búlida), kept at 10°C. Films permeabilities: ($\bigcirc$), 80 000; ($\triangledown$), 40 000; ($\square$), 20 000 cm³/m².24h.atm O_2/CO_2.

Figure 8.1 shows the evolution of CO_2, O_2 and C_2H_4 for 'Búlida' apricot sealed in three films of different non-selective permeability to CO_2 and O_2 (80 000; 40 000; 20 000 cm³/m².24 h.atm). The CO_2 and O_2 concentrations at equilibrium depend on the permeability. In all cases equilibrium was reached after the sixth day. Ethylene evolution shows a partial inhibition which is more pronounced in the less permeable films. This is sufficient to delay maturation. After 17 days at 10°C the fruit shows a firm texture, a high titratable acidity and a low development of colours and soluble solids. Sensory analysis has shown that MA storage can be beneficial for apricot marketing as long as CO_2 levels do not exceed 15%.[7]

The major effect of pre-packing of asparagus in different polymeric films to slow down deterioration was a reduction in moisture loss. Spear's quality was affected by MA. It was found that higher than normal levels of CO_2 in the internal atmosphere retard pathogen growth and tip opening. Best results were obtained using PVC and 0.3% perforated polypropylene.[63] Asparagus spears (cv. Regal) were better stored in non-perforated films, where the equilibrated atmosphere reached 11% CO_2 and less than 2% O_2. In the mean time lignin content increased about

100% as compared to the 400% and 200% increase of the control sample and that in microperforated packs.[64] The use of barrier films is not recommended for this product since concentrations of 28% CO_2 and 2% O_2 are reached after 40 hours, which considerably affects the quality by a putrid odour and a general breakdown due to anaerobic respiration.[65]

The use of non-perforated flexible polyvinyl chloride (PVC) to store broccoli for 3 weeks at 5°C contributed to reduce the weight loss, to preserve the quality with respect to unpacked controls and to decrease the respiration rate.[66] Cytoquinin treatment extends the post-harvest life of broccoli florets kept for 6 days at 16°C in perforated polyethylene bags. Chlorophyll degradation was delayed and the respiration rate fell although ethylene emission increased 4% over the control.[67] MA was also applied to other products, such as white cabbage,[68] lettuce[69] and potato,[70] and improvements in post-harvest shelf-life was achieved.

8.8 Minimally processed fruits and vegetables

Minimally processed fruits and vegetables include those submitted to minimum preparation like peeling or cutting. They reach the consumer in a fresh-like state.[71,72] These conditions mean that the product shows an increased respiration rate and is more open to microbial contamination, which necessitates increased sanitary precautions in preparation and packaging.[73] In cut fruits and vegetables, the respiratory rate increases in proportion to the degree of tissue breakdown. This increase is relatively low in cut vegetables, 50% in endive leaves[74] and 40% in broccoli florets,[81] but it is much higher in grated vegetables and cut fruits: 600 to 700% in carrots[82] and 400% in kiwi.[75,76]

The elaboration of minimally processed products also results in an increase in ethylene production compared with the whole fruit, as has been observed in sliced kiwi.[83] To prolong the shelf-life of these products the correct concentration of gases must be established inside the packaging. As with the packaging of whole fruits and vegetables, several factors should be taken into consideration: the vegetable produce itself, the film used, storage temperature and relative humidity inside the package. For minimally processed salads an equilibrium atmosphere of 2–5% O_2 and 3–8% CO_2 is recommended.[84] However, an atmosphere of 15–20% CO_2 and 1–2% O_2 under a 40 μm thickness polypropylene film permitted mixtures of different endives to be kept 16 days at 10°C with no loss of quality.[77]

One of the factors which most influences the quality of minimally processed products is tissue browning which results from phenolic compounds oxidation catalysed by polyphenoloxidases (PPO) and peroxidases (POD). In endive, the intensity of *in vitro* browning depends on the concentration of phenolic compounds and the preferential substrates of the oxidation reactions are dicaffeic tartaric acid and o-diphenol.[78] In minimally prepared pomegranate anthocians decreased 50% after 4 days at 6°C when a polypropylene film of 25 μm thickness was used.[79] The correct combination of MA with low O_2 concentrations (1%), low temperatures and certain compounds such as ascorbic acid, citric acid, sodium chloride and sodium metabisulphite can reduce the enzymatic browning of sliced peach and apple.[80]

8.9 Microbiological safety

Fruits and vegetables have a characteristic microflora predominated by bacteria, among which are enterobacteria and lactic acid bacteria, yeasts and moulds.[81] Enclosing fresh products in a film tends to increase the relative humidity, which favours the growth of microorganisms, including pathogenic ones. In 'Somatine' tomato packed in different types of plastic at 12.5°C with a relative humidity of 65–73% in the external atmosphere, the development of fungi depends on the internal relative humidity of the pack. Thus, polyethylene packs of 25 μm thickness result in a greater incidence of decay than when PVC packs are used. Despite the equal composition of gases, internal relative humidity is higher under PE.[82] On the other hand cabbage, brussels sprouts, Chinese cabbage and celery show less decay at 98–100% relative humidity than at 90–95%.[83]

Temperature also affects microbial development since higher temperatures result in increased respiration and higher rates of microorganism growth. Iceberg lettuce inoculated with *Listeria monocytogenes* and stored at 5°C shows no increase in microbial growth after 8 days, while at 10°C there is a significant growth after the third day.[84] Nevertheless, the low temperatures that vegetable products can support without damage are not always sufficiently low to prevent the growth of some microorganisms. Certain bacteria responsible for vegetable spoiling are still active at −10°C and fungi like *Alternaria*, *Botrytis* and *Cladosporium* grow at 0°C. Peach and apple *Penicillium* still show activity between 0° and 12°C.

Enclosing fruits and vegetables in a modified atmosphere produces a decrease in O_2 levels inside the pack, which favours the growth of anaerobic microorganisms such as *Clostridium botulinum*. This microorganism grows in conditions of high relative humidity and salinity and low acidity, low O_2 concentrations and temperatures in the region of 3–5°C.[1] It has been estimated that 53% of cases of botulism are associated with vegetables.[85] Similar conditions to those which favour the above microorganism are also favourable to *Yersinia enterocolitica*, *E. coli*, *Listeria monocytogenes*, *Aeromonas hydrophilia* and other toxic bacteria like *Staphilococcus aureus* and *Salmonella*.[77]

The microorganism most commonly found in minimally prepared products are: *Flavobacteria, Pseudomonas, Lactobacillus, Actinobacter* and *Corineformes*.[77] Most of these bacteria are found on the external parts of the vegetables, although they can also be present in internal tissues. However, neither *Salmonella* nor *Staphilococcus aureus* has been detected in this type of product, although *Pseudomonas marginalis* was isolated in minimally processed salad, producing black stains and decay.[86] *Pseudomonas aureaginosa* at 100 cfu/g and *Yersinia enterocolitica* have also been detected.[87]

The microbiological soundness of minimally processed products depends on numerous factors, among which the maintenance of a suitable temperature during manipulation is particularly important. In salads stored at 4, 8 and 12°C, seven times more faecal coliform bacteria were found at the highest temperature. However, the multiplication of faecal coliforms is less than that of mesophilic aerobic bacteria. This last group is principally represented by *Pseudomonas*, which is less sensitive to the cold than faecal coliforms. The average increase in storage period is 1 day at 12°C, 3–4 days at 8°C and 7 days at 4°C.[88]

An atmosphere rich in CO_2 and poor in O_2 may have a selective effect on microflora and the metabolism of the microorganisms themselves can contribute to modifications in the pack's atmosphere. If anoxia conditions are reached or even nearly reached, anaerobic microorganisms growth is favoured. In grated carrots packed in an O_2 concentration below 3% and CO_2 concentrations exceeding 20%, there is a proliferation of lactic bacteria, particularly *Leuconostoc mesenteroides*, and an increase in intermediate metabolites (lactic and acetic acid and ethanol), which produce a slightly acidic taste. To avoid this carrots should be packed [89] in more permeable films, below 6°C.

8.10 External factors affecting MA preservation

8.10.1 *Temperature*

Low temperatures contribute to decrease the respiration rate, to control the microorganism growth and to retard the metabolic activity of plant tissues. A decrease of 10°C usually slows down the respiration rate 2 or 3 times and a fall in temperature from 30 to 0°C can reduce the respiration to 1/27 of the original.[90] For packed products to be preserved as long as possible exposure to low temperatures should last the longest time possible. The optimal temperature can be described as that which delays senescence and maintains quality without causing damage by cold or freezing.[1] This temperature will depend on the type of fruit, the permeability of the plastic used and the tolerance of the fruit to different gaseous concentrations.

During the marketing of products packed in MA the optimum temperature must be kept constant to prevent deterioration. Fluctuations in temperature may produce condensation of water vapour inside the pack, modify the plastic's permeability and increase the respiration rate of the product. This would modify the gaseous composition of the equilibrated atmosphere and the possibility of change to anaerobic respiration. The optimal temperature also depends on the gaseous composition inside the packaging, since high CO_2 levels reduce damage caused by cold.[91] CO_2 is a bacteriostatic and fungistatic agent but its activity depends on temperature [92] and it is more effective at low than at high temperatures. This increase in inhibitory effect is due to a decrease in the product's pH as a consequence of the greater capacity to dissolve CO_2 at low temperatures in the aqueous phase of the product.[93,94]

8.10.2 *Relative humidity*

The plastics used in MAP are of low permeability to water vapour and so water is accumulated inside the pack, producing a rise in relative humidity. High relative humidity decreases weight loss and maintains the produce's firmness over longer periods of time. In tomato, weight loss is inversely related to the relative humidity inside the film and directly related to its permeability to water vapour.[55] Losses in water by evaporation are determined by the external area of the product per unit of weight and the permeability of its epidermis to water. For this reason, some species (lettuce, chicory) dehydrate rapidly and show serious damage during storage. Three possible solutions were suggested to prevent excessive water loss through evaporation:

1. to decrease the water vapour pressure deficit in the atmosphere;
2. to reduce the volume of the atmosphere surrounding the product in the package;
3. to wax the product's surface.[90]

On the other hand, high degrees of humidity favour the growth of microorganisms, which increases the possibilities of product deterioration. To control relative humidity inside the package the use of water-absorbing chemicals has been suggested, such as $CaSiO_4$, KCl, NaCl, xylitol and sorbitol [95] or the inclusion of surfactants in the polymeric formulation of the film to act as antimisting agents.[96]

8.10.3 *Light*

The effect of light on plastic-wrapped products has been little studied. Green vegetables kept in illuminated refrigerators photosynthesize. This modifies the gaseous composition of the pack differently from what might be expected from its permeability. At equilibrium, the CO_2 concentration is lower and the O_2 concentration higher than in dark, as was observed in minimally processed salad.[97] Cut lettuce leaves kept under fluorescent light are slightly lighter in colour than those stored in darkness.[77] Some products can be negatively affected by light and by photosynthesis. For example 'regreening' causes a loss of quality in potato and endive and so opaque packaging is recommended.[1]

References

1. Zagory, D. and Kader, A.A. (1988) *Food Technology*, **42**(9), 70–77.
2. Kader, A.A., Zagory, D. and Kerbel, E.L. (1989) *Critical Reviews in Food Science and Nutrition*, **28**, 1–30.
3. Henning, Y.S. (1975) In: *Post-harvest Biology and Handling of Fruits and Vegetables*, AVI Publishing Company, Inc., Westport, CT. pp. 144–152.
4. Karel, M. (1975) In: *Physical principles of food preservation*, Marcel Dekker, New York. p. 399.
5. Rij, R.E. and Mackey, B. (1986) In: *Int. Hort. Congress*, Davis, C.A., Aug. 15. p. 1286.
6. Kader, A.A. (1986) *Food Technology*, **40**(5), 99–104.
7. Pretel, M.T., Serrano, M.. Martinez, G., Riquelme, F., Romojaro, F. (1993) *Lebensm. Wiss.-u-Technol.* **26**, 8–13.
8. Ke, D., Gorsel, H. and Kader, A.A. (1990) *J. Amer. Soc. Hort. Sci.*, **115**(3), 435–439.
9. Knee, M. (1980) *Ann. Appl. Biol.*, **96**, 243–253.
10. Blakman, F.F. (1928) *Proc. R. Soc.* London, **103**, 412–445.
11. Boersing, M.R., Kader, A.A. and Romani, R.J. (1988) *J. Amer. Soc. Hort. Sci.*, **113**(6), 869–873.
12. Makhlouf, J., Willemot, C., Arul, J., Castaigne, F. and Emond, J.P. (1989) *J. Amer. Soc. Hort. Sci.*, **114**(6), 955–958.
13. Smilanick, J.L. and Fouse, D.C. (1989) *J. Amer. Soc. Hort. Sci.*, **114**(3), 431–436.
14. Ke, D., Rodriguez-Sinobas, L. and Kader, A.A. (1991) *Scientia Horticulturae*, **47**, 295–303.
15. Ke. D. and Kader, A.A. (1990) *J. Amer. Soc. Hort. Sci.*, **115**(5), 779 – 783.
16. Ke, D., Rodriguez-Sinobas, L. and Kader, A.A. (1991) *J. Amer. Soc. Hort. Sci.*, **116**(2), 253–260.
17. Chaves, A.R. and Tomás, J.O. (1984) *Plant Physiol.*, **76**, 88–91.
18. Kerbel, E.L., Kader, A.A. and Romani, R.J. (1988) *Plant Physiol.*, **86**, 1205–1209.
19. Li, C. and Kader, A.A. (1989) *J. Amer. Soc. Hort. Sci.*, **114**(4), 629–634.
20. Kubo, Y., Inaba, A. and Namura, R. (1989) *J. Japan. Soc. Hort. Sci.*, **58**(3), 731–736.
21. Kubo, Y., Inaba, A. and Namura, R. (1990) *J. Amer. Soc. Hort. Sci.*, **115**(6), 975–978.
22. Adams, D.O. and Yang, S.F. (1979) *Proc. Natl. Acad. Sci.*, USA, **76**, 170–174.
23. Beyer, E.M. (1975) *Plant Physiol.*, **56**, 273–278.
24. Hall, M.A., Smith, A.R., Thomas, C.I.R. and Howarth, C.J. (1984) In: *Ethylene, Biochemical, Physiological and Applied Aspects*, Nijhoff Dr. W. Junk, The Hague, pp. 55–64.
25. Bufler, G. and Bangerth, F. (1983) *Physiol. Plant.*, **58**, 486–492.
26. Knee, M. (1980) *Ann. Appl. Bio.*, **96**, 243–253.
27. Bufler, G. and Streif, J. (1986) *Scientia Horticulturae*, **30**, 137–185.
28. Plich, H. (1987) *Fruit Science Reports*, **14**(2), 45–56.
29. Smilanick, J.L. and Fouseo, D.C. (1989) *J. Amer. Soc. Hort. Sci.*, **114**(3), 431–436.
30. Makhlouf, J., Castaigne, F., Arul, J., Willemont, C. and Gosslein, A. (1989) *HortScience*, **24**(4), 637–639.
31. Wang, C.Y. and Ji, Z.L. (1988) *J. Amer. Soc. Hort. Sci.*, **113**(6), 881–883.
32. Burg, S.P. and Burg, E.A. (1969) *Qual. Plant. Mater. Veg.*, **14**, 185–200.
33. Sisler, E.C. and Wood, C. (1988) *Physiol. Plant.*, **73**, 440–444.

34. Zhen-Guo, L., Yu, L., Jian-Guo, D., Rong-Jiang, X. and Mei-Zhen, Z. (1983) *J. Plant Growth Regul.*, **2**, 81–87.
35. Bufler, G. and Streif, J. (1986) *Sci. Horticulturae*, **30**, 177–185.
36. Cheverry, J.L., Sy, M.O., Pouliqueen, J. and Marcellin, P. (1988) *Physiol. Plant.*, **72**, 535–540.
37. Marcellin, P. and Chaves, A.R. (1983) *Acta Hortic.*, **138**, 155–163.
38. Colelli, G., Mitchell, F.G. and Kader, A.A. (1991) *HortScience*, **26**(9), 1193–1195.
39. Lau, O.L., Liu, Y. and Yang, S.F. (1984) *HortScience*, **19**, 425–426.
40. Plich, H. (1989) *Fruit Science Reports*, **16**(3), 173–184.
41. Zamponi, R., Chaves, A. and Añon, M.C. (1990) *Sci. Aliments*, **10**, 141–150.
42. Prince, T.A. (1989) In: *Controlled/Modified Atmosphere Vacuum Packaging of Foods*, Food and Nutrition Press, INC. Trumbell, Connecticut, USA, pp. 67–100.
43. Ben-Yehoshua, S. (1987) In: *Post-harvest Physiology of Vegetables*, weich. Sci., Marcel Dekker, New York, p. 113.
44. Little, G.R. and Peffie, I.D. (1987) *HortScience*, **22**(5), 783–790.
45. Lougheed, E.C. (1987) *HortScience*, **22**(5), 791–794.
46. Siriphanich, J. and Kader, A.A. (1985) *J. Amer. Soc. Hort. Sci.*, **110**, 249–253.
47. Siriphanich, J. and Kader, A.A. (1986) *J. Amer. Soc. Hort. Sci.*, **111**, 73–77.
48. Knee, M. and Hatfield, S.G.S. (1981) *J. Sci. Food Agric.*, **32**, 593–600.
49. Kanellis, A.K., Solomos, T. (1989) *Plant Physiol.*, **90**, 257–266.
50. Ke, D., Goldstein, L., O'Mahony, M. and Kader, A.A. (1991) *J. Food Sci.*, **56**, 50–54.
51. Baxter, L. and Waters, L. (1990) *HortScience*, **25**(1), 92–95.
52. Ontai, S.L., Paull, R.E. and Saltveit, M.E. (1992) *HortScience*, **27**(1), 39–51.
53. Ke, D. and Saltveit, M.E. (1989) *J. Amer. Soc. Hort. Sci.*, **114**(5), 789–794.
54. Powrie, W.D. and Skura, B.S. (1991) In: *Modified Atmosphere Packaging of Fruits*, B. Oraikul, Ed., Ellis Horwood Limited, Chichester, West Sussex, England.
55. Geeson, J.D., Genge, P.M., Smith, S.M. and Sharpless, R.O. (1991) *Inter. J. Food Science and Technology*, **26**(2), 215–223.
56. Geeson, J.D., Genge, P.M., Sharpless, R.O. and Smith, S.M. (1991) *Inter. J. Food Science and Technology*, **26**(2), 225–231.
57. Nakhasi, S., Schlimme, D. and Solomos, T. (1991) *J. Food Sci.*, **56**(1), 55–59.
58. Seseni, E., Rizzolo, A. and Sarlos, S. (1991) *Italian J. Food Science*, **3**(3), 209–218.
59. Yahia, E.M. and Rivera, M. (1992) *Lebensmittel-Wissenschaft Und-Technology*, **25**(1), 38–42.
60. Ben-Arie, R., Zutkhi, Y., Sonego, L. and Klein, J. (1991) *Post-harvest Biology and Technologie*, **1**(2), 169–179.
61. Cohen, E., Luries, S., Shapiro, B., Ben-Yehoshua, S., Shalom, Y. and Posen Berger, I. (1990) *J. Amer. Soc. Hort. Sci.*, **115**(2), 251–255.
62. Lin, T.Y., Koehler, P.E. and Shewfelt, R.L. (1989) *J. Food Science*, **54**(2), 405–407.
63. Tomkins, R.B. and Comming, B.A. (1988) *Scientia Horticulturae*, **36**(1/2), 25–35.
64. Eversom, H.P., Waldron, K.W., Geeson, J.D. and Browne, K.M. (1992) *Int. J. Food Science and Technology*, **27**(2), 187–199.
65. Baxter, L. and Waters, L. (1991) *HortScience*, **26**(4), 399–402.
66. Forney, V.F., Rij, R.E. and Ross, S.R. (1989) *HortScience*, **24**(1), 111–113.
67. Rushing, J.W. (1990) *HortScience*, **25**(1), 88–90.
68. Ponomarev, P.F., Bolyanovskaya, D.S., Batutina, A.P. and Panchenko, G.N. (1991) *Tovarovedenie*, **24**, 15–18.
69. Jeong, J.G., Par, K.W. and Yang, Y.J. (1990) *J. Korean Soc. Hort. Sci.*, **31**(3), 219–225.
70. Shetty, K.K., Dwelle, R.B., Fellman, J.K. and Patterson, M.E. (1991) *Potato Research*, **34**(4), 253–260.
71. King, A.D. and Bolin, H.R. (1989) *Food Technology*, **43**(2), 132–135.
72. Huxsoll, C.C. and Bolin, H.R. (1989) *Food Technology*, **43**(2), 124–128.
73. Ronk, R.J., Carson, K.L. and Thompson, P. (1989) *Food Technology*, **43**(2), 136–139.
74. Chambroy, Y. (1989) *Revue Generale du Froid*, **3**, 78–81.
75. Ballantyne, A. (1987) Technical memorandum no. 464. Campden Food Preservation Research Association, Chipping Campden, UK.
76. McLachlan, A. and Start, R. (1985) Technical memorandum, no. 412. Campden Food Preservation Research Association. Chipping Campden, UK.

77. Saracino, M., Pensa, M. and Spiezie, R. (1990) *Agro-Industry*, Hi-Tech, pp. 11–15.
78. Goupy, P. (1987) Journee 4eme Gamme. Ind. Nat. Rech. Agron. Montfavet (France).
79. Pretel, M.T., Amoros, A., Serrano, M. and Riquelme, F. (1990) II Simp Nac. Maduración y Post-recolección S.E.F.V. Lérida (Spain).
80. Varoquaux, P. (1987) Journee 4eme Gamme. Inst. Nat. Rech. Agron. Montfavet (France).
81. Bracket, R.E. (1990) *J. Food Protect.*, **53**(3), 255–257, 261.
82. Geeson, J.D., Brone, K.M., Maddison, K., Shepherd, J. and Guaraldi, F. (1985) *J. Food Technology*, **20**, 339–349.
83. Fournaud, J. and Laville. M.E. (1981) *Revue Generale du Froid*, **3**, 145–154.
84. Beuchat, L.R. and Brackett, R.E. (1990) *J. Food Sci.*, **55**(3), 755–758, 870.
85. Pierson, M.D. and Reddy, N.R. (1988) *Food Technology*, **42**(4), 196–198.
86. Nguyen-The, C. and Prunier, J.P. (1989) *Int. J. Food Sci.*, **24**, 47–58.
87. Galli, A., Franzetti, L., Bonora, S. and Ponticelli, G. (1990) *Annali Microb. ed Enzimol.*, **44**, 30.
88. Scandella, D. (1988) *Infos-Ctifl*, **39**, 46–48.
89. Carlin, F. (1989) Thèse de Docteur Ingénieur, Inst. Nat. Agronomique. Paris, Grignon.
90. Rizvi, S.S.H. (1981) *Critical Reviews in Food Science and Nutrition*, **14**, 111–134.
91. Lyons, J.M. and Breidenbach, R.W. (1987) In: *Post-harvest Physiology of Vegetables*. Weichman, J. ed., Marcel Dekker, NY, p. 305.
92. Smith, J.P., Ramaswany, H.C. and Simpson, D.K. (1990) *Packaging Science and Technology*, **11**, 111–118.
93. Gill, C.O. and Tan, K.H. (1980) *Appl. Environ. Microbiol.*, **39**, 317–324.
94. Brody, A.L. (1989) In: *Controlled Modified Atmosphere/Vacuum Packaging of Food*, Brody, A.L. ed., Food and Nutrition Press, Inc., Trumbull, Connecticut, USA.
95. Shirazi, A. and Cameron, A.C. (1987) *HortScience*, **22**(5), 1055.
96. Shore, W.S. (1978) Plant package. US. Patent 4,118,890.
97. Day, B.P.F. (1989) CEDRA. Technical Memorandum no. 524–Aug.

9 Bio-packaging: technology and properties of edible and/or biodegradable material of agricultural origin*

N. GONTARD and S. GUILBERT

Abstract

Edible films, coatings and biodegradable packagings produced from biological materials offer numerous advantages over other conventional synthetic packaging materials. Potential applications of edible films are numerous (internal moisture or solute barriers of heterogeneous foods, individual protection of food pieces, encapsulation of food additives, etc.). Advantages, types, formation and properties of edible films with examples are reviewed in detail. Biodegradable packaging, made from entirely renewable natural polymers could contribute to solving environmental pollution and creating new markets for agricultural products. Different approaches are discussed (physical mixing of starch or co-processing of more than 50% starch with synthetic polymers, thermoplastic extruded starch, etc.).

9.1 Introduction

Formulations of packaging films, food or pharmaceutical coatings, etc. must include at least one component able to form an adequately cohesive and continuous matrix. These components are macromolecules that are either synthetic, from which most current packaging is produced, or natural biopolymers. The performances and applications of synthetic packaging are fully controlled and utilized, whereas the two ends of the production line (raw materials and the fate of the packaging after use) are not. Environmental problems can thus result from using non-renewable raw materials and accumulation of such non-biodegradable packaging. As a solution to this dilemma, biopolymers could be used to formulate biodegradable packaging, e.g. to replace short shelf-life plastics, in addi-

*Partially presented at the IFTEC symposium: 'Food Packaging Interactions and Packaging Disposability'. The Hague, 15–18 November 1992. Additional contribution by Pr. Jean-Louis Cuq (Université de Montpellier II).

tion to adoption of plastic recycling programs. These biopolymers can be classified in four general categories: polysaccharides, proteins, lipids and polyesters (obtained by controlled vegetal or bacterial biosynthesis). Films primarily composed of polysaccharides (cellulose and derivatives, starch and derivatives, gums, etc.) or proteins (gelatin, zein, gluten, etc.) have suitable overall mechanical and optical properties, but are highly sensitive to moisture and show poor water vapour barrier properties. In contrast, films composed of lipids (waxes, lipids or derivatives) or polyesters (poly-D-β-hydroxybutyrate, polylactic acid, etc.) have good water vapour barrier properties, but are usually opaque and relatively inflexible. Lipid films could be also quite fragile and unstable (rancidity).

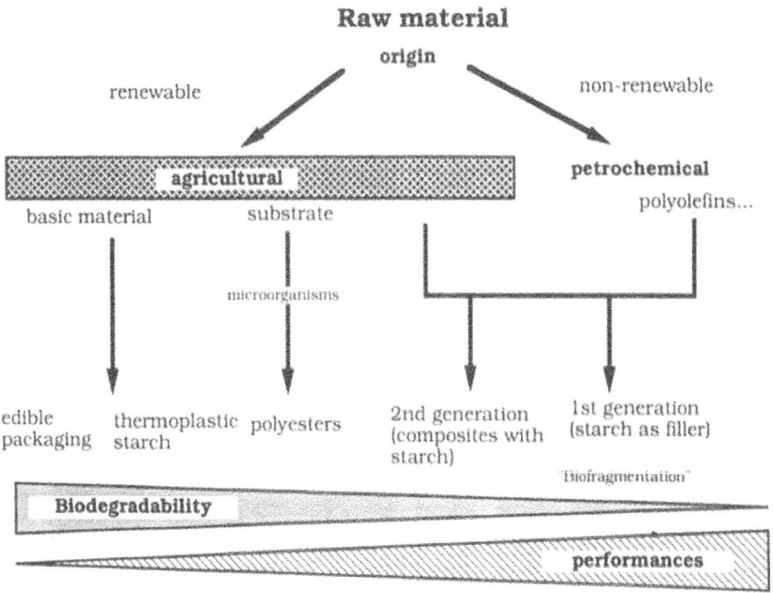

Figure 9.1 Different approaches to make 'bio-packaging' from agricultural raw materials.

Three different techniques using agricultural raw materials (fully renewable raw materials) to make bio-packaging are proposed (Figure 9.1): synthetic polymer/biopolymer mixtures (first and second generations), agricultural materials used as fermentation substrates to produce microbial polymers and finally agricultural polymers used directly as basic packaging material.

Starch is the most commonly used agricultural raw material, especially in the first two techniques, since it is inexpensive, widely available and relatively easy to handle.

9.2   Use of biopolymers in packaging

9.2.1   *Synthetic polymer/biopolymer mixtures*

Synthetic polymers become more susceptible to microbial attack when biopolymers are incorporated, i.e. biodegradability of the synthetic polymer is accelerated by adding components that can be assimilated by microorganisms. Presently, the main marketed products of this type are starch-based. Other types of biopolymers such as cellulose, lipids and vegetable proteins are not widely used and some have been investigated only recently (e.g. cellulose/polyurethane mixtures,[1] gluten/synthetic resin mixtures,[2] vegetable protein/vinylic compound mixtures,[3] casein or lipid/ synthetic polymer mixtures[4]).

9.2.1.1   Filled material (first generation).   The first commercial 'bio-degradable' plastics were developed using a technique involving extrusion mixing of granular native starch (5–20%) and prooxidative and auto-oxidative additives with the synthetic polymer. This technique has been marketed by several firms: the St Lawrence Group (Canada) under the 'Ecostar' trademark; Archer Daniels Midland (USA) as 'Polyclean'; Polychim (France) as 'Ecopolym', and Amylum as 'Amyplast'. Starch granules are uniformly dispersed in the polyethylene matrix without chemical interaction. Microbial enzyme-induced biodegradation of starch reduces the mechanical properties of the material and increases the interface between the polymer and the surrounding atmosphere (oxygen, water, etc.). This stimulates profound chemical degradation (auto-oxidation) of the synthetic phase. Poor starch/polyethylene compatibility weakens the mechanical properties of the material, thus limiting the percentage of starch that can be added. This compatibility has been enhanced by silylation (increased hydrophobicity) of the surface of starch granules. The formed film can therefore contain up to 43% starch (St-Lawrence Corn Starch Company, Canada; Spartech, USA). Biodegradability of these materials is highly controversial,[5–7] and their behaviour is now classified as 'biofragmentation', i.e. fragmentation into small molecules. It takes 3–5 years to degrade this type of product into dust.

9.2.1.2   Composite material (second generation).   A fine molecular mixture of synthetic polymers and starch-based polymers can be made by this technique. These materials are composed of gelatinized starch (up to 40–75%, by destructuring the starch granule with ammonia and water at high temperature), hydrophobic synthetic polymers (polyethylene, etc.) and hydrophilic co-polymers. The latter compounds act as compatibility agents providing an interface between the starch and the synthetic polymer.

Starch is thus not restricted to the dispersed phase, it is able to interact with the synthetic polymer. Compatibility agents can be synthetic (ethylene/acrylic acid, vinyl alcohol, or acrylic ester co-polymers, polyvinyl-alcohol, vinyl acetate, etc.), or obtained by grafting polystyrene chains to amylose chains or amylopectin. These types of materials are marketed by: Ferruzzi (Italy), under the 'Mater-Bi' trademark: Ampacet (USA), as 'Poly-grade II' and Agri-Tech Industries (USA). Their prices (about 25 FF/kg for 'Mater-bi') are still higher than that of standard synthetic packaging films (5 to 10 FF/kg for polyethylene or PVC). This aspect has been studied by Otey and Westhoff,[8-12] Lenk and Merrall,[13] and Narayan and co-workers.[14-16] Full biodegradability of these materials, as claimed by the manufacturers, is still a topic of discussion. Indeed, few standard, strictly-controlled, exhaustive comparative tests on biodegradability have been published. Complete degradation of starch takes 40 days and degradation of the entire film requires a minimum of 2–3 years, as compared to 200 years estimated for entirely synthetic polymers.[17]

9.2.2 Microbial polymers (polyesters)

Polyesters are theoretically biodegradable since ester linkages within polymer chains are potential targets of chemical or enzymatic (microbial) hydrolysis. Polyesters are excreted or stored by microorganisms. Starch hydrolysates are generally used as fermentation feedstock, for the microorganisms that produce polyesters. Isolation and purification costs are very high for these products which are obtained from complex mixtures. They are completely biodegradable[6,7] and recyclable.

Poly (hydroxybutyrate/valerate) (PHB/PHV), marketed under the 'Biopol' trademark by ICI (UK), is produced by an *Alcaligenes eutrophus* strain. The sugar to polymer conversion yield is about 33%. It is thermoplastic and can be formed by the same techniques as those used for synthetic polymers. PHV or polycaprolacton are plasticizers which enhance PHB flexibility. This product is limited to use with products of high value (cosmetic, surgical products, etc.) because of its relatively high expense (50–150 FF/kg).

Polylactic and polyglycolic acids are mainly produced by CCA Biochem (Netherlands) by chemical polymerization of lactic acid and glycolic acid obtained by *Lactobacillus* fermentation of carbohydrates (sucrose, glucose, maltose and lactose). Applications (surgical products) are extremely limited by the prohibitively high price (4000–20000 FF/kg).

Polycaprolactons and chitosans (combined with cellulose) are produced by Union Carbide (USA) as 'Tone', and Aisero Chemical (Japan), but are only used to a small extent in packaging materials.

9.2.3 *Packaging composed of polymers of agricultural origin*

Films composed of polymers of agricultural origin (in a natural state or fractionated, e.g. whole grains, flours, proteins, starch, fractions, etc.) are often much less sophisticated than the above described packaging. These products are economical due to the low cost of raw materials (3 FF/kg for starch, 4–5 FF/kg for gluten, as compared to 4–5 FF/kg for polyethylene as raw material). They are completely biodegradable, and edible when no non food-grade additives are used.

9.2.3.1 *Biodegradable packaging made from thermoplastic biopolymers.* 'All-starch' packaging has been developed mainly by the Warner-Lambert Company (USA), under the 'Novon' trademark, utilizing the thermoplastic properties of corn starch. The controlled presence of water or other plasticizers (glycerol, sorbitol, etc.) lowers the glass transition temperature of starch and films can be formed below the breakdown temperature (molecular degradation) of this polymer. Standard techniques used for forming synthetic polymer films can thus be used (extrusion, injection moulding). Fabrication costs are the same as for synthetic polymers.

This type of material can be used in relatively rigid packaging with a short shelf-life (egg containers, fast food packaging), medical applications, non edible packaging and agricultural mulching.

Raw agricultural starch-based materials have been developed by Future Pop and Alexander Fruit & Trading Co. (USA), with foam starch chips replacing non-degradable polystyrene chips as filling material for packaging of shock sensitive goods. These starch chips cost 20% less than synthetic chips and are biodegradable.

Moreover, Biograna (Switzerland) and Taïwan Sugu (Taïwan) produce injection-molded dishware composed of various raw cereals devoid of other additives.

9.2.3.2 *Edible packaging.* Edible films or coatings have long been used empirically to protect food products. A few examples of such applications to improve product appearance or conservation include sugar and chocolate coatings for candies, wax coating for fruits and traditional lipoprotein 'skins' ('Yuba' obtained by drying the skin formed after boiling soya milk). Solid lipids and oils are also commonly used to cover or coat foods. Edible films are an interesting and often essential complementary parameter to control the quality and stability of many foods.

There are several reviews on the formulation technology and application of edible films.[18–21] Certain basic aspects of the mechanisms and techniques for forming and applying films, coatings and microencapsulation are described in studies aimed at pharmaceutical applications.[22–24] The simple or mixed use of different carbohydrate, protein or lipid materials in various forms (coatings, single-layer, bilayer or multilayer films), has thus been proposed for the formulation of edible films and coatings.

Coatings are applied and formed directly on the food product, whereas films are structures which are applied after being formed separately. They can be superficial coatings or continuous layers between compartments of the same food product. Edible films may be arbitrarily defined as thin layers of material which can be eaten by the consumer as part of the whole food product. The composition of edible films or coatings must therefore conform to the regulations that apply to the food product concerned.

Edible films and coatings can be formed by the following mechanisms:[19]

– Simple coacervation: where a hydrocolloid dispersed in water is precipitated or undergoes a phase change after solvent evaporation (drying), after the addition of a hydrosoluble non-electrolyte in which the hydrocolloid is insoluble (e.g. ethanol), after pH adjustment or the addition of an electrolyte which induces salting out or cross-linking.
– Complex coacervation: where two hydrocolloid solutions with opposite electron charges are mixed, thus causing interaction and precipitation of the polymer complex.
– Gelation or thermal coagulation: where heating of the macromolecule, which leads to its denaturation, is followed by gelation (e.g. proteins such as ovalbumin) or precipitation, or even cooling of a hydrocolloid dispersion causing gelation (e.g. gelatin or agar).

Film coating materials such as waxes or lipids and derivatives can be applied either as a sable emulsion or micro-emulsion with water or by direct application while still melted.[18,19,25]

Direct application and distribution of the film coating material in a liquid form can be obtained by hand-spreading with a paint brush, spraying, falling film enrobing, dipping and subsequent dripping, distributing in a revolving pan (pan coating), bed fluidizing or air-brushing, etc.[19] Suitable food coating and adhesion, which is sometimes difficult to obtain when, for instance, hydrophobic materials are used to protect hydrophilic fillings (or vice versa), can be obtained by hot applications, coating the support with a surfactant or adding it to the film-forming solution. Another way is to apply a preparatory precoating with a material that can adhere to all filling components (e.g. starch precoating on raisins before a wax coating, or cocoa powder precoating on peanuts before a sugar coating).

Full descriptions of numerous industrial production lines are noted in some patents and publications for raisin coatings,[26] for use of the commercial 'Lepak' film-forming preparation,[27] the application of starch films[28] and pectin films,[29] the coating of ice-cream cones[30] and fruit pieces,[31] for a compressed air spraying apparatus capable of homogeneously coating red variety meats,[32] etc.

Free, self-supporting film can be obtained by standard techniques, e.g. extrusion, moulding or rolling mill procedures, which have been developed for non-edible films. Films (or packaging material) are most commonly formed by drying a film-forming solution on a drum dryer, thermoforming (of pulp to make ice-cream cones, French fry and convenience food containers, etc.) or hot extrusion (for thermoplastic biopolymers).

9.3 Properties and applications of 'bio-packaging'

Edible and biodegradable films must meet with a number of specific functional requirements (moisture barrier, solute and/or gas barrier; water or lipid solubility; colour and appearance; mechanical and rheological characteristics; non-toxicity; etc.). These properties are dependent on the type of material used, its formation and application. Plasticizers, cross-linking agents, anti-microbial, anti-oxygen agents, texture agents, etc. can be added to enhance the functional properties of the film. In any polymeric packaging film or coating, two sets of forces are involved: between the film-forming polymer molecules for all polymeric films or coatings (cohesion), and between the film and the substrate for coatings only (adhesion). The degree of cohesion affects film properties such as resistance, flexibility, permeability, etc. Strong cohesion reduces flexibility, gas and solute barrier properties and increases porosity.[22] The degree of cohesion depends on the biopolymer structure and chemistry, the fabrication procedure and parameters (temperature, pressure, solvent type and dilution, application technique, solvent evaporation technique, etc.), the presence of plasticizers and cross-linking additives and on the final thickness of the film. Film cohesion is favoured by high chain order polymers. Excessive solvent evaporation or cooling, which is generally required for industrial reasons, may sometimes produce non-cohesive films due to premature immobilization of the polymer molecule.

9.3.1 Organoleptic properties

Edible films and coatings must have organoleptic properties that are as neutral as possible (clear, transparent, odorless, tasteless, etc.) so as not to be detected when eaten. Enhancement of the surface appearance (e.g. brilliance) and tactile characteristics (e.g. reduced stickiness) could be required. Hydrocolloid based films are generally more neutral than those formed from lipids or derivatives and waxes, which are often opaque, slippery and waxy tasting. It is possible to obtain materials with ideal organoleptic properties, but they must be compatible with the food filling, e.g. sugar coatings, chocolate layers (or imitation chocolate,[33,34]) and starch films for candies, biscuits, some cakes and ice-cream products (wafer coating), etc.

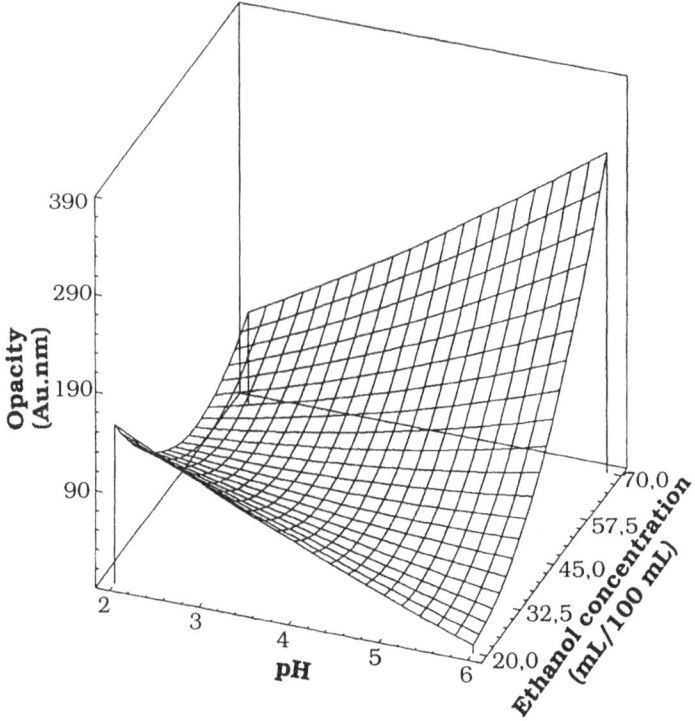

Figure 9.2 Effect of pH and ethanol concentration of the film-forming solution on wheat gluten film opacity.[36]

Films and coatings can also help to maintain desirable colouring, flavour, spiciness, acidity, sweetness, saltiness, etc.[18,19] Some commercial films, especially Japanese pullulane-based films (marketed by Colorcon Ltd., under the 'Opadry' trademark),[35] are thus available in several colours, or with spices and seasonings. This procedure could be used to provide nutritional improvement without destroying the integrity of the food product (edible films and coatings enriched with vitamins and various nutrients).

Optical properties of biopolymeric film depend on the film formulation and fabrication procedure, e.g. opacity of wheat gluten films is highly dependent on film-forming conditions[36] (Figure 9.2). Opacity of low density polyethylene/starch films increases as starch concentration and granule diameter increase.[37]

9.3.2 *Mechanical properties*

Films must be generally resistant to breakage and abrasion (to strengthen the structure of a food filling, to protect it, and ease handling) and flexible (enough plasticity to adapt to possible deformation of the filling without breaking).

The mechanical properties of edible films and coatings depend on the type of film-forming material and especially on its structural cohesion. Cohesion is the result of a polymer's ability to form strong and/or numerous molecular bonds between polymeric chains, thus hindering their separation. This ability depends on the structure of the polymer and especially its molecular length, geometry, molecular weight distribution and the type of position of its lateral groups. The mechanical properties are also linked with the film-forming conditions (type of process and solvent, cooling or evaporation rate, etc.) and the coating technique (spraying, spreading, etc.). The puncture strength of gluten films is strongly dependent on the gluten concentration and pH of the film-forming solution.[36,38] A resistant film can therefore be obtained by using a film-forming solution with high gluten content (12.5%) at about pH 5 (Figure 9.3).

The mechanical properties of biodegradable packaging made from synthetic polymer/starch mixtures depend on the starch content, compatibility (between hydrophobic synthetic polymers and hydrophilic starches)

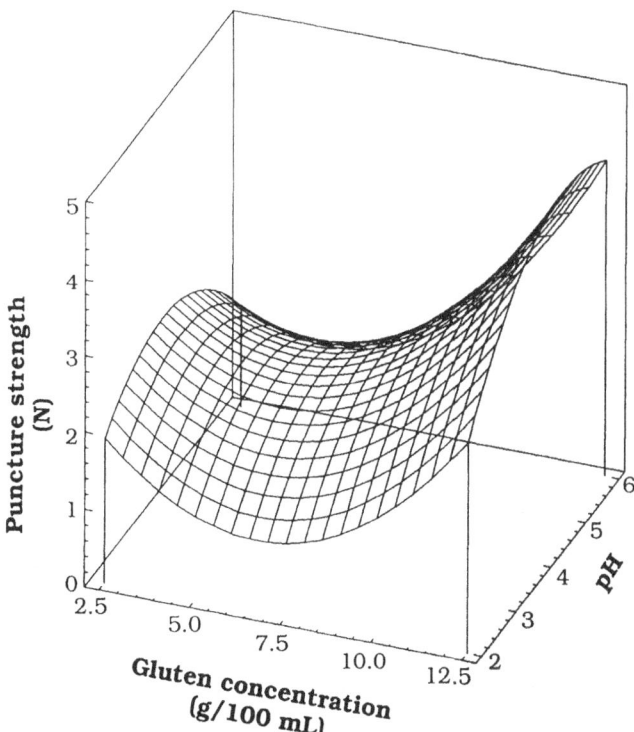

Figure 9.3 Effect of gluten concentration and pH of the film-forming solution on wheat gluten film puncture strength.[36]

and treatments to enhance this parameter (addition of compatibilization agents). For first generation packaging, increasing the percentage of starch reduces puncture strength and extensibility.[37,39]

The mechanical properties of amorphous materials are seriously modified when temperatures of these compounds rise above the glass transition temperature (T_g). The glass transition phenomenon separates materials into two domains according to clear structural and property differences, thus dictating their potential applications. Below T_g the material is rigid, and above it becomes viscoelastic or even liquid. Indeed, below this critical threshold only weak, non-cooperative local vibration and rotation movements are possible. Film relaxation relative to temperature follows an Arrhenius time course. Above the T_g threshold, strong, cooperative movement of whole molecules and polymer segments can be observed. These are cooperative structural rearrangement movements. In the T_g (T_g + 100°C) temperature range, these movements are given by the following equation of Williams, Landel and Ferry:[40]

$$\log a_T = C_1 (T - T_g)/C_2 + (T - T_g)$$

where a_T represents the ratio between values for specific mechanical characteristics and temperatures T and T_g and C_1 and C_2 are constants whose values ($C_1 = 17.4$ and $C_2 = 51.6$) are virtually universal for a wide range of materials.

This glass transition phenomenon has been demonstrated in gluten films by differential scanning calorimetry and confirmed through dynamical, mechanical and thermal analysis by Gontard et al.[41] The phenomenon seems to be a crucial physico-chemical parameter for understanding and predicting the behaviour of edible films such as that formed with gluten.

The mechanical properties of films can be enhanced by plasticization of the polymeric network. There are two different plasticizing effects. Internal plasticization is obtained by modifying the chemical structure of the polymer, e.g. by co-polymerization, selective hydrogenation or transesterification when edible lipid or derivative materials are used. External plasticization is obtained by adding agents which modify the organization and energy involved in the tridimensional structure of film-forming polymers.[22] Reduction of the intermolecular forces between polymer chains, consequently the overall cohesion, facilitates extensibility of the film (less brittle, more pliable) and reduces its T_g. However, this also results in reducing the gas, vapour and solute barrier properties of the film.[22,23,42]

Another plasticization technique involves adding relatively inert solids (filling agents which reduce molecular exchange and cohesion of the final film). Particle sizes and distributions are important. Microcrystalline cellulose, various protein isolates and cocoa powders have thus been used as plasticizers, particularly for lipid films.[43,44]

Water is the most common plasticizer and is very difficult to control in biopolymers which are generally more or less hydrophilic. Plasticization of biopolymeric films is thus dependent on the usage conditions, especially relative humidity (of environment and packaged products). In isothermic conditions, the addition of plasticizers such as water has theoretically the same effect as increased temperature on molecular mobility.

Water causes a substantial drop in the T_g.[45] This effect has been demonstrated in gluten films[41,46] and with other biopolymers such as elastin,[47] gelatin,[48] starch,[49,50] hemicellulose and lignin[51] and with low molecular weight sugar.[52-54] This phenomenon could be utilized to reduce the glass transition temperature of biopolymers to below the decomposition temperature threshold. Standard techniques for synthetic polymer films, such as extrusion or injection moulding, could thus be used. However, the drawback of this phenomenon is that it makes biopolymer packaging moisture-sensitive. Their mechanical properties are generally greatly modified by high temperature and/or moisture (ambient or from the packaged product).[46,55] Some treatments (especially the use of hydrophobic compounds such as octenyl succinate starch, oxidized polyethylene and fatty acids) are reported to reduce these negative effects.[55-57]

Standard techniques to evaluate the mechanical properties of packaging materials can be applied to 'bio-packaging'. Puncture strength, extensibility to puncture, torsion resistance, elasticity, etc. can be evaluated using texturometers and traction/compression testing equipment. However, edible films are more sensitive to ambient physical conditions, i.e. temperature and relative humidity, which must be carefully controlled.

9.3.3 Water and lipid solubility

Solubility or insolubility in water or lipids could be required for films in some specific applications. Edible small bags or capsules[20,58,59] can be used to package premeasured portions of additives for potential dispersion in food mixtures (e.g. emulsifiers for the preparation of cake and bread batters) or for instant dried food preparations (e.g. individual drinks or soups).

Generally, most edible hydrocolloid films and coatings are water soluble, unless a cross-linking or tanning treatment has been carried out or denaturing conditions are used. Gontard et al.[36] demonstrated the influence of film-forming conditions (pH and % ethanol) on water solubility of gluten films after 24 h immersion (Figure 9.4).

When developing films that are effective moisture barriers under a broad range of relative humidities, it is often necessary to use materials that are almost or entirely insoluble in water so as to avoid loss of the film qualities through swelling or disintegration upon contact with the food.[18] In such

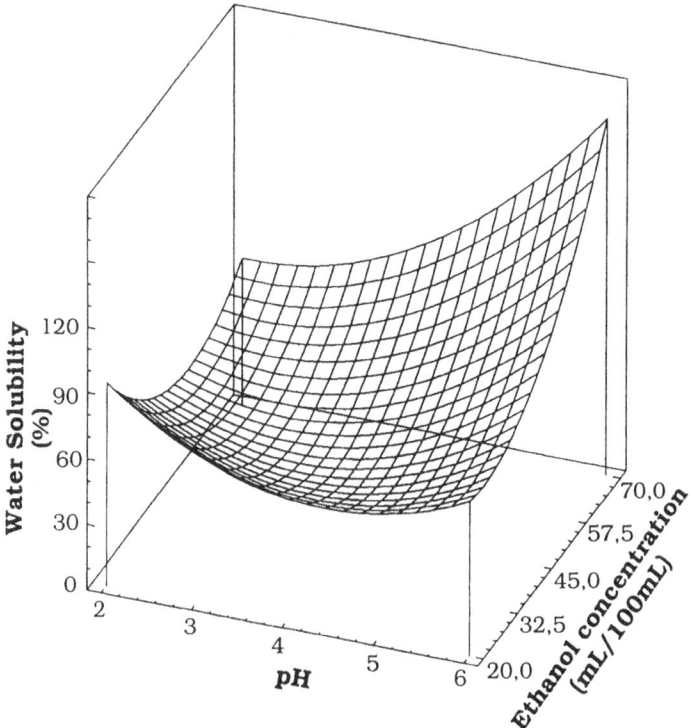

Figure 9.4 Effect of pH and ethanol concentration of the film-forming solution of wheat gluten film solubility and disintegration in water after 24 h immersion and agitation.[36]

instances, the use of lipids or almost insoluble proteins such as zein or gluten is recommended.

9.3.4 Gas, solute, lipid and water vapour permeability

9.3.4.1 *The permeability concept.* Permeability is defined as a state which permits the transmission of permeants through materials.[60] When there are no pores, faults or membrane punctures, permeability (P) is equal to the product of the diffusion coefficient (D), representing the mobility of permeant molecules in the polymer, and of a solubility coefficient (S), representing the permeant concentration in the film in equilibrium with the external pressure:

$$P = D.S$$

In practice, P is determined by steady state measurements:

$$P = \Delta W.x/ \Delta t.A.\Delta p$$

where ΔW is the permeant weight that passes through a film of thickness x and area A; and where t is the time and Δp the differential partial pressure across the film.

The diffusion coefficient can be obtained by taking measurements before the steady state is reached. The solubility coefficient can either be calculated from P and D, or measured in a separate experiment (sorption isotherms).

The diffusion and solubility of permeants are affected by temperature and the size, shape and polarity of the diffused molecule. Moreover, these two parameters depend on film characteristics, including the type of forces influencing molecules of the film matrix, the degree of cross-linking between molecules, crystallinity, the presence of plasticizers or additives, etc.[42,46,61,62]

Permeability is only a general feature of films or coatings when D and S are not influenced by permeant content, thus when Fick's and Henry's laws apply. In practice, for most edible films the permeant interests with the film and D and S are dependent on the differential partial pressure. For instance, concerning the water vapour permeability of hydrophilic polymer films, the water solubility and diffusion coefficients increase when the water vapour differential partial pressure increases because of the moisture affinity of the film (nonlinear sorption isotherm) and increased plasticization of the film due to water absorption.[46,62] The film thickness can also influence permeability when using film-forming materials that do not behave ideally.

Hence, it should be noted that the permeability of biopackaging is a property of the film-permeant complex under defined ambient conditions (temperature and humidity).

Using sorption curves, it is quite easy to determine the effect of temperature and relative humidity on solubility of the permeant in the material. In contrast, it is more difficult to determine the nature of the functional relationship between diffusivity and temperature or water content.[62] Various theories, including the free volume theory, have been put forth to explain this functional relationship.[63] According to the free volume theory,[64,65] molecular diffusion results from redistribution of the free volume in the material. This diffusion is only valid if the size of the free volume faults is greater than the critical value defined by the size of the diffused molecule. This value can be reached after a temperature increase. At $T < T_g$, mobility is controlled by the pre-existence of pores in the glassy material; at $T > T_g$, polymer relaxation and porosity influence mobility.[66] The free volume theory allows one to qualitatively predict variations of D relative to the difference between the temperature and the glass transition temperature (T_g). This theory is especially valid in the T_g to $T_g + 100°C$ range.

There is usually a difference between water vapour and gas permeability (CO_2, O_2) of the same film. According to Banker[22] and Kester and Fennema,[67] gas diffusion is crucial for gas permeability, whereas both sorption and diffusion are essential for moisture transfer.

Gravimetric methods are generally used to measure water vapour permeability.[68,69] According to these techniques, the film is attached to a cell containing a desiccant and the differential partial pressure of the water vapour is kept constant by placing the cell in an extremely humid atmosphere. The opposite configuration is also possible but may lead to differences in the results, as is the case with chocolate films.[43] The time course of the weight gain of the cell allows calculation of the water vapour transmission coefficient (WVTC in $g/m^2.day$) and the 'apparent' water vapour permeability constant (P_{app} in $g.mm/m^2.day.mmHg$).

The moisture barrier properties can also be monitored directly via variations in the water content (or water activity) of the different components. When this technique becomes too complex or difficult to carry out or interpret, food system models can be used.[43,70]

Gas permeability (air oxygen, carbon dioxide, nitrogen, etc.) can be measured with air porosity meters or specialized cells for specific measurements.[71,72]

The sorption isotherms of water and gas must be measured in order to calculate their solubilities in the material and evaluate the effect of humidity on film performance.

9.3.4.2 *Gas barrier properties.* Materials with suitable oxygen barrier properties are required to protect oxidizable foods (to reduce rancidity and vitamin loss), but some permeability to oxygen, and especially to CO_2, is essential for fresh fruit and vegetable coatings. Some biopolymer-based packaging has impressive gas barrier properties, especially against oxygen. The oxygen permeabilities of various edible and non-edible films are given in Table 9.1.

Hydrocolloid films have good oxygen barrier properties when they are not moist. Gelatin films can be used to protect frozen meats from rancidity,[78] to coat candies and dried products[59] and to microencapsulate flavors.[79] When moisture is present, the macromolecule chains become more mobile which leads to a substantial increase in oxygen permeability.[42]

Lipids, which are very often used to delay water transfer as described later, also have significant oxygen barrier properties.

According to Blank,[80,81] lipids made up of linear saturated fatty acids, fatty alcohols and fatty acid esters containing 16 or more carbon atoms have the best oxygen barrier characteristics. An increase in the degree of unsaturation or branching and reduction in the length of the carbon chain lowers oxygen permeability. The following barrier efficiency order was observed by Kester and Fennema:[82] stearic alcohol > tristearine > beeswax > acetylated monoglycerides > stearic acid > alkanes. These differences can be explained by the presence of pores or cracks, the lipid content, homogeneity of the composition, density of the network, which is

Table 9.1 Oxygen permeability of various films

Film	T (°C)	X (mm)	Oxygen permeability 10^{12}.g.cm./cm^2.sec.mmHg.
Polyethylene (low density)	25	0.025	0.50
Starch	24	0.79	0.42
Polyethylene (high density)	25	0.025	0.125
Gliadins and glycerol	23	0.10	0.089
Acetostearin	26	0.17	0.030
Amylomaize	25	0.05	0.0325
MC/HPMC and beeswax	25	0.051	0.021
Gluten and glycerol	23	0.11	0.016
Beeswax and C_{18}-C_{16} MC/HPMC	25	0.045	0.007
Waxed paper	25	–	0.005–0.075
Cellophane	25	–	0.001

According to references: 38, 71–77.
x is film thickness, t is temperature.
HPMC is hydroxypropylmethyl cellulose, C_{18} and C_{16} are stearic and palmitic acids, MC is methyl cellulose.
–, Data not available in literature cited.

dependent on the polymorphic shape and orientation of the chains and morphological differences in the lipid layers (as viewed by electron microscopy).[67,82]

The development of edible films and coatings with selective gas permeability could be very promising for controlling respiratory exchange and improving the conservation of fresh fruits and vegetables. Composite films or carboxymethyl cellulose and fatty acid sucroesters appear to have suitable oxygen barrier properties while remaining relatively permeable to CO_2.[83,84] This type of film was applied to refrigerated bananas and caused a 5-fold reduction in oxygen transfer, whereas CO_2 exchange was only reduced by about half.[85,86] However, this effect is dependent on the type and variety of fruit and temperature.[30,87]

The selectivity coefficient of a film relative to two gases is defined as the ratio of the respective permeabilities of these gases under the experimental conditions. This selectivity is the basis for gas permeation procedures. Concerning oxygen and carbon dioxide, the relative solubility of CO_2 in water explains its high permeability in hydrophilic materials.[62] The selectivity coefficient relative to these two gases is therefore dependent on the moisture content of the film.[62]

9.3.4.3 *Lipid and solute barrier properties.* Oil penetration into foods to be fried (breaded fish or meats) or dried by frying (e.g. potato chips or fruits such as bananas) can be reduced by precoating these foods with lipid-resistant hydrophilic materials.[58]

Moreover, solute penetration during freezing of meats and seafoods in brine, or during osmotic dehydration of fruit and vegetable pieces, which is a considerable limitation in these processes, can be reduced by prior application of films that are resistant to the solutes in question.[18]

Certain specific additives (antimicrobial agents, antioxidants, nutritional additives, flavours, colouring, etc.) can be incorporated in edible films to obtain localized functional effects (usually on the surface of the food) at very low mean additive concentrations. Torres et al.[88] used a film composed of zein, aceto-monoglycerides and glycerol to maintain a high surface concentration of sorbic acid on an intermediate moisture cheese analogue. The coating very significantly slowed migration of the sorbic acid. Diffusion of sorbic acid in this film was at $3-7 \ 10^{-11} \ m^2.s^{-1}$, which represented a 150- to 300-fold lower level than that in the food mass. The increased microbiological stability on the surface of this intermediate moisture cheese was confirmed by counts after surface-seeding of *Staphylococcus aureus*.[89] Moticka and Nairn[90] and Vojdani and Torres[91] studied the diffusion of benzoic acid and sorbic acid salts in lipid coatings (stearic alcohol, stearic acid and waxes) and composite films (cellulose derivatives and fatty acids). These materials significantly decreased the permeability to benzoic acid and sorbic acid. They could thus be recommended for maintaining high surface concentrations of antimicrobial agents.

Guilbert[92] developed a coating composed of a casein film containing sorbic acid and treated with a cross-linking agent. When this film was used to coat a food product with a moisture activity (A_w) of 0.95, there was more than 30% surface retention of sorbic acid after 35 days storage. When the same protein film was used to coat dried fruits, with A_w 0.85, the shelf-life was increased from only a few days to over 40 days after contamination with osmiophilic yeasts and moulds. Guilbert[92] also measured the retention of a tocopherol in gelatin films used on the surface of margarine. After 50 days storage, no migration was observed when the film was pretreated with a cross-linking agent (tannic acid), whereas without a film, a tocopherol diffusion was as found to be as high as $10-30 \ 10^{-11} \ m^2.s^{u1}$.

9.3.4.4 *Moisture barrier properties.*   Films with suitable moisture barrier properties are required for a great number of applications. Indeed, control of the moisture content and activity of heterogeneous food components or of the elements of a mixture influences the microbiological, physico-chemical and organoleptic characteristics of the food.

Surface drying of certain fresh or frozen foods or, conversely, moisture regain of dry or intermediate moisture foods, can be reduced by using films that are good barriers to moisture migration.

In order to conserve the different crunchy and soft textures, it is essential to reduce moisture exchange between compartments of heterogeneous foods (e.g. pizza, quiche, cakes or biscuits) or between the components of mixtures (e.g. aperitif mixes, breakfast mixes with dried fruit and/or cereals) with different water activity levels. In many cases, the only realistic technique to halt moisture exchange involves using edible barrier layers at the interfaces and between compartments or elements of mixtures.[18,19,93,94]

Vegetal and microbial gums, starches, soluble cellulose derivatives and many proteins are used to form films which often have poor moisture barrier properties especially at high relative humidity. The use of such films as protective layers against moisture exchange (although described in many studies and patents) is limited to providing short-term protection for dried foods such as dried fruit.[29,95]

Sugar coatings applied by centrifugal turbine action have interesting moisture barrier properties. Most hydrophilic groups of the sucrose molecule are thus involved in intermolecular linkage of the tridimensional crystal arrangement. Breakage of this structure can only occur above a certain energy level, i.e. above a certain water activity level at a given temperature (about 0.80 at 25°C).[96] At lower A_w levels, water diffusion and especially the water solubility coefficient are very weak.

Moreover, according to its sorption isotherm,[43] the moisture content of chocolate is very low within the intermediate moisture range. This explains the suitable moisture barrier properties of chocolate, which have long been used even to coat liquids (e.g. liqueur chocolates).

Many lipid compounds, such as animal and vegetable fats, aceto-glycerides, surfactants and waxes,[44,77,97-99] have been used in the formation of edible films and coatings because of their excellent moisture barrier properties (Table 9.2).

Waxy coatings on fresh fruit and vegetables thus reduce weight loss due to dehydration during storage by 40–75%.[102-106]

Some edible lipid or derivative films with good moisture barrier characteristics have been patented.[33,34,58,107-109]

Due to the weak polarity of lipid compounds and their ability to form dense organized molecular networks after cooling, the water remains relatively immobile and insoluble (quite straight sorption isotherm up to high relative humidities). The moisture barrier capacities of different films can be classified in decreasing order of efficiency, as follows: waxes > lipids and solid fatty acids > lecithin, aceto-glycerides > liquid oils. This efficiency order was confirmed by Kester and Fennema[110] in a study on the resistance of various lipids, heated and absorbed into filter papers, to water vapour transmission.

The composition, fusion, solidification range and crystalline structure (polymorphic shape) of lipids and derivatives, in addition to the interactions with water, oxygen and other components of the food product, influence the physico-chemical, functional and organoleptic properties of lipid films.[67,82]

Moisture permeability rises substantially when the proportion of liquid lipids increases. Solidification of lipids (especially saturated) in a densely

Table 9.2 Water vapour permeability of various films

Film	T (°C)	Δp (mmHg)	x (mm)	Permeability (g.mm/m².mmHg.day)
Starch, cellulose acetate	37.7	49.2–15.7	1.19	29.3
Sweet milk chocolate	26.7	26.3–0	1.91	9.94
Pectin	25	19.2–7.3	0.036	8.2
Casein-gelatin treated with lactic acid	30	28.9–18.5	0.25	7.1
Hydrogenated soya and coton oils, HPMC	25	20.0–0	0.12	1.74
Gliadins and glycerol	30	32.2–0	0.05	1.36
Acetylated glycerol monostearate	21.1	18.8–14.1	1.75	1.215
Gluten and glycerol	30	32.2–0	0.05	1.05
Gluten, DATEM and glycerol	30	32.2–0	0.05	0.55
Glutenins and glycerol	30	32.2–0	0.05	0.75
Zeine and oleic acid	37.8	44.7–0	0.04	0.30
Tempered cocoa butter	26.7	26.3–0	1.61	0.288
Palmitic and stearic acid on HPMC	25	23–15.4	0.04	0.253
Dark chocolate	20	14.1–0	0.61	0.14
Cellulose acetate	37.7	44.3–0	0.025	0.113
C_{18}–C_{16} MC/HPMC	25	20.0–0	0.02	0.035
Gluten and monoglyceride	30	32.2–0	0.11	0.024
Polyethylene (low density)	37.7	44.3–0	0.025	0.010
C_{18}–C_{16} MC/PEG and beeswax	25	20.2–0	0.056	0.0075
C_{18}–C_{16} MC/HPMC and beeswax	25	20.2–0	0.051	0.0075
C_{18}–C_{16} HPMC/PEG	25	32.2–0	0.09	0.0048
Beeswax	25	20.0–0	0.12	0.0025
Waxed paper	37.8	46.7–0	–	0.0016–0.125
Paraffin wax	25	20.0–0	0.66	0.0002
Aluminium foil	37.7	44.3–0	0.025	0.00006

According to references: 18, 25, 36, 38, 43, 46, 71, 73, 74, 76, 77, 93, 94, 99, 100, 101
x is film thickness, T is temperature, Δp is water vapor pressure gradient.
DATEM is Diacetyl Tartaric Ester of Monoglyceride, HPMC is hydroxypropylmethyl cellulose, C_{18} and
–, data not available in literature cited.

organized crystalline structure results in a very significant reduction in moisture permeability.[99,100,111,112]

Some surfactants, when applied as a thin surface layer, effectively inhibit water evaporation. They reduce water activity on the surface (surface A_w) of the food product, thus substantially slowing the rate of water evaporation. This effect depends on the structure of the surfactant/water system, i.e. on the temperature, surfactant content, length of the carbon chain and degree of saturation. Saturated fatty alcohols with 16–18 carbons and the corresponding monoglycerides (glycerol monopalmitate and monostearate) are the most effective, with fatty acids and unsaturated monoglycerides being less effective.[113]

Lipids and derivatives can form good barriers to moisture transmission, but these compounds have certain drawbacks with respect to application, mechanical and chemical stability and/or organoleptic quality. Hence,

lipid-hydrocolloid   associations   have   been   investigated
often,[19,25,26,38,58,70,73,74,93,94,100,101,114–117]   and   patented.[118–124]   These
films, which can constitute a better water vapour barrier than synthetic
material such as low density polyethylene (Table 9.2), can be applied as
emulsions, suspensions and dispersions of non-miscible compounds, or in
successive layers (films and multilayer coatings) or in a solvent solution.
The coating operation affects the barrier properties of the film. According
to Schultz et al.[117] and Gontard et al.,[115,116] who investigated the moisture
permeability of films composed of demethoxylated pectins or gluten and
various lipids (waxes, fatty acids, etc.), it is better to form two successive
layers than to apply a dispersion in solvent. Kamper and Fennema[100,101]
carried out detailed studies of films composed of soluble esters, cellulose
and a mixture of palmitic and stearic acids and demonstrated that application of solvent solutions (ethanol/water) resulted in reducing moisture
permeability by 10-fold relative to bilayer systems. Variations in homogeneity and/or structure (size, form and orientation of the crystals) of the
lipid layer affect film permeability and could be related to the coating
operation.

Guilbert[25] developed a multicomponent film composed of gelatin or
casein and carnauba wax and glycerol monopalmitate and monostearate.
This film, which was applied as an emulsion and then acidified with lactic
acid after drying, showed good water vapour barrier properties. Composite
films of casein and aceto-glycerides or wax applied as an emulsion were
also investigated by Krochta et al.[125] They showed that lowering the pH
of the film to the isoelectric point of casein reduced water vapour transmission by half.

A composite film (a mixture of hydroxypropylmethylcellulose and fatty
acids) developed by Kamper and Fennema[101] was tested as a moisture
barrier between two compartments of a heterogeneous food product
(pizza-type foods), and the results indicated an increase from 20 days to
more than 70 days in the shelf-life of the frozen product without any loss
in crust crispiness.[70] The water vapour barrier properties of this film were
excellent up to 90% relative humidity, above this level hydration caused
structural loss. To improve the moisture barrier properties at high humidities, Kester and Fennema[19,93] coated this film with a thin layer of beeswax.
The water vapour permeability of this film was close to that of polyvinyl
chloride and low density polyethylene films (Table 9.2), and it remained
constant at all relative humidity levels (up to 97%). Kester and Fennema[93]
were able to correlate permeability with the morphology of the lipid
components by electron microscopic analysis. The film was translucent,
had suitable mechanical properties, and was an effective barrier to water
transmission when tested with pizza-type frozen foods.[94] The film was not
discernible when the food was eaten hot (melting point of the film: 80–
85°C).

Using the falling film technique, Clark and Shirk[27] studied the semi-industrial application of a cellulose acetoglyceride and acetobutyrate coating (marketed under the 'LEPAK' trademark by the American Cyanamid Co.) for surface protection of frozen fish and meats. They found that this film provided excellent protection (against discoloration, dehydration, microbial development, etc.), which was close to that obtained with non-edible films.

9.4 Conclusion

This research provides evidence of the multiple advantages of using edible and biodegradable packaging made from biopolymers. Investigations on this type of packaging call on the use of biochemistry, food science and synthetic polymer technology techniques. The studies presented here have demonstrated a number of characteristics of food macromolecules that make them suitable for the formation of different types of wrappings and films. The use of these properties and their ability to be modified and controlled thus opens a new field of application for these macromolecules in a non-food sector, for the manufacture of biodegradable packaging. Problems resulting from the disposal of synthetic polymer packaging have caused a boom in research to obtain biodegradable products from biological macromolecules. This biodegradable objective has been partially reached by the addition of synthetic agricultural macromolecules (starch, proteins, etc.). Research and development is required to develop packaging material composed entirely of renewable biodegradable macromolecules from agricultural products that have good performance characteristics and are economical. This is essential for purposes of environmental protection and to create a new outlet for agricultural products.

References

1. Mueller, H.P., Tillman, H. and Gunter, W. (1991) Wolff Walsrode A-G, European P. Appln., EP 449041 A2; EP 91103879 (910314); DE 4009758 (900327).
2. Aranyi, C., Gutfreund, K., Hawrylewicz, E.J. and Wall, J.S. (1970) U.S. Patent 3522197.
3. Friedman, M. (1973) In *Industrial Uses of Cereals*, ed. Pomeranz, Y., Symp. Proceed. A.A.C.C., pp. 237–251.
4. Robey, M.J., Field, G. and Styzinski, M. (1989) *Degradable Plastics. Materials Forum*, **13**, 1–10.
5. Vert, M. (1991) *Caoutchoucs et plastiques*, **706**, 71–85.
6. Krupp, L.R. (1991) *The Biodegradability of Modified Plastic Films in Controlled Biological Environments*. Cornell University Press, Ithaca.
7. Krupp, L.R. and Jewell, W.J. (1992) *Environ. Sci. Technol.*, **26**(1), 193–199.

8. Otey, F., Westoff, R. and Russell, C. (1975) In: Symposium papers, Technical Symposium Int. Nonwoven Disposables Assoc. (INDA) vol. 3, pp. 40–47.
9. Otey, F. and Westoff, R. (1977) *Ind. Eng. Chem. Prod. Res. Dev.*, **16**, 4.
10. Otey, F. and Westoff, R. (1980) *Ind. Eng. Chem. Prod. Res. Dev.*, **19**, 592–595.
11. Otey, F. and Westoff, R. (1984) *Ind. Eng. Chem. Prod. Dev.*, **23**, 284–287.
12. Otey, F.H., Westoff, R.P. and Doane, W.M. (1987) *Ind. Eng. Chem. Res.*, **26**, 1659–1663.
13. Lenk, R.S. and Merrall, E.A. (1981) *Polymer*, **22**, 1279–1282.
14. Biermann, C.J. and Narayan, R. (1987) Polymer preprint, **2**, 240–241.
15. Narayan, R., Lu, Z., Chen, Z. and Stacy, N. (1988) *Polymer Preprints*, **29**, 106–107.
16. Narayan, R., Stacy, N. and Lu, Z. (1989) *Polymer Preprints*, **30**, 105–108.
17. Tomka, I., Muller, R., Innerebner, F., Pukansky, B. and Hollo, J. (1991) In: *International Symposium Proceedings: valorisation industrielle non alimentaire des productions de grandes cultures*, CIIAA, Paris, 20–21 Nov.
18. Guilbert, S. and Biquet, B. (1989) In: *L'emballage des denrées alimentaires de grande consommation*, Eds G. Bureau and J.L. Multon, Tech. et Doc., Lavoisier, Apria, Paris.
19. Kester, J.J. and Fennema, O.R. (1986) *Food Technol.*, **40**(12), 47–59.
20. Kroger, M. and Igoe, R.S. (1971) *Food Prod. Dev.*, November, 74–79.
21. Morgan, B.H. (1971) *Food Prod. Dev.*, June–July, pp. 75–78.
22. Banker, G.S. (1966) *J. Pharm. Sci.*, **55**, 81–92.
23. Deasy, P.B. (1984) *Microencapsulation and Related Drug Processes*, Marcel Dekker, Inc., New York.
24. Kondo, A. and Van Valkenburg, J.W. (1979) *Microcapsule Processing and Technology*, Marcel Dekker, Inc., New York.
25. Guilbert, S. (1986) In: *Food Packaging and Preservation, Theory and Practice*, M. Mathlouthi, Ed. Elsevier Applied Science, London, 371–394.
26. Lowe, E., Durkee, E.L., Hamilton, W.E., Watters, G.G. and Morgan, A.I. (1963) *Food Technol.*, **11**, 109–111.
27. Clark, W.L. and Shirk, R.J. (1965) *Food Technol.*, **19**(10), 1561–1599.
28. Jokay, L., Nelson, G.E. and Powell, E.L. (1967) *Food Technol.*, **21**, 1064–1066.
29. Swenson, H.A., Miers, J.C., Schultz, T.H. and Owens, H.S. (1953) *Food Technol.*, **7**, 232–235.
30. Bank, H.U. and Rubenstein, H.I. (1985) U.S. Patent 4, 505, 220.
31. Shea, R.A. (1970) U.S. Patent 3, 576, 836.
32. Peyron, A. (1991) *Viandes Prod. Carnés*, **12**(4), 123–127.
33. Tresser, D. (1983) U.S. Patent 4, 394, 392.
34. Tresser, D. (1983) U.S. Patent 4, 396, 633.
35. Hannigan, K. (1984) *Food Eng.*, March, 98–99.
36. Gontard, N., Guilbert, S. and Cuq, J.L. (1992) *J. Food Sci.*, **57**(1), 190–195.
37. Lim, S.T., Jane, J.L., Rajagopalan, S. and Seib, P.A. (1992) *Biotechnol. Progress*, **8**, 51–57.
38. Gontard, N. (1991) Films et enrobages comestible: étude et amélioration des propriétés filmogènes du gluten. University of Montpellier II, Fr.
39. Kastner, K.P. and Whitney, T. (1990) In: *ANTEC'90: Plastics in the Environment Held*, pp. 1973–1977.
40. Williams, M.L., Landel, R.F. and Ferry, J.D. (1955) *J. Amer. Chem. Soc.*, **77**, 3701–3706.
41. Gontard, N., Ring, S., Guilbert, S. and Botham, L. (1994) *J. Food Sci.*, to be published.
42. Kumins, C.A. (1965) *J. Polymer Sci.*, Part C, **10**, 1–9.
43. Biquet, B. and Labuza, T.P. (1988) *J. Food Sci.*, **53**(4), 989–998.
44. Feuge, R.O. (1955) *Food Technol.*, **9**, 314–318.
45. Simatos, D. and Karel, M. (1988) In *Food Preservation by Moisture Control*, ed. Seow, C.C., Elsevier Applied Science, Barking, pp. 1–56.
46. Gontard, N., Guilbert, S. and Cuq, J.L. (1993) *J. Food Sci.*, **58**(1) 206–211.
47. Kakivaya, S.R. and Hoeve, C.A.J. (1975) *Proc. Nat. Acad. Sci.*, **72**(9), 3505–3507.
48. Marshall, A.S. and Petrie, S.E.B. (1980) *J. Photogr. Sci.*, **28**, 128–136.
49. Maurice, T.J., Slade, L., Sirrett, R.R. and Page, C.M. (1985) In *Influence of Water on Food Quality and Stability*, eds D. Simatos and J.L. Multon, M. Nijhoff, Dordrecht, The Netherlands, pp. 211–222.

50. Zelezniack, K.J. and Hoseney, R.C. (1987) *Cereal Foods World*, **64**, 121–126.
51. Kelley, S.S., Riabs, T.G. and Glaser, W.G. (1988) *J. Materials Sci.*, **22**, 617–623.
52. Blond, G. (1989) *Cryoletters*, **10**, 299–310.
53. Mackenzie, A.P. and Rasmussen, D.H. (1972) In *Water Structure at the Polymer Interface*, ed. Jellinck, N.H.G., Plenum Press, New York, pp. 146–152.
54. Rasmussen, D. and Luyet, B. (1969) *The Technology of Plasticizers*, J. Wiley Interscience, New York.
55. Evangelista, R.L., Nikolov, Z.L., Sung, W., Jane, J. and Gelina, R.J. (1991) *Ind. Eng. Chem. Res.*, **30**, 1841–1846.
56. Griffin, G.J.L. (1977) US Patent 4, 021, 388.
57. Jane, J., Evangelista, R.L., Wang, L., Ramrattan, S., Moore, J.A. and Gelina, R.J. (1990) In: *Corn Utilisation Conference 3 Proceedings*, **4**, pp. 1–5.
58. Daniels, R. (1973) *Edible Coatings and Soluble Packaging*, Noyes Data Corporation, Park Ridge, NJ.
59. Grouber, B. (1983) *Labo. Pharma. Probl. Tech.*, **337**, 909–916.
60. Mannheim, C. and Passy, N. (1985) In: *Properties of Water in Foods*, eds D. Simatos and J.L. Multon, Martinus Nijhoff, Dordrecht, The Netherlands, pp. 375–392.
61. de Leiris, J.P. (1986) Water activity and permeability. In: *Food Packaging and Preservation, Theory and Practice*, M. Mathlouthi, Ed. Applied Science, London, pp. 213–233.
62. Schwartzberg, H.G. (1985) In: *Food Packaging and Preservation*, M. Mathlouthi, Ed. Elsevier Applied Science, New York, pp. 115–135.
63. Lecomte, C. (1988) personal communication.
64. Cohen, M.H. and Turnbull, O. (1959) *J. Chem. Phys.*, **31**, 1164–1169.
65. Turnbull, O. and Cohen, M.H. (1961) *J. Chem. Phys.*, **34**, 120–125.
66. Coutandin, J., Ehlich, D., Sillescu, H. and Wang, C.H. (1985) *Macromolecules*, **18**, 587–589.
67. Kester, J.J. and Fennema, O. (1989) *J. Amer. Oil Chem. Soc.*, **66**(8), 1147–1153.
68. AFNOR (1974) Emballages, Matières en feuille, Détermination du Coefficient de Transfert de Vapeur d'Eau., NF H 00–030.
69. ASTM (1983) Standard Test Methods for Water Vapor Transmission of Materials, American Society for Testing and Materials, 15.09, E96–80.
70. Kamper, S.L. and Fennema, O.R. (1985) *J. Food Sci.*, **50**, 382–384.
71. Allen, L., Nelson, A.I., Steinberg, M.P. and McGill, J.N. (1963) *Food Technol.*, **17**, 1437–1441.
72. Mark, A.M., Roth, W.B., Mehltretter, C.L. and Rist, C.E. (1966) *Food Technol.*, **20**(1), 75–77.
73. Greener, I.K. and Fennema, O. (1989) *J. Food Sci.*, **54**(6), 1393–1399.
74. Greener, I.K. and Fennema, O. (1989) *J. Food Sci.*, **54**(6), 1400–1406.
75. Karel, M., Issenberg, P. and Jurin, V. (1963) *Food Technol.*, **17**, 327–330.
76. Karel, M. (1975) In: *Principles of Food Science Part II. Physical Principles of Food Preservation*, Ed. O.R. Fennema, Marcel Dekker, New York, pp. 400–410.
77. Lovegren, N.V. and Feuge, R.O. (1954) *J. Agric. Food Chem.*, **2**, 558–563.
78. Klose, A., Mecchi, E.P. and Hanson, H.L. (1952) *Food Technol.*, **6**, 308–315.
79. Anandaraman, S. and Reineccius, G. (1980) *Microencapsulation of Flavour*, FIPP., May, 22–25.
80. Blank, M. (1962) In: *Retardation of Evaporation by monolayers: Transport Processes*, Ed. V.K. LaMer, Academic Press, New York, pp. 75–79.
81. Blank, M. (1972) In: *Techniques of Surface and Colloidal Chemistry and Physics*, Eds. R.J. Good, R.R. Stromberg and R.L. Patrick, Marcel Dekker, New York, pp. 41–88.
82. Kester, J.J. and Fennema, O. (1989) *J. Amer. Oil Chem. Soc.*, **66**(8), 1129–1138.
83. Drake, S.R., Fellman, J.K. and Nelson, J.W. (1987) *J. Food Sci.*, **52**, 1283–1285.
84. Lowings, P.H. and Cutts, D.F. (1982) *Proc. Inst. Food Sci. Tech. Ann. Symp.*, July, Nottingham, UK, pp. 52–67.
85. Banks, N.H. (1983) *J. Expert. Bot.*, **34**, 871–876.
86. Banks, N.H. (1984) *J. Expert. Bot.*, **35**, 127–134.
87. Smith, S.M. and Stow, J.R. (1984) *Ann. Appl. Biol.*, **104**, 383–391.
88. Torres, J.A., Motoki, M. and Karel, M. (1985) *J. Food Proc. Pres.*, **9**, 75–79.
89. Torres, J.A. and Karel, M. (1985) *J. Food Proc. Pres.*, **9**, 107–119.

90. Moticka, S. and Nairn, J.G. (1978) *J. Pharm. Sci.*, **67**, 500–521.
91. Vojdani, F. and Torres, J.A. (1990) *J. Food Sci.*, **55**(3), 841–846.
92. Guilbert, S. (1988) In: *Foods Preservation by Moisture Control*, C.C. Seow, Ed. Elsevier Applied Science, London.
93. Kester, J.J. and Fennema, O. (1989) *J. Food Sci.*, **54**(6), 1383–1389.
94. Kester, J.J. and Fennema, O. (1989) *J. Food Sci.*, **54**(6), 1390–1406.
95. Forkner, J.H. (1958) U.S. Patent 2, 821, 477.
96. Bussiere, G. and Serpelloni, M. (1985) In *Properties of Water in Foods*, Eds D. Simatos and J.L. Multon, Martinus Nijhoff, Dordrecht, The Netherlands, pp. 627–645.
97. Alfin-Slater, R.B., Coleman, R.D., Feuge, R.O. and Altschul, A.M. (1958) *J. Amer. Oil Chem. Soc.*, **35**, 122–127.
98. Feuge, R.O., Vicknair, E.J. and Lovegren, N.V. (1953) *J. Amer. Oil Chem. Soc.*, **30**, 283–296.
99. Landman, W., Lovegren, N.V. and Feuge, R.O. (1960) *J. Amer. Oil Chem. Soc.*, **37**, 1–4.
100. Kamper, S.L. and Fennema, O.R. (1984) *J. Food Sci.*, **49**, 1478–1481.
101. Kamper, S.L. and Fennema, O.R. (1984) *J. Food Sci.*, **49**, 1482–1485.
102. Brusewitz, G.H. and Singh, R.P. (1985) *J. Food Proc. Press.*, **9**, 1–9.
103. Espelie, K.E., Carvalho, S.C. and Kolattukudy, P.E. (1982) *Hort Sci.*, **17**(5), 779–780.
104. Hardenburg, R.E. (1967) *Wax and Related Coatings for Horticultural Products. A Biography*. Agricultural Research Service Bulletin 51–15, United States Department of Agriculture, Washington, D.C.
105. Kaplan, H.J. (1986) In *Fresh Citrus Fruits*, Eds W.F. Wardoski, S. Nagy and W. Grierson, The AVI Publishing Company, Westport, CT., p. 379.
106. Paredez-Lopez, O., Carmargo-Rubio, E. and Gallardo-Navarro, Y. (1974) *J. Sci. Food Agric.*, **25**, 1207–1210.
107. D'Atri, J.J., Swidler, R., Colwell, J.J. and Parks, T.R. (1980) U.S. Patent 4, 207, 347.
108. Watters, G.G. and Brekke, J.E. (1959) U.S. Patent 2, 909, 435.
109. Werbin, S., Rubenstein, I.H., Weinstein, S. and Weinstein, D. (1970) U.S. Patent 3, 526, 515.
110. Kester, J.J. and Fennema, O.R. (1986) *J. Amer. Oil Chem. Soc.*, **66**(8), 1139–1146.
111. Kester, J.J. and Fennema, O. (1989) *J. Amer. Oil Chem. Soc.*, **66**(7), 1154–1157.
112. Watters, G.G. and Brekke, J.E. (1961) *Food Technol.*, **5**, 236–238.
113. Roth, T. and Loncin, M. (1984) In *Engineering and Food*, Volume 1, Ed. B.M. McKenna, Elsevier Applied Science, New York, pp. 433–444.
114. Chu, C.L. (1986) *Hort Science*, **21**(2), 267–280.
115. Gontard, N., Duchez, C., Cuq, J.L. and Guilbert, S. (1994) *Int. J. Food Sci. Technol.*, to be published.
116. Gontard, N., Marchesseau, S., Cuq, J.L. and Guilbert, S. (1994) *Int. J. Food Sci. Technol.*, to be published.
117. Schultz, T.H., Miers, J.C., Owens, H.S. and Maclay, W.D. (1949) *J. Phys. Colloid Chem.*, **53**, 1320–1330.
118. Cole, S.M. (1969) U.S. Patent 3, 479, 191.
119. Cosler, H.B. (1957) U.S. Patent 2, 791, 509.
120. Hamdy, M.M. and White, S.H. (1969) U.S. Patent 3, 471, 304.
121. Harris, N.E. and Lee, F.H. (1974) U.S. Patent 3, 794, 742.
122. Seaborne, J. and Egberg, D.C. (1987) U.S. Patent 4, 661, 359.
123. Silva, R., Ash, D.J. and Scheible, C.E. (1981) U.S. Patent 4, 293, 572.
124. Ukai, N., Ishibashi, S., Tsutsumi, T. and Marakami, K. (1976) U.S. Patent 3, 997, 674.
125. Krochta, J.M., Pavlath, A.E. and Goodman, N. (1989) International Congress on Engineering and Food 5.

10 Bacterial poly(hydroxyalkanoates)

G. EGGINK, J. SMEGEN, G. ONGEN-BAYSAL and
G.N.M. HUIJBERTS

Abstract

Poly(3-hydroxyalkanoates) (PHAs), of which poly(3-hydroxybutyrate)
(PHB) is the most common, can be accumulated by a large number of
bacteria as energy and carbon reserves. Due to their biodegradability and
biocompatibility these optically active polyesters may find industrial applications. The properties of PHAs are dependent on their monomeric
composition and therefore it is of great interest that recent research has
revealed that, in addition to PHB, a variety of PHAs can be synthesized
with bacteria. The monomeric composition of PHAs depends on the
nature of the carbon source and the microorganism used. PHB is a typical
highly crystalline thermoplastic whereas long side chain PHAs are elastomers with low melting points.

In this chapter a general overview is given on: the biosynthetic pathways
leading to the formation of PHAs, the fermentative production of PHAs,
the recovery of PHA from biomass, the polymeric properties and biodegradability of PHAs, and some (potential) applications of PHAs. It is
concluded that due to the increasing interest in the use of biodegradable
polymers, e.g. in packaging materials, PHAs may be widely used in the
future. A prerequisite for the large production and use of PHAs is,
however, that the current high costs of production should be lowered by
increasing the productivity of PHA fermentations to a level similar to other
large scale fermentation processes. Also the possibility of production of
PHAs in transgenic plants is considered.

10.1 Introduction

The Japanese Science and Technology Agency has recently conducted a
survey among scientists concerning their expectations of breakthroughs
in science and technology with the next generation.[1] One of the remarkable outcomes was that scientists in Japan predicted that by the year
2003 biodegradable packaging materials will be widely used. A variety of
truly biodegradable polymers is available on the market, such as poly-
(lactic acid), thermoplastic starch and polyhydroxybutyrate/valerate. Com-

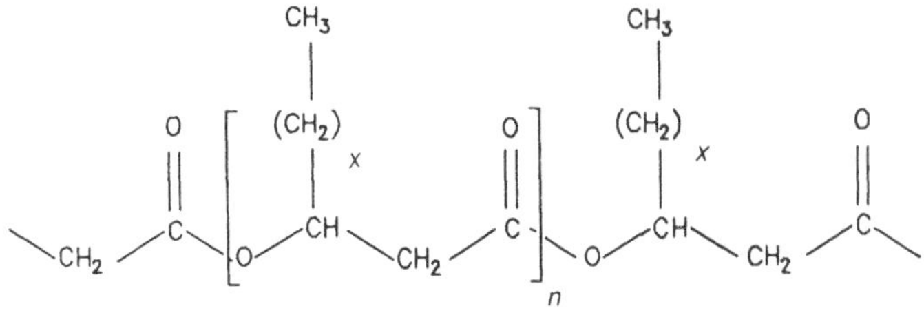

Figure 10.1 General structural formula of poly(3-hydroxyalkanoates): n = 1000–15 000; x = 0–10.

pared to commodity plastics, these biodegradable polymers are relatively expensive and therefore only produced on a small scale. However, it can be anticipated that due to increasing academic and industrial interest, considerable progress will be made with respect to the production costs and applicability of these bioplastics.

The poly(hydroxyalkanoates) (PHAs) (Figure 10.1) are a very common class of bacterial reserve materials, that have attracted considerable industrial attention. These microbial polyesters are biodegradable and biocompatible thermoplastics with physical properties dependent on their monometric composition. The production of PHAs is a typical biotechnological process which requires for development the involvement of several scientific disciplines, i.e. genetics, biochemistry, microbiology, bioprocess engineering, polymer chemistry and polymer engineering. A number of specialized reviews[1-4] on bacterial PHAs have appeared describing, in particular, the rapid progress made in each of these areas of research. In this chapter a general overview is given on the biosynthesis, production, downstream processing, characterization and application of bacterial PHAs.

10.2   Biosynthesis of poly(3-hydroxybutyrate)

Poly(3-hydroxyalkanoates) (PHAs), of which poly(3-hydroxybutyrate) (PHB) is the most common, can be accumulated by a large number of different bacteria. These polymers are stored in intracellular granules and function as both a carbon and energy reserve (Figure 10.2). Synthesis of PHA occurs under conditions of excess carbon source and limitation in one of the other nutrients, e.g. nitrogen or phosphorous.

PHB synthesis is extensively studied in *Alcaligenes eutrophus*, which is currently used for the industrial production of PHB. Genetic and biochemical studies have shown that PHB is synthesized from acetyl-CoA via a three step biosynthetic pathway (Figure 10.3). 3-Ketothiolase catalyses the condensation of two molecules of acetyl-CoA to produce acetoacetyl-

Figure 10.2 Freeze fracture electron micrograph of a recombinant *P. oleovorans* containing both genes for PHA and PHB synthesis.[8] The cell shows both needle type deformations of granules as well as mushroom type deformation of granules. The needle type structures are typical for PHB and the mushroom type structures are ascribed to long side chain PHA. The bar represents 0.5 μm. (Courtesy of Dr Witholt.)

CoA; this is subsequently reduced by an NADPH dependent acetoacetyl CoA reductase.[2] The product of this reaction is substrate for PHB polymerase. Molecular cloning and nucleotide sequence analysis revealed that the PHB biosynthetic genes are clustered and organized in a single operon.[3,5] Introduction of this cloned PHB operon in *Escherichia coli*, which normally does not produce PHB, resulted in a recombinant strain capable of accumulating PHB to an astonishing 90% of the cell dry weight.[6] The PHB operon was also introduced in *Pseudomonas* strains capable of accumulating PHAs comprised of medium chain length monomers (see below). Though it was aimed to produce a copolymer of PHB with medium chain length monomers, it appeared that two distinct types of granules were formed in the cell, one consisting of PHB and another consisting of PHA containing exclusively medium chain length monomers[7,8] (see Figure 10.2).

In addition to 3-hydroxybutyrate a limited number of other monomers (such as 3-hydroxyvalerate, 4-hydroxybutyrate and 5-hydroxyvalerate) can be incorporated into the polymer by the PHB polymerizing system. This, however, can only be achieved when precursors of these monomers, i.e. valeric acid or propionic acid, 4-hydroxybutyrate and 5-chlorovaleric acid, are respectively included in the medium.[9,10-12] In this respect, it is impor-

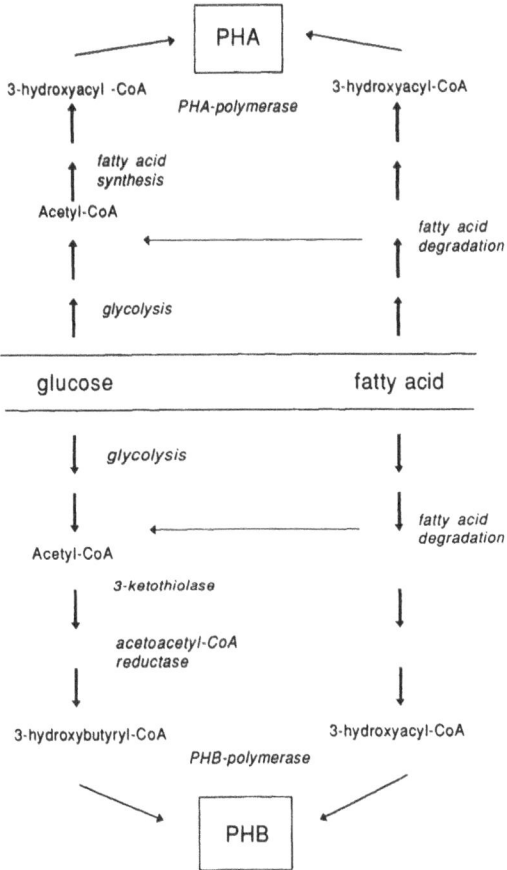

Figure 10.3 Schematic representation of metabolic pathways leading to PHB synthesis in *A. eutrophus*[2] and PHA synthesis in *Pseudomonas*.[52]

tant to notice that a number of bacteria have recently been isolated which are capable of synthesizing the copolymer PHB/HV from one single carbon source, like glucose or sucrose.[13,14]

10.3 Biosynthesis of poly(3-hydroxyalkanoates)

In the early 1980s a second class of bacterial PHAs was discovered. Witholt and coworkers found that *Pseudomonas* strains accumulate PHA during cultivation on medium chain length alkanes, alkenes and fatty acids.[15–17] These so-called long side chain PHAs consist of monomers which are, with respect to the carbon chain length, either directly derived from the substrate or shortened by one or more C2 units. Therefore it has been proposed that the (S)-3-hydroxyacyl-CoA intermediates of beta-oxidation

Table 10.1 Composition of PHA synthesized by *P putida* KT2442 during growth on different carbon sources[19,22]

Substrate	%PHA	Relative amount of monomers in purified PHA (%, wt/wt)									
		C6	C8	C10	C12:1^5	C12:1^6	C12	C14:2	C14:1^5	C14:1^7	C14
Glucose	16.9	tr	6.9	74.3	8.8	–	7.7	–	–	1.6	tr
Decanoate	27.6	5.3	52.3	42.3	–ᴶ	–	–	–	–	–	–
Oleate	37.2	4.4	33.5	32.2	–	–	14.4	–	15.5	–	–
Linoleate	17.8	5.6	38.9	22.7	–	15.9	–	16.9	–	–	–

C6, 3-hydroxyhexanoate; C8, 3-hydroxyoctanoate; C10, 3-hydroxydecanoate; C12:1^5-3-hydroxy-5-*cis*-dodecenoate; C12, 3-hydroxydodecanoate, C14:1^5, 3-hydroxy-5-*cis*-tetradecenoate; C14:1^7-*cis*-tetradecenoate; C14, 3-hydroxytetradecanoate; C14:2, 3-hydroxy-5-*cis*-8-*cis*-tetradecadienoate.
tr, trace amounts (<0.1% wt/wt).

are incorporated into PHA after inversion to the (*R*)-enantiomer by the action of an epimerase (Figure 10.3)/[16,18]

Long chain fatty acids can also be used as substrates for PHA synthesis. In this case, however, the fatty acids are degraded via the beta-oxidation cycle until medium chain length intermediates are formed. This implies that the composition of PHA synthesized from long chain fatty acids is determined by the specificity of the PHA synthesizing system, the fatty acid degradation pathway and the structure of the fatty acid. The latter is of special interest since a variety of different long chain fatty acids is available, allowing the synthesis of a series of different PHAs[19] (see Table 10.1).

It was recently found that fluorescent *Pseudomonas* strains also accumulate PHAs during unbalanced growth on unrelated substrates such as glucose.[20,21] Haywood *et al.*[20] found that these polymers predominantly consist of 3-hydroxydecanoate. In addition to this monomer, we have identified six other saturated and unsaturated monomers (see Table 10.1). The chemical structure of these constituents are identical to the structures of the acyl moieties of the 3-hydroxy-acyl carrier protein (ACP) intermediate of the fatty acid biosynthetic pathway.[22] We therefore concluded that during growth on non-related substrates, (*R*)-3-hydroxyacyl-ACP intermediates of the *de novo* fatty acid biosynthetic pathway are diverted to the PHA polymerizing system (see Figure 10.3).

Genetic studies revealed that the genes responsible for PHA synthesis from fatty acids are also responsible for PHA formation from glucose.[22] The molecular characterization of the *P. oleovorans pha* locus was found to encode for two PHA polymerase genes which are flanking a PHA depolymerase gene.[23] So far no clear difference in function of specificity has been found for the two polymerases; in fact the substrate range of PHA polymerases appears to be very broad. Alkanoic acids that contain functional groups such as bromide, chloride, nitrile and phenyl can be applied[24] as a co-substrate and will be incorporated into PHA with retention of the functional group.

10.4   Production of poly(3-hydroxyalkanoates)

10.4.1   PHB fermentations

In the past decade several industries have explored the possibility of commercial PHB production (Figure 10.4). In 1990, Imperial Chemical Industries (ICI) started the production of the copolymer poly(3-hydroxybutyrate-co-valerate) using *Alcaligenes eutrophus* in a two stage fed-batch process with glucose and propionic acid as feedstocks.[9] The ratio of propionic acid to glucose determines the 3-hydroxyvalerate content of the polymer. The Austrian company Btf has developed a pilot plant for the production of PHB by the microorganism *Alcaligenes latus* using sucrose as a carbon source.[25]

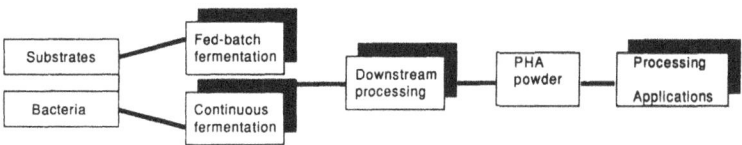

Figure 10.4 Scheme of poly(hydroxyalkanoate) production and recovery process.

In general, the efficiency of a fermentation is assessed by three factors: product yield, production rate and final concentration of the product. The published data from ICI show that their large scale process has a production rate of 0.7 g per litre per hour. The concentration of the product is approximately 75 gram per litre. Btf claims that their process has a productivity of 1 ton PHB per week in a 15 m³ reactor. This is equivalent to a production rate of 0.4 gram PHB per litre per hour. Compared to other biotechnological processes these productivities are rather low, and since the yield (0.3 kg PHB per kg sugar) is also low and the downstream processing costs are high, PHB is relatively expensive. For this reason, attempts are being made to lower the production costs by using very cheap raw materials. Suzuki *et al.*[26] have shown that with *Protomonas extorquens* growing on methanol, extremely high cell densities with a high PHB content can be obtained. However, the PHB accumulated by these microorganisms is very difficult to isolate and has a low molecular weight. Molasses has also been considered as a substrate for the large scale production of PHB. Page[27] has isolated an *Azotobacter vinelandii* UWD strain which produces relatively large amounts of PHB during growth of molasses. However, molasses is disadvantageous for its low sugar content. About 8 to 10 kg of molasses are required to produce 1 kg of PHB. Such dilute feedstocks are highly impractical and could make a large scale high cell density fermentation uneconomic.

A different category of substrates was recently examined as potential feedstock for PHB fermentations. Akiyama et al.[28] and Eggink et al.[29] demonstrated that several *Alcaligenes* species grow very well on vegetable oils or long chain fatty acids. Moreover, the PHB yield with *A. eutrophus* cultivated on oleic acid was found to be almost twice the yield obtained with carbohydrates, with less than 2 kg of LCFAs being required to produce one kg of PHB. Thus, LCFAs have potential as feedstocks for PHB fermentations. However, further research is required since the use of LCFAs in fermentations brings specific problems such as high oxygen demand and inadequate mass transfer due to the two-liquid phase system. Results in our laboratory have shown that cell dry weight of 60 g/l with 65% PHB can easily be obtained with *A. eutrophus* growing on oleic acid.[29]

10.4.2 *PHA fermentations*

In a previous section we described that long side chain PHAs can be synthesized from aliphatic compounds as well as carbohydrates. From an economical point of view, however, it seems unlikely that PHAs will be produced on a large commercial scale from sugars. So far no data on product yield are available, but based on published data on single cell oil it can be anticipated that approximately 5 kg of glucose is required to synthesize 1 kg of PHA. Highly reduced aliphatic compounds, such as alkanes and fatty acids, will therefore be more suitable substrates for PHA production with *Pseudomonas*.

Witholt and coworkers[30,31] explored the possibility of both continuous and fed-batch fermentation for the production of PHA. In a fed-batch fermentation with *P. oleovorans* and n-octane as carbon source a maximum cell density of 37 grams dry weight per litre was reached. The polymer content of the cell mass was 35%. This is equivalent to a volumetric productivity of 0.25 gram per litre per hour.[30] In high cell density continuous fermentations using octane as carbon source a considerably higher productivity was obtained. At a dilution rate of 0.2 a cell density of 12 gram dry weight per litre could be obtained containing 23% PHA. The continuous culture of the PHA producing *P. oleovorans* was maintained successfully for one month.[31] The average volumetric productivity during this period was 0.5 gram PHA per litre per hour, which is still insufficient for the large scale commercial production of PHA. However, further improvements of the production process can be expected from the use of less toxic and less reduced substrates (ie LCFAs), increase of mass and oxygen transfer in the bioreactor, strain selection, and genetic alteration with regard to regulation of PHA synthesis.

10.4.3 PHA synthesis in plants

A very recent development is the production of PHAs in transgenic plants. The cost of PHA produced by means of fermentation is, and will always be, higher than the cost of biopolymers, such as starch, produced by plants. For this reason several research groups are now trying to introduce genes responsible for PHA synthesis into plants. Suitable crops would be oil or starch accumulating crops. Through genetic engineering of these plants the carbon flux should be diverted from lipid or starch synthesis to PHA synthesis.

As a first step Poirier et al.[32] have succeeded in producing PHB in Arabidopsis thaliana which was transformed with PHB synthesis genes from A. eutrophus. Transgenic plant lines accumulated PHB granules in the cytoplasm, nucleus and vacuole. The size and appearance of these granules were similar to the PHB granules formed in bacteria. The amount of PHB, however, was very low, approximately 0.1% of the cell dry weight of leaves.

10.5 Downstream processing

Downstream processing is an important step in the production of PHAs since the method of isolation will affect the purity and quality of the polymers. Several methods for the isolation and purification of PHAs have been developed: (1) solvent extraction; (2) chemical digestion; (3) selective enzymatic digestion.

1. PHAs can be extracted from the biomass with suitable organic solvents.[33] Chlorinated solvents, like chloroform, methylene chloride and 1,2-dichloroethane, are often used on a small scale. Propylene carbonate is also an appropriate solvent. Once the polymer is extracted from the biomass, it is separated from other dissolved cell components by precipitating it with a non-solvent such as methanol or ethanol. Very pure polymers can be obtained with these procedures. Although Btf has applied this method successfully on a pilot scale,[25] the use of solvents is too impractical and expensive for application on an industrial scale.[9]
2. With the chemical treatment, the biomass is selectively disrupted and digested. Usually an alkaline hypochlorite solution is used.[34] A major disadvantage of this method is that it results in a serious degradation of the polymer. Berger et al.,[35] however, have improved the procedure in order to minimize degradation.

3. In the third method PHA granules are isolated from any biomass by an enzymatic treatment in which all cell materials with exception of PHA are solubilized. The enzymatic treatment is usually preceded by some kind of shock disruption, e.g. heat or ultrasound and the enzymes used are lipases and proteases. The aqueous suspension of the PHA granules can be dried by spray drying. ICI is currently using this method for PHB/HV isolation and a purity of 95% is obtained.[9]

The purification methods 2 and 3 have been specifically developed for PHB or its copolymers. The applicability of these procedures for the purification of long side chain PHAs has to be determined and probably be optimized.

10.6   Properties of PHB and PHA

10.6.1   *Molecular weight*

The mechanisms which in the bacterial cell affect and determine the molecular weights of PHAs are not well understood. Molecular weights of one million or more have been reported for PHB.[9,11] However the molecular weight of PHB may vary greatly depending on the microorganism and the feedstock used. Molecular weights of long side chain PHAs with saturated or unsaturated pendant groups are relatively low and usually in the range from 160 000 to 360 000[36,37] (see Table 10.2).

10.6.2   *Thermal properties*

The melting temperature (T_m) and glass transition temperature (T_g) of several PHA variants have been summarized in Table 10.2. Incorporation

Table 10.2 Physical characteristics of some saturated and unsaturated PHA types

	$Mw^b \times 10^{-3}$	D (Mw/Mn)[a]	T_g °C	T_m °C
PHA from *A. eutrophus*[11,38]				
PHB	1400	1.8	9	177
PHB/HV (15 mol %HV)	982	2.5	−1	140
PHB/HV (28 mol %HV)	930	4.0	−8	102
PHA from *P. oleovorans*[37] grown on:				
n-hexane	330	1.8	−25.8	–
n-octane	178	1.8	−36.5	58.5
n-decane	225	2.0	−38.4	47.6
1-octene	242	2.4	−36.6	–
1-decene	260	2.2	−43.1	–

[a]Mn = number averaged molecular weight.
[b]Mw = weight average molecular weight.

of comonomers into PHB decreases the T_m as well as the T_g. The lower T_m make PHB/HV a more attractive polymer than PHB, since a major disadvantage of the homopolymer PHB is that it melts and degrades to crotonic acid when kept above 180°C. This makes PHB difficult to process by extrusion or injection moulding.

Compared to PHB long side chain PHAs have low melting points (in the range of 45 to 60°C). The T_g values of the PHAs decreases as the average length of the side chain group increases.[38]

10.6.3 Crystallinity/morphology

PHB is an optically active polyester with complete linear chains. As a result PHB can achieve a relatively high degree of crystallinity,[39] usually between 55–75%. Copolyesters of PHB and PHB have approximately the same high degree of crystallinity. They show a minimum in their melting point versus composition curve, which is explained by the phenomenon of isodimorphism.[39]

Some differences in crystallinity for long side chain PHAs were found.[36,37] Although the saturated PHAs appeared to be able to crystallize they showed a lower degree of crystallinity than PHB. The unsaturated PHAs were amorphous implying that unsaturated endgroups somehow prevent polymeric chains from crystallizing. All PHAs seem to crystallize in a similar layered packing order, forming thin lamellar crystals.[38]

It was interesting to note[40] that PHA and PHB granules are completely amorphous *in vivo*. Harrison *et al.*[41] suggested that water and/or lipids act as plasticizers but this has been contradicted by de Koning *et al.*[42] who argue that the granules found in microbes are simply too small to crystallize. A typical time for a granule of 0.5 μm across to nucleate for crystallization would be approximately 30 000 years at 20°C.

10.6.4 Mechanical properties

The mechanical properties of PHB and its copolymers have been extensively studied.[43,44] PHB resembles isotactic polypropylene with respect to Young's modulus (3.5 Gpa) and the tensile strength (40 Mpa). However, the elongation needed to break (6% for PHB) is markedly lower than that of polypropylene (400%) which makes PHB stiffer and more brittle than PP.

Incorporation of comonomers (3HV or 4HB) produces dramatic changes in the mechanical properties: the stiffness (Young's modulus) decreases while the toughness (elongation to break) increases.[11] The ageing of PHB, and to a lesser extent its copolymers, is of great importance. Moulded PHB materials are tough, but over a period of several days

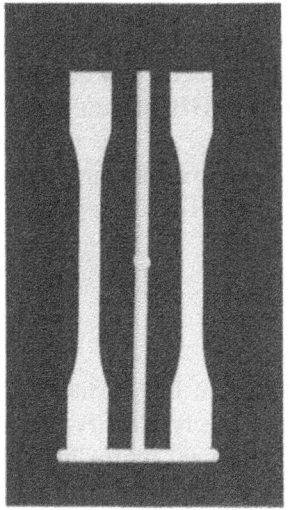 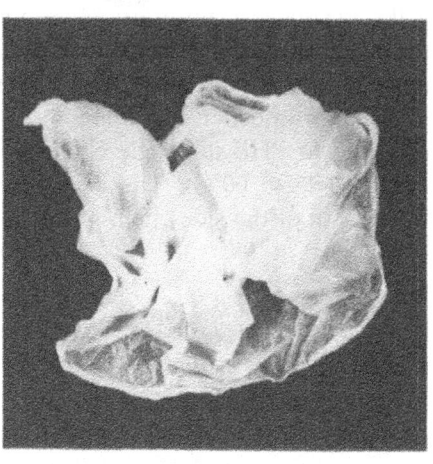

Figure 10.5 On the left, a sample of the copolymer PHB/HV processed via injection moulding. On the right, a sample of long side chain PHA made by solution casting.

at ambient conditions they slowly become brittle. This detrimental ageing process is a major drawback for the commercial use of the homopolymer.

Long side chain PHAs, unlike PHB, behave as elastomers (Figure 10.5) with crystals acting as physical crosslinks. Marchessault *et al.*[38] found an elongation to break of 250%–350%, and a Young's modulus of 17 Mpa which is about 10–20 times greater than found for a typical crosslinked rubber. In any case, however, it should be considered that the thermal history of the material greatly affects the mechanical properties. This implies that thermal treatment, i.e. annealing, can dramatically change the properties.[45]

10.6.5 *Biodegradability*

One of the most important features of PHAs is that they are fully biodegradable (Figure 10.6). PHAs are naturally occurring polymers and many different microorganisms excrete extracellular enzymes (depolymerases) which catalyse the hydrolysis of PHAs.[46] The PHA fragments are utilized as carbon source by these microorganisms resulting in the formation of biomass, water and carbon dioxide. Also PHA polymers which are processed into fibres or films are degraded in soil, seawater or in a waste water treatment system. The rate of biodegradation (ranging from days to years) is affected by a number of factors, such as thickness, surface properties, composition and physical state of the polymer, and of course the environmental conditions.[47] In general degradation and assimilation of

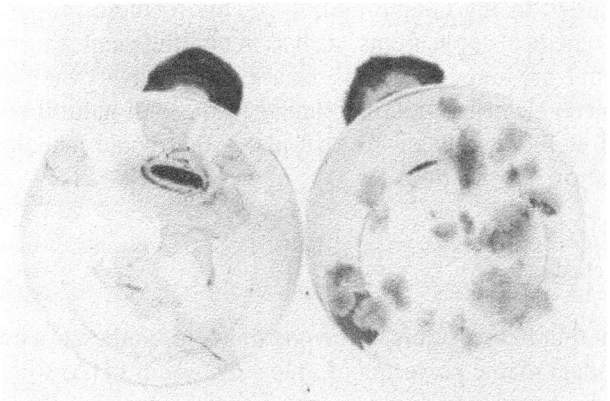

Figure 10.6 Biodegradation of pieces of long side chain PHA tested in two Erlenmeyer flasks containing a sterile minimal salt medium without additional carbon source. The flask on the right was inoculated with a PHA degrading fungus (*Epicoccum nigrum*) isolated in our laboratory. Within two weeks the pieces of PHA were completely consumed by this fungus.

PHAs will be slower in environments with low temperatures and low microbial activity.

10.7   Applications of PHB and PHA

The variation in physical and mechanical properties of different PHA-types offers a wide range of potential applications. So far commercial applications have been developed only for PHB/HV.[9,48] This polymer is produced by ICI and has been brought to the market under the trade mark of Biopol®. This polymer can be blended with (natural) fillers, pigments and modifiers, and extrusion and injection moulding of Biopol is possible on conventional equipment. PHB/HV has been processed into bottles, mouldings, films, fibres, non-woven fabric and coated onto paper. Currently, Biopol® is used for the manufacture of bottles for hair care products and biodegradable motor-oil.

In general due to their biodegradability PHAs can be used for all sorts of biodegradable packaging materials, including composting bags and food packaging. A particular property of PHB films that could be of potential interest for its use for food packaging is the relatively low oxygen diffusivity.[33] Also the use of PHAs in disposable sanitary articles like diapers is considered. Other applications are in agriculture, i.e. mulching films and biodegradable matrices for the sustained release of plant growth factors or pesticides.[49]

PHAs are biocompatible with human and mammalian tissue and are resorbed at slow rate, being hydrolysed to normal mammalian metabolites when implanted in the human body.[11,33] Due to these properties PHAs may find biomedical applications, such as in multifilament sutures, vascular grafts, wound dressings and slow release drug delivery systems. Finally PHB has piezo-electric properties similar to those of natural bone, giving it potential as biodegradable fixative plates that could actually promote bone formation and healing.[33]

10.8 Conclusions

It is evident that nature offers us the possibility to synthesize a large variety of truly biodegradable polyesters. Large scale use of PHAs will depend on the production costs and applicability of these polymers. The production costs can only be lowered when the productivity is increased to 2 grams per litre per hour which is similar to the productivity of other large scale commercial fermentation processes.[50] Furthermore the product yield should be improved by using feedstocks other than glucose, e.g. vegetable oils. Production of PHAs in plants may also lower the production cost. However, at present, fermentation will probably continue to be the means of production because an extraordinary range of PHAs can be synthesized by bacteria and genetic modification of metabolic pathways in plants is laborious and time consuming work.[51]

The mechanical and physical properties of PHAs can be regulated by variation of the monomeric composition. In the future, specific modification of the substrate range of PHA polymerases by protein engineering will further increase the possibilities for controlled synthesis of well defined PHAs. The properties of PHAs can also be varied and improved by blending PHA with other biodegradable polymers, such as poly(vinyl alcohol), starch, cellulose, etc.[4] An additional advantage could be the price of blends which may be lower than that of PHA.

References

1. Anon. (1993) Science, 259, 461.
2. Anderson, A.J. and Dawes, E.A. (1990) Microbiol. Rev., 54,450–472.
3. Steinbüchel, A. and Schlegel, H.G. (1991) Mol. Microbiol., 5, 535–542.
4. Inoue, Y. and Yoshie, M. (1992) Prog. Polym. Sci., 17, 571–610.
5. Peoples, O.P. and Sinskey, A.J. (1989) J. Biol. Chem., 264, 15298–15303.
6. Slater, S., Voige, W. and Dennis, D. (1988) J. Bacteriol., 170, 5837–5847.
7. Timm, A., Byrom, D. and Steinbüchel, A. (1990) Appl. Microbiol. Biotechnol., 33, 296–301.
8. Preusting, H., Kingma, J., Huisman, G.W., Steinbüchel, A. and Witholt, B. (1992) J. Environ. Polym. Degrad., 1, 11–21.
9. Byrom, D. (1987) TIBTECH., 5, 246–250.
10. Doi, Y., Segawa, A. and Kunioka, M. (1990) Int. J. Biol. Macromol., 12, 106–111.
11. Doi, Y. (1990) Microbial Polyesters, VCH Publishers, Inc., Weinheim.

12. Doi, Y., Tamaki, A., Kunioka, M. and Soga, K. (1987) *Macromol. Chem. Rapid Commun.*, **8**, 631–635.
13. Kim, G.J., Yun, K.Y., Bae, K.S. and Rhee, Y.H. (1992) *Biotechn. Letters*, **14**, 27–32.
14. Haywood, G.W., Anderson, A.J., Williams, O.R. and Dawes, A.E. (1991) *Int. J. Biol. Macromol.*, **13**, 83–88.
15. de Smet, M.J., Eggink, G., Witholt, B., Kingma, J. and Wijnberg, H. (1983) *J. Bacteriol.*, **154**, 870–878.
16. Lageveen, R.G., Huisman, G.W., Preusting, H., Ketelaar, P., Eggink, G. and Witholt, B. (1988) *Appl. Environ. Microbiol.*, **54**, 2924–2932.
17. Huisman, G.W., De Leeuw, O., Eggink, G. and Witholt, B. (1989) *Appl. Environ. Microbiol.*, **55**, 1949–1954.
18. Brandl, H., Gross, R.A., Lenz, R.W. and Fuller, R.C. (1988) *Appl. Environ. Microbiol.*, **54**, 1977–1982.
19. de Waard, P., Van der Wal, H., Huijberts, G.N.M. and Eggink, G. (1993) *J. Biol. Chem.*, **268**, 315–319.
20. Haywood, G.W., Anderson, A.J., Ewing, D.F. and Dawes, E.A. (1990) *Appl. Environ. Microbial.*, **56**, 3360–3367.
21. Timm, A. and Steinbüchel, A. (1990) *Appl. Environ. Microbiol.*, **56**, 3360–3367.
22. Huijberts, G.N.M., Eggink, G., de Waard, P., Huisman, G.W. and Witholt, B. (1992) *Appl. Environ. Microbiol.*, **58**, 536–544.
23. Huisman, G.W., Wonink, E., Meima, R., Kazemier, B., Terpstra, P. and Witholt, B. (1991) *J. Biol. Chem.*, **266**, 2191–2198.
24. Kim, Y.B., Lenz, R.W. and Fuller, R.C. (1992) *Macromolecules*, **25**, 1852–1857.
25. Hänggi, U.J. (1990) *Novel Biodegradable Microbial Polymers*, E.A. Dawes (ed.), NATO ASI series, Kluwer Academic publishers, Dordrecht, The Netherlands, pp. 65–70.
26. Suzuki, T., Yamane, T. and Shimizu, S. (1986) *Appl. Microbiol. Biotechnol.*, **24**, 370–374.
27. Page, W.J. (1989) *Appl. Microbiol. Biotechnol.*, **31**, 329–333.
28. Akiyama, M., Taima, Y. and Doi, Y. (1992) *Appl. Microbiol. Biotechnol.*, **37**, 698–701.
29. Eggink, G., Van der Wal, H., Huijberts, G.N.M. and de Waard, P. (1992) *Industrial Crops and Products*, **1**, 157–163.
30. Preusting, H., Van Houten, R., Hoefs, A., Kool Van Langenberghe, E., Favre-Bulle, O. and Witholt, B. (1993) *Biotechnol. Bioengineering*, **41**, 550–556.
31. Preusting, H., Hazenberg, W. and Witholt, B. (1993) *Enzyme Microb. Technol.*, **15**, 311–316.
32. Poirier, Y., Dennis, E., Klomparens, K. and Sommerville, C. (1992) *Science*, **256**, 520–523.
33. Lafferty, R.M., Korsatko, B. and Korsatko, W. (1988) in H.J. Rehm and G. Reed (eds) *Biotechnology*, vol. 6b, Verlagsgesellschaft, Weinheim, pp. 136–176.
34. Williamson, D.H. and Wilkinson, J.F. (1958) *J. Gen. Microbiol.*, **19**, 198–209.
35. Berger, E., Ramsay, B.A., Ramsay, J.A. and Chavarie, C. (1989) *Biotechnol. Techn.*, **3**, 227.
36. Gross, R.A., de Mello, C., Lenz, R.W., Brandl, H. and Fuller, R.C. (1989) *Macromolecules*, **22**, 1106–1115.
37. Preusting, H., Nijenhuis, A. and Witholt, B. (1990) *Macromolecules*, **23**, 4220–4224.
38. Marchessault, R.H., Monasterios, C.J., Morin, F.G. and Sundararajan, P.R. (1990) *Int. J. Biol. Macromol.*, **12**, 158–165.
39. Bluhm, T.L., Hamer, G.K., Marchessault, R.H., Fyfe, C.A. and Veregin, R.P. (1986) *Macromolecules*, **19**, 2871–2876.
40. Calvert, P. (1992) *Nature*, **360**, 535.
41. Harrison, S.T.L., Chase, H.A., Amor, S.R., Bonthorne, K.M. and Sanders, J.K.M. (1992) *Int. J. Biol. Macromol.*, **14**, 50–56.
42. de Koning, G.J.M. and Lemstra, P.J. (1992) *Polymer*, **33**, 3292–3294.
43. Scandola, M., Cecorilli, G. and Doi, Y. (1990) *Int. J. Biol. Macromol.*, **12**, 112–117.
44. Nahamura, S., Doi, Y. and Scandola, M. (1992) *Macromolecules*, **25**, 4237–4241.
45. Gagnon, K.D., Lenz, R.W., Tanis, R.J. and Fuller, R.C. (1992) *Macromolecules*, **25**, 3723–3728.
46. Chowdhury, A.A. (1963) *Arch. Microbiol.*, **47**, 167–200.

47. Kunioka, M., Kawaguchi, Y. and Doi, Y. (1989) *Appl. Microbiol. Biotechnol.*, **30**, 569–573.
48. Holmes, P.A. (1985) *Phys. Technol.*, **16**, 32–36.
49. White, B.G. (1983) *Brit. Pat. Appl.*, 8221567.
50. Witholt, B., de Smet, M.J., Kingma, J., Van Beilen, J.B., Kok, M., Lageveen, R.G. and Eggink, G. (1990) *TIBT.*, **8**, 46–52.
51. Poirier, Y., Dennis, D., Klomparens, K., Nawrath, C. and Sommerville, C. (1992) *FEMS Microbiol. Reviews*, **103**, 237–246.
52. Eggink, G., de Waard, P. and Huijberts, G.N.M. (1992) *FEMS Microbiol. Reviews*, **103**, 159–164.

11 NMR imaging of packaged foods

J.J. WRIGHT and L.D.HALL

Abstract

Nuclear magnetic resonance (NMR) imaging (MRI) is a technique for producing non-invasive images of heterogeneous systems based on the NMR properties of nuclei within those systems. In foods this corresponds to spatial mapping of water and fat molecules. It is envisaged that the non-invasive nature of MRI will allow serial studies on the stability and composition of packaged foods. However, certain packaging materials may interact with the MRI experiment and prevent these measurements being made. A selection of foods contained in different types of packaging materials was studied and it was established that certain types of packaging will prevent successful serial study by NMR imaging.

11.1 Introduction

Nuclear magnetic resonance (NMR) imaging (MRI) is a relatively new technique; its growth since its conception by Paul Lauterbur in 1973[1] has been dominated by its applications to many areas of medicine, especially in neurology and orthopaedics.[2] The *in vivo* diagnostic images of anatomical features such as the brain and joints exhibit soft tissue contrast which exceeds that produced by other imaging techniques such as X-ray computed tomography (CT).[3] In contrast to those well developed medical applications, the use of MRI to investigate the structure of 'materials' is still in its infancy. The main reason for this is that since the development of NMR imaging instrumentation has been largely directed towards clinical imaging of man. Most instrumental specifications do not match those required for non-medical applications; furthermore the cost is far too high for most industrial applications. Fortunately, this imbalance is now being redressed and the potential of NMR imaging for investigating the physical and chemical properties of many non-medical problems is being evaluated. Industries which are actively involved in this area include the oil,[4] polymer and polymer composite,[5] rubber[6] and pharmaceutical industries.[7] The food industry has also recognized the potential of MRI and a few preliminary studies have already confirmed the expectation that the enormous benefits already enjoyed by medical science can also be obtained when the technique of MRI is applied to food systems. Studies already reported include the evaluation of quality in fruit and vegetables,[8] the ripening of tomato fruits,[9] the imaging of chocolate confectionery and spatial detection of polymorphic states of cocoa butter,[10] the study of cheese structure,[11] the freezing and thawing of courgettes[12] and the drying[13] and steeping[14] of corn. Those preliminary studies have demonstrated that the abundance of water and fat in food systems makes them ideally suited to study by MRI. It is also becoming increasingly obvious that MRI can be a diagnostic probe to investigate the microstructures and dynamics of the constituents in food systems, with a view to establishing better structure–function relationships. The non-invasive nature of MRI, while at first seeming less relevant than in medicine, is particularly beneficial since many physical and chemical changes which occur when a food undergoes some process would be perturbed by invasive methods, and hence not observed. Extensions of the methodology developed for medical imaging can provide quantitative information on water and fat distribution in foods and their changes in response to external influences; importantly it is also possible to study the mobilities and diffusivities of water and fat molecules as functions of position and time.

The stability and hence shelf-life of foods, is a critical factor in determining not only the hygiene and safety of the food, but also its quality as perceived by the consumer. Consequently a massive effort is expended to optimize the preservation of foods; food packaging plays a decisive role in this function, and almost all food items are now packaged before sale.[15] Clearly if NMR imaging has a role to play in the evaluation of the quality of packaged foods then it is important to know which forms of food containment are incompatible with the NMR hardware and data acquisition methodology. Given that MRI systems use a powerful magnet, one can anticipate that ferrous metals, such as cans, will be prohibited; however, it is not obvious if small iron components, such as bottle caps, are similarly precluded. In principle, aluminium is acceptable, but again it is not obvious if eddy currents induced in aluminium objects by rapid switching of the magnetic field gradients used in imaging will cause local image distortion. Given that the sample must be bathed evenly with radio-waves of the correct frequency, complete enclosure within a container that prevents entry of radio-waves will also preclude MRI; hence sealed aluminium cans, or aluminium foil or aluminium-coated plastic may disallow measurement. On the positive side there seems to be no reason why containers and packaging made of glass, paper, wood, cardboard or plastic should interfere unduly with an MRI examination.

This chapter will attempt to answer some of these questions by showing images of a selection of foods contained in typical food packaging materials. In order to fully appreciate the experimental procedure and results obtained, a brief introduction to the NMR imaging experiment is described, followed by some practical considerations. The suitability of the technique for studying different food types is also discussed, and examples of 'easy' and 'difficult' foods are given.

11.2 Principles of NMR imaging

NMR imaging measurement of a food system provides a spatial map of the concentration and physico-chemical properties of the protons of water and fat in the nuclei of the molecules that are present in the food. The measurement begins by placing the sample in a strong static magnetic field, known as the B_0 field. The protons tend to become aligned by this field and hence the sample acquires a net magnetization vector in the direction of this static field. The application of a second and much smaller magnetic field (B_1), generated transiently by a pulse of radio-frequency energy of the correct frequency and duration, allows the direction of the magnetization vector to be oriented perpendicular to the direction of the static field. That vector then precesses about B_0 and induces an alternating current in the receiver coil of the NMR probe which surrounds the sample. The magnetization vector rotates with frequency ω_0, which depends on the strength of the main field, B_0, and has the form:

$$\omega_0 = \gamma B_0 \qquad (11.1)$$

where γ is constant for the hydrogen nucleus.

Nuclei excited by the oscillating magnetic field B_1 are in an unstable, high energy state and spontaneously revert to their original orientation; as they do, so too does the intensity of their magnetization vector. Consequently the observed NMR signal decays with a characteristic time constant T_2 known as the spin–spin relaxation time. The magnetization vector in the direction of the main field, B_0, also returns to its original value, again exponentially; this time constant, T_1, is termed spin-lattice relaxation. Numerical values of T_1, and T_2 are greatly influenced by the physico-chemical environment of the nuclei.

From equation (11.1) we note that the frequency of NMR precession, ω_0, is proportional to the magnetic field strength. In an NMR imaging experiment the magnitude of the homogeneous B field is made spatially dependent by the application of a linear magnetic field gradient. The local resonance frequency of a proton is now dependent on its spatial position within the sample, and the induced NMR signal now contains a range of resonance frequencies which reflect this distribution. As a result, Fourier transformation of the NMR signal measured in the presence of a gradient results in a one-dimensional (1-D) spatial profile of the sample; effectively the NMR signal has been spread along an axis defined by the direction of a gradient field.

There are several ways of generating two- or three-dimensional NMR images of a sample. The first method developed, by Lauterbur[1] used reconstruction from a set of 1-D profiles obtained by varying the angle between the sample and the direction of the applied gradient. In the measurements presented in this chapter, the 2-D spin-warp imaging experiment[16] is used; now by varying the magnitude of a gradient in one direction, whilst leaving a second gradient on continually but applied at right angles to the first, it is possible to line-scan through the sample, somewhat akin to the line-scan of a TV picture. The resultant 2-D matrix is then subjected to a 2-D Fourier transformation to produce the image. In practice, a spatially encoded spin-echo is formed at time T_E after the initial excitation of the sample by the radiofrequency pulse. Spin–spin relaxation imposes a limit on how long a T_E can be used, since the NMR signal at the time of the spin-echo decays according to $S(t) \propto \exp(-T_E/T_2)$. In most medical applications the T_2 of the mobile water and lipid is typically several hundreds of milliseconds, which allows plenty of time for the hardware manipulations to be carried out to enable the NMR to be spatially encoded. However, in many material science applications, including food, the T_2 value of the samples can lie in the range 100 microseconds to 100 milliseconds, which necessitates fast timing and switching requirements on the hardware used to generate the NMR image. It is for that reason that many foods cannot be studied using commercially produced clinical MRI scanners.

Two-dimensional (2-D) images are characterized by their slice thickness and the data matrix size within the plane of the image; the latter is typically (128×128) or (256×256) pixels. The in-plane resolution in an image is given by the field-of-view (FOV), or size of the image, divided by the number of pixel rows; thus the resolution for a typical packaged food image would be approximately 0.25 mm. The slice thickness of an image is a function of the duration of a selective radiofrequency pulse used to define the slice and the strength of the slice select gradient which allows selective excitation of a plane of spins through a sample; typically this will be several millimetres thick. The spatial resolution of an image may be improved either by increasing the data matrix size or by reducing the field of view of an image; however, there is a finite limit to the spatial resolution which can be obtained in NMR imaging which reflects both physical and technical constraints. Physical limitations involve the interactions of the nuclear spins within the sample such as spin relaxation and molecular diffusion.[17] Technical constraints involve the overall performance of the NMR spectrometer, primarily the achievable signal-to-noise (S/N) for a given volume element; thus clearly it is necessary for the signal from one voxel to be distinguishable from those of the surrounding voxels, and is of primary importance in determining both the overall measurement time and resolution. Consequently every possible effort must be made to optimize

S/N sensitivity, e.g. by using the smallest possible radio-frequency coil for the object to be studied. In most practical applications, time is the dominant factor that limits the resolution attainable for a given experiment. The imaging time for a typical 2-D spin warp image is of the order of several minutes; this depends on the data matrix size and the time between successive increments of the magnetic field gradients. It should be noted however that other NMR imaging techniques exist which enable an image to be obtained in less than a second;[18,19] however, these involve compromises in both spatial resolution and S/N, and are not used here.

Although it is theoretically possible to obtain an image from the response of any NMR active nucleus the vast majority of work reported in the literature is of proton (^{1}H) NMR imaging; other magnetic nuclei such as sodium-23 or phosphorus-31 are not as sensitive to the NMR experiment and are generally not present at high enough concentration to allow image acquisition in a reasonable length of time. Food products which have high water and/or fat contents can be considered relatively easy to image because of their high proton density, and the long nuclear relaxation times; these include fruit and vegetables, meat and meat products and dairy produce including milk, yoghurt and cheese.

NMR images of the internal structure of fruit and vegetables display good spatial resolution and contain impressive morphological detail since the NMR relaxation times of water and/or fat molecules vary with the precise morphology and physico-chemistry of the structures which contain them. NMR imaging can be used to study the ripening of fruit, to evaluate the damage to plant tissue during the freezing and thawing of vegetables or as a quality control for bruises, pest damage, etc. NMR signals arising from the protons of lipid and water in dairy products have slightly different resonance frequencies and this can be used as the basis for a 'chemical filter' to produce images that display separately the spatial distribution of just one component, i.e. lipid-only or water-only images.[20] Meat and meat products are similarly suitable for study by NMR imaging since meat has both high water and fat content. As the water and fat content of foods is reduced it becomes increasingly more difficult to obtain an image simply because the signal intensity is low. Thus bakery products, including bread, cakes and biscuits, which have relatively low moisture contents and large amounts of air have low proton density and hence give low S/N images. The problem is compounded by the fact that as the water content of the food system is reduced, the transverse relaxation time (T_2) of its water protons falls rapidly; consequently the available magnetization must be sampled very quickly before it has decayed to zero, which is extremely difficult to achieve in practice. Using the conventional slice-selective spin warp imaging sequence a minimum T_E of approximately 2 milliseconds is achievable; this limits NMR imaging to studies of samples with water contents above approximately 20% moisture. Fortunately the transverse relaxation times of the protons of lipids are less dependent on the concentration of fat, and are sufficiently long in cakes and biscuits to enable suitable images to be obtained. It should, however, be noted that only 'liquid' fats can be observed using this technique, the signal arising from 'solid' fat decays far too rapidly to be measured. This property can be used to determine the polymorphic state of cocoa butter and examine the thermal history of chocolate.

11.3   Materials and methods

11.3.1   Sample preparation

Samples for the food packaging experiments were purchased from a local supermarket and are summarized in Table 11.1. The intact package was placed in the instrument, subjected to the NMR imaging experiment, then removed.

Table 11.1 Food products imaged and their packaging materials

Food product	Type of packaging
Black pudding	Shrink wrapped low density polyethylene
Tomatoes	Polyethylene bag
Pork pie	Greaseproof paper
Hazelnut yohurt	Polypropylene carton with an aluminium foil top
Camembert cheese	Paper with aluminium foil coating
Peanut butter	Glass jar with polypropylene top
Raspberry jam	Glass jar with ferrous metal top

11.3.2   NMR imaging measurements

The NMR images were obtained using an Oxford Research Systems Biospec 1 spectrometer operating at 83.7 MHz for protons, connected to an Oxford Instruments 31 cm horizontal bore, 2 Tesla superconducting magnet. The linear magnetic field gradients were generated using home built gradient coils;[21] the samples were studied using a home-built birdcage resonator probe[22] with an inner diameter of 10 cm. A standard spin warp imaging pulse sequence[16] was used to obtain 5 mm thick, single slices through the samples; the total echo time (T_E) was 16 ms and the recycle delay (T_R) was 1 second; no signal averaging was required and all experiments were performed at room temperature (23°C).

11.4   Results and discussion

The image of the black pudding is displayed in Figure 11.1. Black pudding is a type of sausage made from pigs' blood and suet. It is shrink wrapped

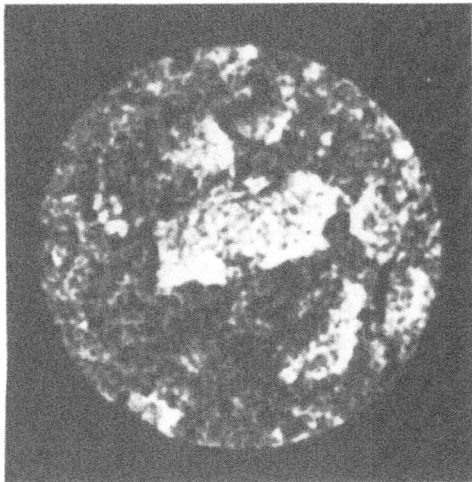

Figure 11.1 Black pudding, shrink wrapped in low density polyethylene.

in low density polyethylene film; the image shows voids as regions of zero signal intensity (black), whilst the very bright regions in the image are from pieces of fat. The packaging does not interfere with the production of this image, indeed all plastic packaging materials are translucent to the NMR imaging experiment. Plastics are easily penetrated by the radiofrequency field and do not cause any disruption to the homogeneity of the B_0 field. Although the polymers used in plastics do contain protons those protons are in a solid environment; consequently their transverse magnetization decays very rapidly, on the order of microseconds, and hence it is not sampled by an NMR imaging experiment.

An example of how foods in plastic packaging can be monitored over a period of time is given in Figure 11.2. A fresh tomato was placed in a clear polyethylene bag which is typical of those used to collect fresh produce in a supermarket, and left on a shelf for 3 weeks; a second tomato was purchased on the day of the imaging experiment and was also placed in a clear polyethylene bag. The two tomatoes were then imaged simultaneously. Figure 11.2a is the 3-week-old tomato and Figure 11.2b the fresh tomato. The morphological features of the two tomatoes are quite similar, but there is clearly a difference in contrast in the central region of the fruits. The image was acquired with a T_E of 300 ms, so the contrast is due to differences in transverse relaxation times (T_2). Transverse relaxation times are greatly influenced by the physico-chemical environment of the nuclei, hence temporal changes in internal fruit structures can be monitored non-invasively by NMR imaging.

Wood pulp is the base for many widely used packaging materials and appears ranging from cardboard to tissue paper. Paper used in food packaging may be modified by treatment or by coating in order to tailor its properties; for example greaseproof paper, used largely for baked

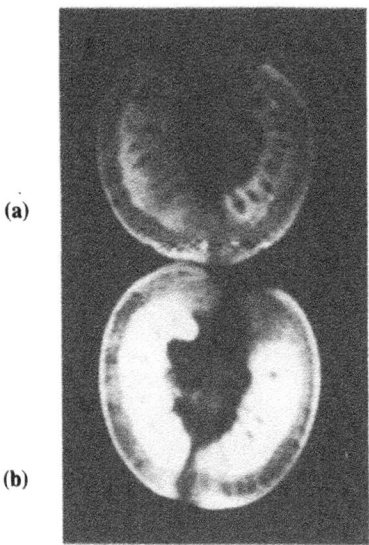

(a)

(b)

Figure 11.2 Tomatoes packaged in plastic bags. (a) Imaged 3 weeks after purchase. (b) Imaged on day of purchase.

goods. This is demonstrated for the pork pie shown in Figure 11.3; the pasty crust and meat filling of the pie are clearly distinguished and, as with the black pudding, the very intense regions in the meat are from pieces of fat. The cellulose fibres do not disrupt the imaging measurement since in common with plastics, the T_2 relaxation of their protons is so rapid that their magnetization cannot be detected by most NMR imaging sequences.

The image shown in Figure 11.4 is of hazelnut yoghurt contained in a polypropylene carton with an aluminium foil top; again the image shows

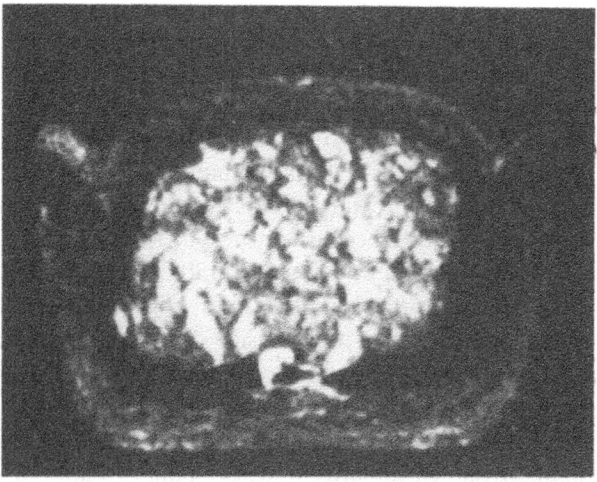

Figure 11.3 Pork pie, wrapped in greaseproof paper.

Figure 11.4 Hazelnut yoghurt in a polypropylene carton with an aluminium foil top.

good spatial resolution and pieces of hazelnut are distinguished as regions of high spatial intensity, the darker regions are air bubbles. Although one could predict that the plastic packaging would not disturb the NMR image, there are two possible ways in which the aluminium foil top could disrupt the experiment. Firstly, as the magnetic field gradient changes during the imaging experiment the changing flux lines passing through an aluminium surface induce 'eddy' currents which can persist for many milliseconds; these eddy currents cause additional unwanted magnetic fields which can cause image artefacts. The image of the hazelnut yoghurt is not subject to such artefacts caused by unwanted magnetic fields. This implies that relatively small areas of the aluminium foil do not generate sufficiently large fields to cause image artefacts. Secondly the radiofrequency field cannot penetrate metallic materials; hence it is possible that the aluminium top could act as a 'shield' and prevent the RF from exciting all parts of the sample. This would manifest itself as a dark region in an image. This does not appear to happen with the aluminium foil top because the surface of the yoghurt closest to the top is not obviously darkened due to poor excitation. This is because the radiofrequency field can penetrate the sample from below and to the side; hence covering only one side of the sample does not prevent image formation due to radiofrequency shielding. This conclusion is supported by the images in Figure 11.5 of Camembert cheese. Figure 11.5a shows the cheese placed on packaging material made of paper with an aluminium foil coating; the wrapping material covers only one side of the cheese and an undistorted image is obtained. In contrast, the image shown in Figure 11.5b was produced when the wrapping was folded over one half of the cheese; now the loss of signal from part of the image can be attributed to the radiofrequency shielding effect of the aluminium foil coated paper, which prevents the radiofrequency field from reaching that part of the sample. When the cheese was completely enclosed

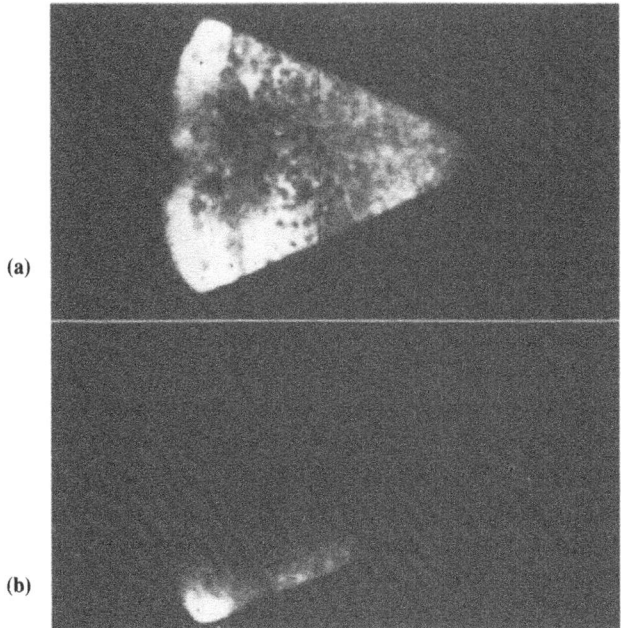

(a)

(b)

Figure 11.5 Camembert cheese packaged in aluminium foil coated paper. (a) The cheese sitting on its wrapping paper. (b) The paper is folded over one half of the cheese.

in its wrapping material, no signal whatsoever was observed; now no radiofrequency can reach the sample, thus no image is observed. It follows that MRI cannot be applied to any food which is enclosed in any aluminized-paper or -plastic packaging materials nor to the wide variety of drinks that are sold in aluminium cans.

An example of a glass packaged food is given in Figure 11.6, which is an image of a glass jar of peanut butter with a polypropylene screw top. The image is clear and undistorted, showing small pieces of peanut as bright regions in the image. The chemical composition of glass, being largely silicon dioxide, ensures its suitability to the NMR imaging experiment; it contains no protons, does not prevent radiofrequency penetration and does not disrupt the static magnetic field homogeneity. In contrast to this, Figure 11.7 demonstrates the effect of a ferrous metal cap on a glass jar. The first image, Figure 11.7a, was acquired with the ferrous metal cap removed, and an acceptable image is obtained. However, when the cap is replaced and the measurement repeated the resultant image is very distorted; now a large area of the sample has been excluded from the image, and only the contents of a small portion at the bottom of the jar remains visible. This is because the cap of the jar, being made of ferrous based metal and hence magnetic, causes a gross change in the size of the static

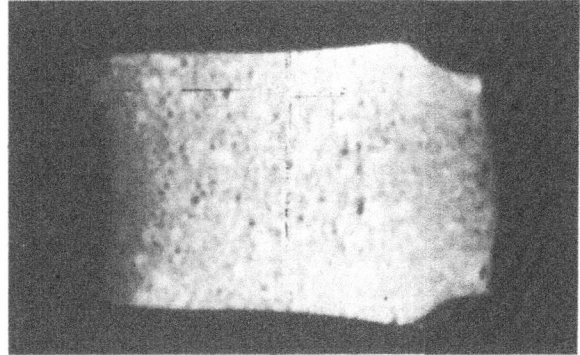

Figure 11.6 Peanut butter, in a glass jar with a polypropylene screw top.

magnetic field (B_0) homogeneity. Consequently protons in those parts of the sample where the B_0 field has changed magnitude now resonate at such a different Larmor frequency that they are not excited by the radiofrequency pulse, so no signal results from that part of the image. Thus foods whose packaging materials contain only small amounts of ferrous metal are excluded from imaging studies, unless they are removed from their packaging. Indeed, it must be stressed that it would be quite dangerous to attempt to place a large ferrous metal object such as a tin can inside the magnet of an NMR imaging spectrometer; the object would become violently attracted when taken within a few feet of the magnet.

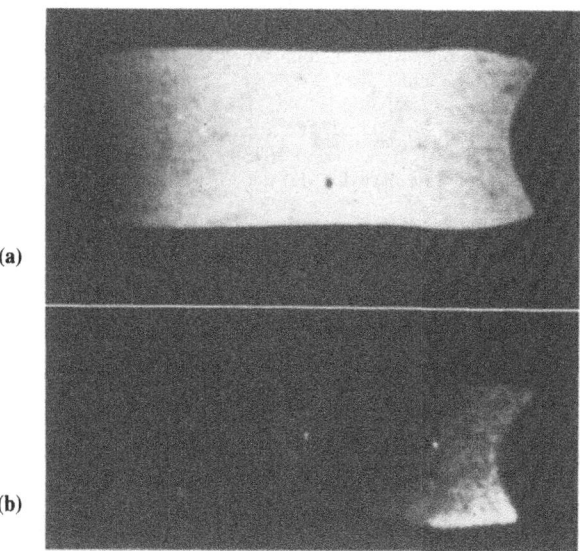

Figure 11.7 Raspberry jam, in a glass jar. (a) With the cap removed. (b) With the ferrous metal cap replaced.

11.5 Conclusion

It has been demonstrated that commonly used food packaging materials made of plastic, paper or glass do not interfere with the acquisition of NMR images of food contained within those types of packaging. However, ferrous materials cause considerable image distortion due to their interaction with the magnetic field. It has also been established that although aluminium foil does not disrupt the magnetic field, it can preclude NMR images being obtained from within a sealed vessel because it shields the sample from the radiofrequency waves. It follows from these demonstrations that MRI can be used to make sequential studies of many foodstuffs in the form they are packaged, either for storage, for distribution, or for retail sale; in principle then MRI can be used to evaluate methods for improving the quality of the food as received by the consumer.

Acknowledgement

It is a pleasure to thank Dr Herchel Smith for a munificent benefaction (L.D.H.) and for a research studentship (J.J.W.).

References

1. Lauterbur, P.C. (1973) *Nature*, **242**, 190–194.
2. Andrew, E.R. (1984) *British Medical Bulletin*, **40**(2), 115–119.
3. Andrew, E.R., Bydder, G., Griffiths, J., Iles, R. and Styles, P. (1991) *Clinical Magnetic Resonance Imaging and Spectroscopy*, John Wiley, London.
4. Horsfield, M.A., Hall, C. and Hall, L.D. (1990) *J. Magn. Reson.*, **87**, 319–330.
5. Jezzard, P., Wiggins, C.J., Carpenter, T.A., Hall, L.D., Barnes, J.A., Jackson, P. and Clayden, N.J. (1992) *J. Mater. Sci.*, **27**, 6365–6370.
6. Chang, C. and Komorowski, R.A. (1989) *Macromol.*, **22**, 600–607.
7. Checkley, D., Johnstone, D., Taylor, K. and Waterton, J.C. (1989) *Magn. Reson. Medicine*, **11**, 221–235.
8. Chen, P., McCarthy, M.J. and Kauten, R. (1989) *Trans. ASAE*, **32**, 1747–1753.
9. Ishida, N., Kobayashi, T., Koizuma, M. and Kano, H. (1989) *Agric. Biol. Chem.*, **53**, 2363–2367.
10. Duce, S.L., Carpenter, T.A. and Hall, L.D. (1990) *Leben.-Wiss.u.-Tech.*, **23**, 545–543.
11. Rosenberg, M., McCarthy, M.J. and Kauten, R. (1991) *Food Structure*, **10**, 185–192.
12. Duce, S.L., Carpenter, T.A. and Hall, L.D. (1992) *J. Food. Eng.*, **16**, 165–172.
13. Song, H. and Litchfield, J.B. (1990) *Cereal Chem.*, **67**(6), 580–584.
14. Ruan, R. and Litchfield, J.B. (1992) *Cereal Chem.*, **69**(1), 13–17.
15. Paine, F.A. and Paine, H.Y. (1983) *A Handbook of Food Packaging*, Leonard Hill, London.
16. Edelstein, W.A., Hutchinson, J.M.S., Johnson, G. and Redpath, T. (1980) *Phys. Med. Biol.*, **25**, 751–758.
17. Callaghan, P.T. (1991) *Principles of Nuclear Magnetic Resonance Microscopy*, Clarendon Press, Oxford.
18. Hasse, A. (1990) *Magn. Reson. Med.*, **13**, 77–83.

19. Mansfield, P. (1977) *J. Phys. C.*, **10**, L55.
20. Dixon, W.T. (1984) *Radiology*, **153**, 189–194.
21. Carpenter, T.A., Hall, L.D. and Jezzard, P. (1989) *J. Magn. Reson.*, **84**, 383–390.
22. Hayes, C.E., Edelstein, W.A., Schenck, J.F., Mueller, O.M. and Eash, M. (1985) *J. Magn. Reson.*, **63**, 622–628.

12 The recycling, reuse and disposal of food packaging materials: a UK perspective

A.J. CAMPBELL

Abstract

The environmental issues associated with food packaging are of concern to the consumer. Pressure groups within the UK are influential and large supermarket chains are becoming involved in the issues of recycling by the placement of collection containers outside their stores. However, as has been observed with certain materials, collection costs can be far greater than the potential market for these materials.

The consumer perception of what packaging does for the product and how it can be recycled varies depending on the understanding of the environmental problem by the consumer.

Within the UK there have been a large number of public initiatives relating to the recycling, reuse and disposal of packaging, such as RECOUP, Save-A-Can and others. Food packaging contributes approximately 66% to the waste stream and potentially creates the biggest problem regarding recycling.

Legislation is being proposed both on a national basis and on an EC-wide basis which will have an effect on use, packaging, the environment, and inter-community trade.

The food industry is being pushed into making decisions regarding recycling of food packaging based on insufficient information supplied by different sectors of the packaging industry. The potential of 'life cycle analysis' properly co-ordinated should be able to aid the 'environmental' impact of packaging and guide industry towards making the correct decision.

12.1 Introduction

Packaging has been with us for thousands of years in one form or another. Nature developed food packaging well before man, even to the extent of the outer surface being an indicator to show when the product is ready to eat. Today food packaging is used, or abused, to provide prepared food for the sophisticated consumer.

Modern packaging can be defined as a means of ensuring the safe delivery of a product to the consumer in a sound condition at minimal overall cost. To put packaging materials in perspective, 66% of all packaging is used for food. In developed countries, where packaging is used with a food processing system, food waste is approximately 2–3%. In developing countries, where no such packaging systems exist, food waste may be as high as 30–50%.

Today's consumer wants to buy food which is ready to eat, or needs a minimum of preparation, and is good value for money. To be able to supply these demands requires the use of different types of packaging for different applications. Packaging is also used by the manufacturer to make his product stand out on the shelf in the supermarket so that the consumer will select his product rather than that of his competitor. This means that pack design has an important part to play in the perception of packaging by the consumer.

Packaging becomes waste when its original functions[1] have been fulfilled. These functions are containment, protection, preservation, identification and convenience. Packaging is a lot less objectionable than the organic putrescible matter of which it has taken the place. Packaging can provide hygiene and an insurance against illness and disease. The average consumer is confused by some of the issues regarding packaging. We, as an industry, should attempt to put a balanced view to try to educate the consumer so that misunderstandings and conflicts between packaging materials are minimised. Figures for materials in municipal solid waste in Europe[2] are shown in Table 12.1.

However, when new products packaging in direct contact with food[3] is analysed, then the figures in Table 12.2 appear. Seeing these figures, it is easy to see where consumer misconceptions can arise.

Packaging materials can readily be classified into two categories:

1. renewable resource, and
2. non-renewable resource.

Table 12.1 Materials in municipal solid waste in Europe

Material	Percentage
Textiles	4
Plastics	7
Metal	8
Glass	8
Miscellaneous (ash, etc.)	10
Paper and board	30
Organic products	33

Source: ref. 2.

Table 12.2 New products packaging in direct contact with food

Material	Percentage
Plastics	66.5
Metals	12
Glass	9.5
Paper and board	12

Source: ref. 3.

Basically, anything which is grown (trees, plants, etc) is a renewable resource. The paper and board industry in the UK has an advertisement stating that for every tree cut down at least another tree is planted. Non-renewable resources relate to anything which comes out of the ground, such as steel, tin, oil, etc.

What do we mean by reuse, recycling, and disposal?[4]

12.1.1   Reuse

This simply means using something again after its initial purpose has expired. Subsequent uses can be similar or dissimilar to the original function, and the use of a single product for different purposes is something referred to as multiple use, e.g. returnable bottles are reused by washing and refilling them.

12.1.2   Recycling

This covers a diverse range of processes, including repulping waste paper to make new paper, using broken glass in the production of new bottles, and incineration for the recovery of energy.

12.1.3   Disposal

This basically means taking what you no longer want and putting it somewhere else. Landfill is just one of the many methods of disposal. Ninety per cent of the waste in the UK is landfilled, and the design of these sites means that all matter placed in them is maintained in that state for years. Newspapers which were landfilled 20–30 years ago have been unearthed still intact and can be easily read.

An area of recent confusion in the recycling environmental industry is the level of recycling that is taking place and the amount of recycled material used in a particular type of packaging. Certain material manufacturers have stated that they are using, say, 80% recycled material in their product, but when questioned further it would appear that the majority of recycled material is clean scrap from their own production lines (post-industrial) and only about 5% is genuinely recycled (post-consumer) waste.

12.2 Recycling schemes in the UK

There are a range of material-specific recycling schemes in the UK, as well as schemes covering all materials. The majority of these schemes are small pilot schemes and are based not so much on how much material can be collected but on looking at the economics of collecting the material and how to deal with it afterwards. The problems for food contact use of recycled materials are: (a) initial sorting of waste into material/polymer types, and (b) proving that the collected materials are free from contamination which could prove harmful to the consumer.

12.2.1 Metal

12.2.1.1 *Steel.*   In 1991, 1200 million tinplate cans were recycled in the UK.[5] Three-quarters of all cans sold are tinplate, and in 1991 78% of all cans recycled were tinplate. Steel cans, whether tin coated or tin-free steel, can be collected in one of two ways.

(a) Magnetic extraction. Regular kerbside collections from the house-holder are taken to a transfer site where the refuse is sorted. Steel is removed from the waste by means of large electromagnets.

(b) Save-A-Can. In areas not yet served by magnetic extraction, a British can recycling scheme, Save-A-Can, provides a viable alternative. Save-A-Can will accept all metal cans for recycling and operates through local authorities and supermarket chains. British Steel, which owns Save-A-Can, has more than 720 banks in operation and has pledged to install more than 1000 banks by 1994.

12.2.1.2 *Aluminium.*   Aluminium cans, because they are not magnetic, need to be collected separately from steel, one way of which is to use consumer-aided schemes. The Aluminium Can Recycling Association (ACRA) co-ordinates one such programme whereby the aluminium industry pays for the collected cans. In 1991, the number of aluminium collection points in the UK was over 1150. Aluminium recycling rates, behind other parts of Europe, have risen sharply from 11% to 16%.[6] In specific areas of the UK where extra promotional effort has been made, recycling rates above 40% have been achieved. The 1991 European average of all aluminium cans recycled is about 21%, twice the percentage of 1987.

Recently, to support aluminium can recycling in Europe, British Alcan spent £28 million building a can recycling plant.

12.2.2 *Plastic*

Plastics are an emotive issue. Most consumers would like to replace plastics packaging with some other form of packaging. However, if plastics packaging were to be replaced, then figures[7] show that:

– energy consumption would double;
– raw material consumption would quadruple;
– the volume of waste would increase by 150%;
– the cost of packaging would double.

As stated previously, it is frequently perceived that plastics waste contributes the major part of domestic waste. By weight, the plastics waste is quite small, accounting for approximately 6–7% of all waste.

Post-consumer plastics recycling is in its infancy and it may prove difficult to successfully separate all plastics. Local authorities which have kerbside collections for pre-sorted materials are currently running them on an experimental basis and are sponsored by industry.

However, schemes for recycling plastics exist. Shoppers are able to take their carrier bags back to the supermarket and place them in a collection bin. These bags are then taken away and formed into new bags. J. Sainsbury plc is one supermarket chain that currently has bags made from recycled plastic. Dow Plastics, in conjunction with Mono Containers, has a pilot scheme operating initially in 42 factory and office sites in North London which is collecting used polystyrene cups from vending machines. Called Save-A-Cup, the scheme relies on users to place their used cups into plastic sacks, which are then collected and returned for recycling.

Plastics packaging, e.g. film wrapping on pallets, is collected by large supermarkets and returned to suppliers for recycling. This type of clean plastics material is easily recycled.

12.2.3 *Paper/board*

Paper and board form the basis for 40% of all packaging.[8] These materials have perhaps been seen as the easiest to recycle and possibly the easiest to collect as post-consumer waste. Newspaper was the first material that I can remember being collected for recycling. At the time we were still taking glass bottles back to collect the deposit. The beverage can had not yet arrived. The level of recycled material in each type of paper and board varies according to the end use of the product. High grade papers for direct food contact may be from virgin pulp, but the outer shipping cases, used to protect the products in transit, may have up to 96% recycled material.

Fibre used in the production of paper and board is a renewable resource. However, each time that it is recycled, it loses some of its strength and after about six cycles it needs to be replaced.

In 1991, 2.9 million tonnes of waste paper was used by UK paper mills. This figure has increased by approximately one million tonnes in the last 20 years. The UK waste paper industry, however, has been hit recently by the rapidly increasing European 'waste mountain' resulting from recently introduced recycling legislation. Cheap imports of waste paper have forced the closure of several UK paper machine companies.[9] The users of fibreboard cartons, such as industry and supermarkets, bale their cartons and return them for recycling.

12.2.4 Glass

Glass recycling has been in successful operation in the UK for 15 years and there are now 5600 bottle banks in the UK. In 1989 the national recovery rate was 18% or 300 000 tonnes. This rose to 372 000 tonnes in 1990. Many of the local recycling schemes have collected mixed coloured glass together. The point has now been reached where 68% of cullet is either mixed or green. Mixed glass can only be used to make green bottles. The 'Sutton' model scheme[10] showed that local levels of collection could not only be increased but had the additional benefit that clear, amber and green glass could be collected separately. Glass is a material that in the UK has a very high level of reuse for doorstep deliveries of milk but not for other sectors of glass packaging. Milk bottles have an average trippage rate of 20 cycles, with 40 cycles recorded in some instances. In Scotland 73% of shoppers return their beverage bottles to be refilled.[11] This is possibly due to the differing shopping patterns throughout the UK.

The UK's first national glass recycling company will be launched next year by Rockware Glass and United Glass. This joint company intends to develop the collection of cullet through bottle bank schemes.

12.2.5 Other schemes

Sheffield, designated the UK's first 'recycling city', was established with support from FoE, the BPF and Sheffield City Council. This pilot scheme was based on 3300 households. About 70% of households participate in the scheme.[12] Each house is provided with a blue box into which all recyclable materials (glass, paper, metal and plastic) are deposited. Evidence suggests that there is an improved awareness level of recycling and public support in the collection area compared to the rest of Sheffield. A tremendous amount has been learnt from the first year's operation and modifications have been made to second year collection plans.

Halton Borough Council has agreed a 20 year deal with Biomass Recycling to reuse 98% of its household waste.[13] Aluminium waste will be sold to Japan. It would be sold to the nearby British Alcan recycling site if the Japanese price could be matched. HDPE and LDPE will be sold to Dubai. Other plastics will be used to make growbags, agricultural sheeting, and soundproofing materials. Paper and board will be composted with other organic waste. Gas extracted by compression will be sold to make electricity, and water warmed by the composting will be used to farm prawns and crayfish. The compost will eventually be spread over land to grow Christmas trees and trees for fuel.

The RECOUP scheme in Milton Keynes has extended its kerbside collection scheme to 70 000 homes.[14] Collection levels have now reached eight tonnes per week.

12.3   Environmental pressure groups

12.3.1   *Friends of the Earth (FoE)*

FoE is an environmental pressure group funded by voluntary contributions. It is part of a worldwide federation of similar organisations. It actively pursues campaigns on the control of packaging, protection of endangered species, energy conservation policy and transport, as well as recycling. FoE was campaigning against nonreturnable bottles in 1971, some six years before the first bottle bank to collect glass was installed.

12.3.2   *Women's Environmental Network (WEN)*

WEN is a non-profit organisation, working to educate, inform and empower women who care about the environment. WEN first came to the nation's notice when it started its Campaign for Minimum Packaging. This protest involved women removing what they deemed excess packaging and leaving it in a big pile outside the supermarket. As an extension to that protest, they called upon consumers who objected to the use of non-refillable, one-way packaging to return the packaging, by post, to the manufacturers who use it.

While WEN is striving to reduce the amount of packaging, they do believe that some packaging is necessary, i.e. it protects food and supplies useful product information.

12.4 Industrial bodies

12.4.1 *RECOUP (RECycling Of Used Plastic Containers Ltd)*

Sponsored primarily by plastics bottles packing and manufacturing companies, RECOUP started work in 1989 and is now on target to exceed collection levels of 2000 tonnes during 1992.

12.4.2 *INCPEN (The Industry Committee for Packaging and the Environment)*

INCPEN was founded in 1975 with two main goals:

1. to provide a forum for debate on the economic, social and environmental aspects of packaging;
2. to provide factual information and act as a channel for communication with government, environmentalists and the consumer on the role of packaging in society.

Its members include suppliers of raw materials, packaging manufacturers, producers of packaged goods, wholesalers and retailers.

12.5 Supermarkets

Within the UK we have five very large supermarket chains which supply over 60% of all food sold. These chains each have their own policy on how they are being more environmentally friendly than their competitors. The benefits are that they are trying to educate the consumer into thinking 'green'. To enable recycling of materials to take place, banks are now placed near their stores. Tesco,[15] for instance, has 160 glass bottle banks, 40 steel banks, aluminium banks at most stores, 60 paper banks, and is trialing plastic bottle banks at several Manchester stores. Other supermarkets, such as Safeway, Sainsbury, Gateway and Asda, are all in a similar position.

The buyers for these chains are all trying to reduce the amount of packaging seen by the consumer. This may mean that in some instances the secondary packaging is increased to reduce the primary packaging seen by the consumer.

12.6 Legislation

The UK is part of the EC and the implementation of any directive has a bearing on the type of packaging used. The proposed Council Directive on packaging and packaging waste will attempt to harmonise some of the schemes within Europe and hopefully eliminate the excesses of some schemes in operation.

While there is nothing preventing any company using recycled materials for food packaging in direct contact with food, any such packaging must be safe. The current legislation, both in the EC[16] and the UK, lays down that:

> ... materials and articles, in their finished state, must not transfer their constituents to foodstuffs in quantities which could endanger human health or bring about an unacceptable change in the composition of the foodstuffs.

This puts the onus on the packaging material supplier to ensure that *all* materials for direct food contact are safe. The only way to ensure safety at the moment is to produce food contact materials from virgin materials. The term 'functional barrier' appeared in one of the draft waste proposals but this term was not defined.

German legislation is not generally seen as the correct way to proceed. It is not acceptable to collect materials for recycling when no acceptable methods for recycling exist or where the levels of materials collected exceed the demand for these materials. However, German legislation has rapidly focused industry into looking into ways of making recycling commercially viable.

12.7 Recycling symbols

The use of symbols on packaging (Figures 12.1 and 12.2) to indicate that material is recyclable is not new, but it does seem that the consumer does not fully understand what they mean. Recently at a consumer panel,[17] two questions or statements were made regarding recycling logos. The first query regarding the arrow triangle was: 'Does it mean that it (the material) has already been recycled?', and the second query specific to the plastics recycling code proposed by the SPI was: 'Does the number mean how many times it (the material) can be recycled?'.

The proposed EC Council directive on packaging and packaging waste suggests its own list of numbers for each material and the way that packages should be marked. Different symbols are used for reusable packaging and recoverable packaging, as well as packaging incorporating recycled materials.

If reasonable consumers cannot clearly understand these symbols, how can they understand all the variations of symbols used by the different sectors of industry and even different supermarket chains? The plastics industry within the UK is concerned that certain manufacturers are using the SPI code incorrectly and adding to this consumer confusion.

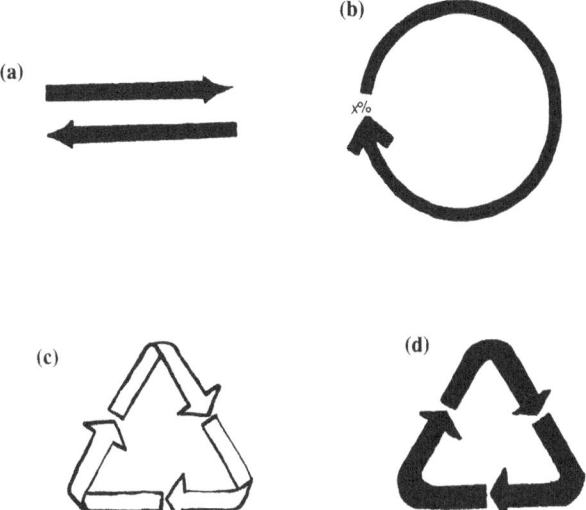

Figure 12.1 Proposed EC logos. (a) Reusable packaging. (b) Packaging made partly or entirely of recycled materials; $x\%$ = the percentage of recycled material used in the manufacturing of the product. (c) and (d) Recoverable packaging.

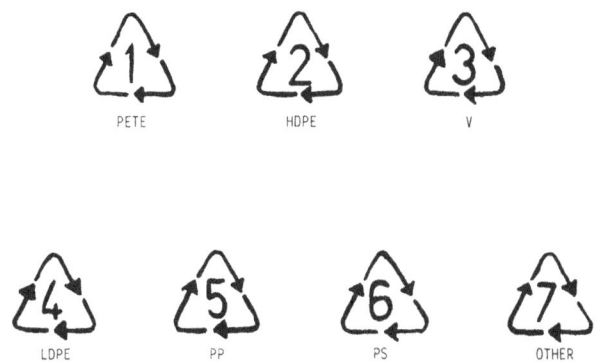

Figure 12.2 Society of Plastic Industries (SPI) codes.

12.8   Life Cycle Analysis (LCA)

LCA or eco-balance, although useful, does not supply the total answer to the food industry. This may be due to the complex issues of using certain materials for specific purposes, i.e. the obtained benefits for the product or, more importantly, the total cycle for the product is not fully documented and understood.

The example which is used by certain groups relates to the use of glass bottles. Is there less cost in using single-trip bottles than refillable bottles?

The single-trip bottle is lighter (it is designed to be used once) and can be easily crushed to reduce its volume to make it easier to return to the glass factory, where it may be remelted and new bottles made. In contrast, refillable glass bottles are designed to be heavier and stronger to withstand the rigours of multi-trip usage, full one way and returned empty. How do the energy costs of returning the bottles to the supplier, sorting and washing them compare with single-trip glass crushed and remelted, both before filling?

Those people who take part in LCA need to fully understand the processes they are studying and preferably be seen to be totally independent of their sponsors. Studies have shown that governments are not credible participants in this debate.

12.9 Conclusions

The consumer, as well as food processors and retailers, is very confused by the whole issue of packaging and the environment. The various packaging sectors which have been claiming that their material is better for the environment than others have in a way confused the issues further. It is only recently that the various sectors of the packaging industry in the UK have joined together to present a united front. This is more a response to government pressure of proposed legislation than pressure from environmental groups.

For the future, the proposed EC packaging directive will ensure that the materials used by manufacturers are the best for the intended application. However, the levels of materials recycled may not reach those intended by the directive due to limited outlets for those materials.

At the end of the day it will be the consumer who decides which packaging suits. He will decide whether to use heavy glass bottles or lightweight flexible pouches, or to use plastic packaging or cartons on the basis that they are only marginally more damaging to the environment. In the process the consumer will determine which packaging concepts survive.

References

1. Paine, F. (1975) What Packaging Means to the Quality of Life. *Packaging Technology and Management*, **21**(137), 4.
2. SEMA Group. (1992) Materials in municipal solid waste in western Europe. *Financial Times*, 28 May.
3. CAMPDEN (1992) NEWS: Packaging. August.
4. Thomas, C. (1974) *Material Gains – Reclamation, Recycling and Reuse*. Friends of the Earth, p. 2.
5. Guise, W. (1992) Packaging in Cans and Tins. *Packaging*, **63**(692), 4.

6. Gooding, K. (1992) UK Aluminium Recycling Rate Increases to 16 Percent. *Financial Times*, 5 August.
7. Gallagher, R. (1990) Recycle, Reuse and Disposal – Solutions. In proceedings of symposium, *Heat Processed Foods in Plastic Containers*, CAMPDEN, 26–27 November.
8. Anon (undated) Paper and Board Packaging – Fact Sheet. Pulp and Paper Information Centre.
9. Thornhill, J. (1992) Paper Recycling Undetermined by Cheap German Imports. *Financial Times*, 10 August.
10. Anon (1990) *The 'Sutton' Model Scheme for Glass Recycling*. Report on Trial Period. United Glass, April.
11. Anon (1992) Bring Back the 'Bring Back'. Friends of the Earth Briefing Sheet, April, p. 4.
12. Birley, D. (1991) Blue Box Annual Review – Sheffield Kerbside Project November 1989 – November 1990. Recycling City Ltd, April.
13. Pidgeon, R. (1992) Waste will be sold to Japan in council deal. *Packaging Week*, **8**(15), 1.
14. Anon (undated) Getting Better All the Time. Recoup News, Issue 4, p. 5.
15. Anon (undated) *Green Choice – Land Pollution and Waste Disposal*. Tesco Stores Ltd.
16. Council Directive 89/109/EEC of 21 December 1988 on the Approximation of the Laws of the Member States Relating to Materials and Articles Intended to Come into Contact with Foodstuffs. Article 2.
17. McEwan, J. and Bond, S. Viewpoint – Packaging and the Consumer. CAMPDEN, unpublished report.

13 Influence of light transmittance of packaging materials on the shelf-life of milk and dairy products – a review*

J.O. BOSSET, P.U. GALLMANN and R. SIEBER

Abstract

This paper summarises the test results on photodegradation of milk and dairy products, particularly yoghurt and butter, obtained by the authors and their colleagues over the past ten years. A large amount of data from relevant literature is also considered in this review. Two major aspects are treated: the extrinsic and intrinsic factors that influence photooxidation and the effects of light on chemical components and physico-chemical parameters of milk and dairy products. The principal extrinsic factors that influence photooxidation are: the spectrum and intensity of the light source, the conditions of light exposure (geometry, duration etc.), the degree of light transmittance and the oxygen permeability of the packaging material as well as the storage temperature of the product. These factors must be selected after careful consideration of the intrinsic factors responsible for the photosensitivity of the product. The latter include the composition (particularly oxidant and antioxidant contents), pH, redox potential and processing of the product.

Light induces loss of vitamins, especially riboflavin (which also acts as a photosensitiser), β-carotene and vitamin C, production or degradation of free amino acids, increase of the peroxide value, formation of sensorially unpleasant volatile compounds (methional, aldehydes and methyl ketones) as well as colour changes. The kinetics of these alterations and the limits of detection of the compounds involved are compared. Some practical recommendations complete this study.

*Elaboration of a lecture held during a FIL/IDF-seminar entitled 'Protein and fat globule modifications by heat treatment, homogenization and other technological means for high quality dairy products' at Munich, August 25th to 28th, 1992 (P.U. Gallmann) as well as a lecture given at the meeting 'Conditionnement alimentaire – 2 défis: Innovation et environnement', in Pouzauges, France, October 7th to 8th, 1992 (J.O. Bosset).

13.1 Introduction

Few foods are consumed immediately after production. Often they are further processed, stored and transported over long distances before reaching the consumer. Food products are packaged in an attempt to prevent chemical, biochemical, physico-chemical, bacteriological or mechanical degradation. The packaging also serves as the product container as well as for the display of product information and for sales promotion. Among all these functions, the protection of foodstuffs against light plays a key role particularly during storage, transport and sales display. For this reason numerous packagings with varying properties are used, e.g. cardboard, paper, glass, metals in the form of cans, composite foils or films (aluminum and plastic), plastic pouches and cups.

The sensitivity to natural and artificial light differs from one foodstuff or beverage to the other.[1] Owing to their composition, milk and dairy products can be used as ideal models for basic research on light sensitivity.[2-8] The light sensitivity of a foodstuff depends above all on its composition, particularly on its content of riboflavin, which acts as a photosensitiser. It is also influenced by the content of sulphur compounds, antioxidants and heavy metals as well as the fat composition.

In opaque foods and beverages the photochemical degradation occurs almost exclusively at the surface since the depth of penetration of the light is very low, particularly when the products have a solid structure. Milk and natural yoghurt are comparable. Flavoured yoghurt however differs considerably because of the varying effects of its pigments, colourants and natural antioxidants.

The objective of this paper is to compile some of the results of various investigations on the effect of light on milk and dairy products obtained by the authors in co-operation with Nestlé in Lausanne and Vevey.[9-24] These results have been supplemented by data from recently published literature. Similar investigations have been carried out on the effects of light on orange juice,[25] beer,[26] tea,[27] champagne,[28] sake[29] and lemon oil.[30]

The ecological aspects of the use of different packaging materials are not dealt with in this study in spite of their increasing importance in our consumer society. However when choosing packaging materials these aspects must be taken into consideration as well as the technical, technological and economic factors.

13.2 Effects of physico-chemical parameters

There is an interdependence between the container and its content. This interdependence is decisive for the choice of the packaging material, i.e. its mechanical, chemical, physical and microbiological properties (Figure

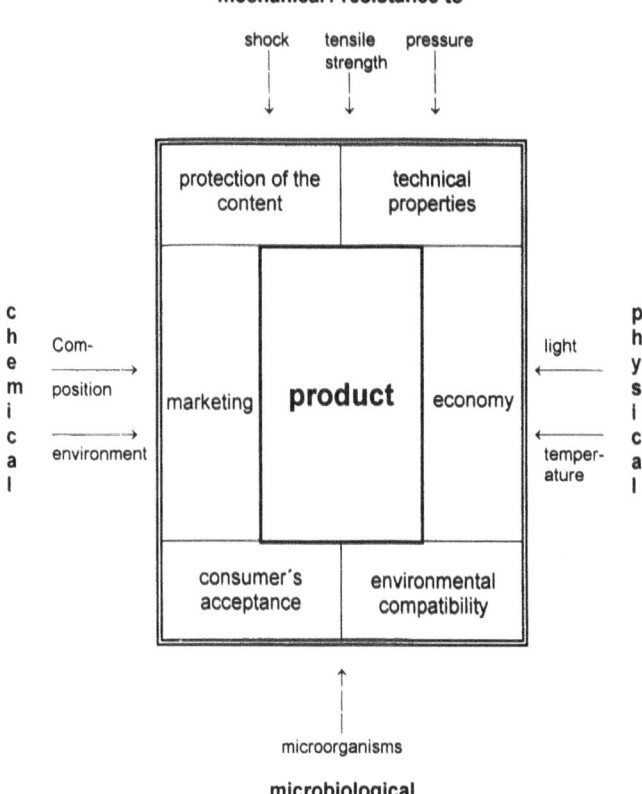

Figure 13.1 Interdependence of packaging and product: influences and requirements.

13.1). The photosensitivity of food products is also affected by further factors, i.e.

- the content of dissolved or free oxygen and the oxygen permeability of the packaging;
- the temperature;
- the duration of light exposure and the intensity and the emission spectrum of the light source;
- the degree of light transmittance of the packaging, which determines the energy level of the photooxidative reactions;
- the spectra of light reflection, transmission and absorption of the products.

Additional factors such as seasonal influences (the foodstuffs are more sensitive in winter), homogenisation of milk as well as cold storage of raw milk and packaging of dairy products can increase the risks of light induced defects.[2]

Table 13.1 Visible spectrum and its colours

Wavelength region (nm)	Perceived colour	Absorbed (complementary) colour
380–440	Violet	Yellow–green
440–480	Blue	Yellow
480–490	Green–blue	Orange
490–500	Blue–green	Red
500–560	Green	Purple
560–580	Yellow–green	Violet
580–600	Yellow	Blue
600–620	Orange	Green
620–650	Red	Blue–green

Visible light covers a wavelength range of 380 to 700 nm (Table 13.1) and ultraviolet (UV) light a range of 200 to 380 nm. Theoretically, this study should consider the whole emission spectrum of the light source, both in the visible and in the ultraviolet. Moreover, the analysis of photodegradative processes induced by UV light should be given priority because of its high energy content, which is capable of splitting certain chemical bonds (Table 13.2). This part of the (emission) spectrum is however almost entirely absorbed by the packaging (glass, polystyrene, polyethylene, polyethylene–terephthalate), except for some highly intensive spectral lines of mercury vapour emitted by fluorescent tubes. This absorption is called the cut-off of the packaging material. In contrast to this, light of the low wavelength range of the visible spectrum (420 to 520 nm) causes substantial problems, especially when the product contains riboflavin (see Figures 13.2–13.6, hatched zones).

Table 13.2 Energy levels of some chemical bonds and of various wavelengths

Chemical bonds in molecules (kcal/mole)				Wavelength region (nm)	Energy (kcal/mole)
C–H	99	C–C	83	750–400	38–74
N–H	93	C=C	146	400	71
O–H	111	C≡C	200	Hg 313	91
S–H	83	C–N	73	285	100
P–H	76	C=N	147	H 254	112
N–N	39	C–O	213	200	142
N=N	100	C=O	176–192		
O–O	35	C–S	65		
S–S	54				
Intermolecular forces					
Hydrogen bonds		3–10			

Source: reference 7.

13.2.1 *Emission spectrum of sunlight*

The emission spectrum of sunlight (natural light) is broad, homogenous and rich in high energies, both in the ultraviolet and in the visible light (Figure 13.2). To avoid photodegradation, milk and milk products should not be exposed at all to direct sunlight. Fluorescent tubes (artificial light) are poorer in energy and therefore less harmful.

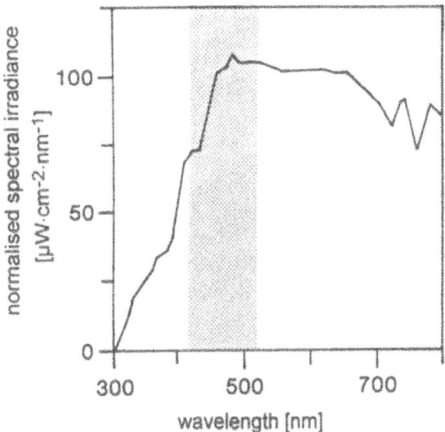

Figure 13.2 Emission spectrum of sunlight.[31]

13.2.2 *Emission spectra of fluorescent tubes*

Figure 13.3 shows the emission spectra of some commonly used commercial fluorescent tubes. Table 13.3 summarises their principal characteristics and fields of application. Fluorescent tubes emitting polychromatic (white) light can be divided into two main groups:

- the so-called 'cool white' fluorescent tubes rich in all colour components of visible light;
- the so-called 'warm white' fluorescent tubes rich in yellow, orange and red components, and poor in violet, blue and green components and therefore poorer in energy.

The high energy emission spectra of cool white fluorescent tubes makes them inappropriate for the illumination of store windows, display cabinets and storage rooms of photosensitive foods such as milk and dairy products. The high emission in the blue–green spectral region (approx. 444 nm) corresponds to the third absorption band of riboflavin. And furthermore, emission spectra of fluorescent tubes such as the Philips 33 exhibit lines of mercury vapour at 366, 405 and 436 nm, which were found[18] identical with the absorption bands of riboflavin (see Figure 13.6).

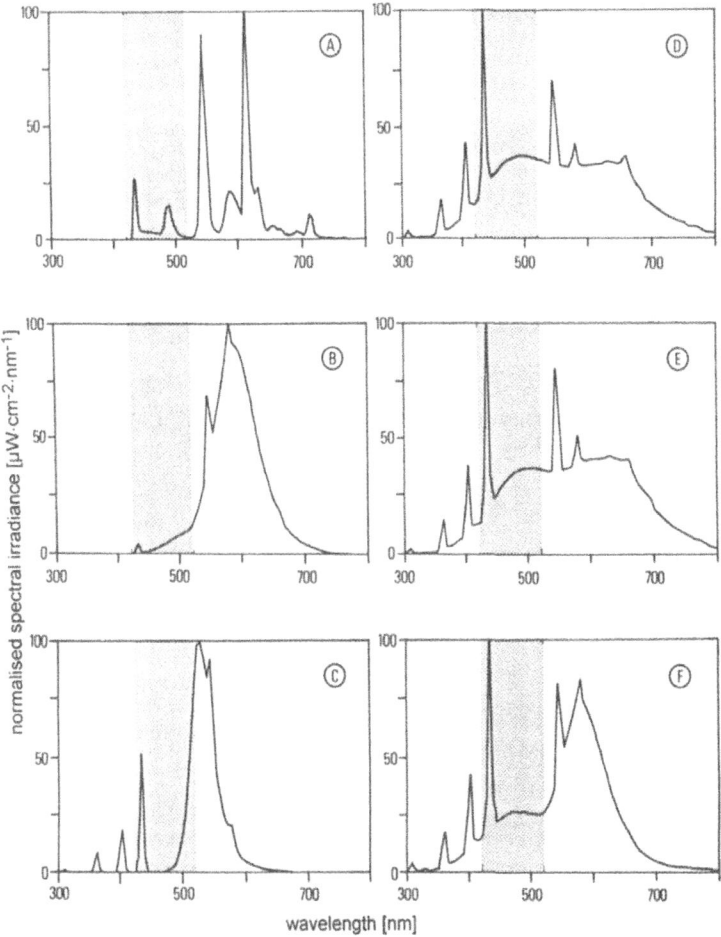

Figure 13.3 Emission spectra of some commercial fluorescent tubes.[10] (A) Philips 83/36 W white; (B) Philips 16/40 W yellow; (C) Philip 17/40 W green; (D) Philips 55/40 W daylight; (E) Osram 19/40 W daylight; (F) Philips 33/40 W white.

13.2.3   Light intensities in display cabinets

In addition to the spectral emission characteristics of fluorescent tubes the actual intensity of light applied must be considered. In several display cabinets used for the presentation of milk and milk products a light intensity of about 2000 lx was measured (authors' unpublished results). This corresponds to the values reported in the relevant literature and has thus been applied in the authors' trials. The average light intensity of the display cabinets of 105 supermarkets was found[5] to be of about 2000 lx in 1974. In three other studies, it was 269–5380 lx in milk displays,[32] 110–4950 lx in 19 retail stores[33] and 129–4304 lx in four further retail stores.[34]

Table 13.3 Technical data on commercial fluorescent tubes

Trade mark, type and power	Normalised irradiance (μW/cm²)*		Colour temperature (K)	Flux (lm)	Colour designation	Colour efficiency	Luminous efficiency	Application	Equivalent commercial product
	400–500 nm	380–700 nm							
Philips 83/36 W	370	2630	3000	3400	Warm white de luxe	Very good	Very good	Offices, sales rooms, restaurants, apartments	Osram 31
Philips 16/40 W	190	6220	–	2000	Yellow	–	–	Decorations	–
Philips 17/40 W	470	4120	–	3300	Green	–	–	Decorations	–
Philips 18/40 W	**	**	–	650	Blue	–	–	Decorations	–
Philips 55/40 W	2200	7300	6500	2020	Daylight	Good	Good	Display cabinets	Osram 10
Osram 19/40 W	2120	8200	5000	1900	Daylight de luxe	Good	Good	Conference halls Colour selection Inappropriate for foods	Philips 47
Philips 33/40 W	1950	7830	4200	3200	White	Good	Good	Offices, industry, outside lighting, stores and warehouses	Osram 20
Philips 29/40 W	**	**	3000	3100	Warm white	Low	High	Outside lighting, heavy industry	Osram 30
Philips 82/36 W	**	**	2550	3250	Warm white extra	–	–	Apartments	–
Thorn NX/40 W	**	**	3600	1750	Natural daylight	Good	–	Sales rooms for flowers, meat, fruits and vegetables	Osram 36
Thorn KR/40 W	**	**	4000	2000	Daylight	Excellent	–	–	–
Thorn TD/40 W	**	**	6500	2410	Tropical daylight	–	–	–	Osram 10

Source: reference 10.
*Rounded off to ten
**Not measured

13.2.4   Light transmittance of packagings

The packaging material should be selected as carefully as the light source. Materials pervious to the blue–green bands should not be used for milk and dairy products in order to avoid absorption by riboflavin due to its third absorption band. If a level of transparency of the package is desired for a more attractive presentation of the product, preference should be given to brown–red coloured materials. Table 13.4 shows some values of transmittance of artificial light by several milk packaging materials. Figure 13.4 presents the transmission spectra of five packaging materials tested in Switzerland for the packaging of pasteurised milk.[24] Four of them are

Table 13.4 Fluorescent light transmittance of milk packaging materials

Packaging materials	Thickness (mm)	Light transmittance (%)
Clear flint glass	3.4	91
Clear polycarbonate	1.5	90
Tinted polycarbonate	1.5	75
Non returnable polyethylene	0.5	70
High density polyethylene	1.7	57
Unprinted fibreboard	0.7	4

Source: reference 5.

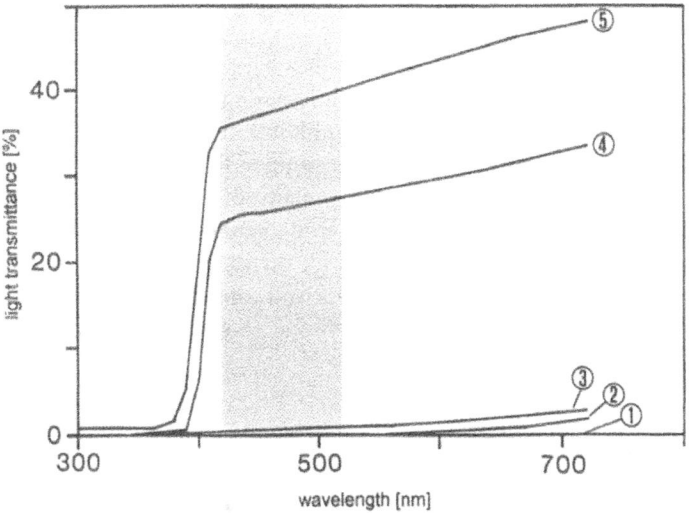

Figure 13.4 Light transmittance of packaging materials used for pasteurised milk (bags).[24] (1) Plastic with light barrier (black); (2) cardboard without aluminum foil; (3) plastic with light barrier (brown); (4) and (5) plastic without additional light barrier. (Spectra measured by EMPA, Dübendorf, on behalf of Toni-MIBA Produktion AG, Zurich, with the kind permission of this company.)

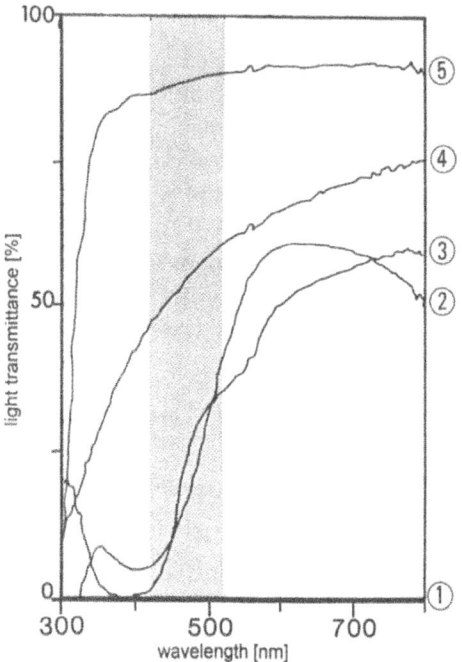

Figure 13.5 Light transmittance of packaging materials used for yoghurt.[12] (1) Cardboard wrapped polystyrene cup; (2) transparent brown-coloured glass jar; (3) transparent brown-coloured polystyrene cup; (4) transparent uncoloured polystyrene cup; (5) transparent uncoloured glass jar.

plastic films, of which two are fitted with a light barrier (brown or black); the fifth consists of cardboard without an aluminum foil. (These packaging materials cannot be used for UHT milk because of their high oxygen permeability.) Figure 13.5 shows the transmission spectra of different yoghurt packaging materials (filling weight of 180 g): transparent un-coloured glass (about 119 g), brown–red coloured glass (about 119 g), transparent uncoloured polystyrene (wall thickness 0.3–0.4 mm), brown–red coloured polystyrene (about 7 g, wall thickness 0.2–0.4 mm) and cardboard wrapped polystyrene (about 8 g, wall thickness 0.1–0.2 mm).[12,13]

13.2.5   Absorption spectrum of riboflavin

Riboflavin plays a key role in all problems related to the photosensitivity and photodegradation of milk and dairy products. Figure 13.6 shows its absorption spectrum. This vitamin has several absorption bands with their maxima at 223, 268 and 359 to 375 nm in the ultraviolet and at 446 and 475 nm in the visible. Within the critical range of 350 to 520 nm, the

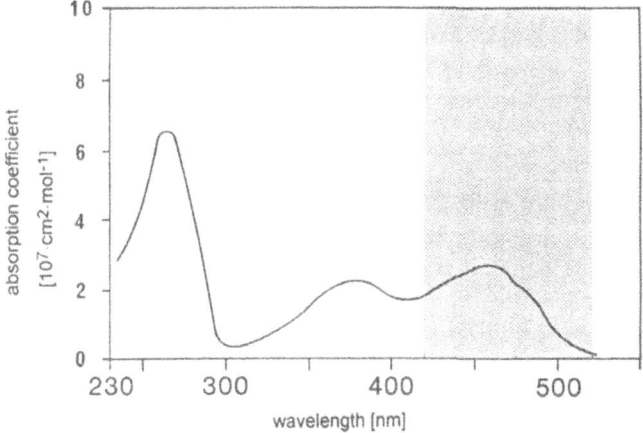

Figure 13.6 Absorption spectrum of riboflavin.[18]

wavelengths between 415 and 455 nm are responsible for the degradation of riboflavin[35] and for a possible light induced off-flavour.[1]

13.3 Test conditions

Trials were carried out under the following conditions (for detailed descriptions refer to the respective publications):

1. milk and milk products in 1 litre Petri dishes of uncoloured glass, exposed to cool white fluorescent tubes Philips TL 33/40 W, 20000 lx, at room temperature for 6 hours;[9]
2. natural yoghurt in 180 g jars, 5 different packaging materials (uncoloured glass, brown–red coloured glass, uncoloured polystyrene, brown–red coloured polystyrene, cardboard wrapped polystyrene), exposed to fluorescent tubes as in (1), 2000 lx, at approx. 8°C for 22 days (12 h a day);[12,13]
3. flavoured yoghurt (strawberry, chocolate, mocha, natural yoghurt as the control) in 180 g jars, 2 packaging materials (brown–red coloured glass and uncoloured polystyrene), exposed to fluorescent tubes as in (1), 2000 lx, at approx. 7°C for 21 days (12 h/day);[20]
4. pasteurised whole milk in white polyethylene bags slightly pervious to light (without a light barrier), exposed to fluorescent tubes as in (1), 2000 lx, at 4°C for 5 days (12 h/day);[21]
5. whole milk homogenised at various pressures and pasteurised at different temperatures in 1 litre uncoloured glass bottles filled with a 250 ml headspace of air, exposed to warm white fluorescent tubes Philips TLD 58/33 W, 750 lx (at 8°C for 20 h);[22]

6. butter in different packaging materials, exposed to cool light fluorescent tubes Philips TL 33/40 W, 2300 lx, at 8°C for 2 weeks (10 h/day) and then stored for 4 weeks in the dark. The controls were stored in the dark at 8°C for 4 weeks, at −18°C for 3 weeks or at 8°C for 1 week.[36]

13.4　Effects of light exposure on milk and dairy products

The influence of light on components and other parameters of milk and dairy products have been dealt with in different reviews herein referred to.[1,4,5,7,37,38] Many nutrients contained in food products, particularly in milk and dairy products, are sensitive to light, e.g. vitamin A and its provitamin β-carotene, vitamin B_2 (riboflavin), B_6, B_{12}, D and K, folic acid, tocopherol, tryptophan and unsaturated fatty acids. The following chapters are particularly concerned with the effects of light on vitamins, free amino acids, peroxides, some volatile flavour-active compounds, colour and several other constituents of milk and dairy products.

13.4.1　*Effects on vitamins*

Since vitamins are essential nutrients, their loss by photodegradation decreases the nutritional value of foodstuffs. The reactions of the vitamins to light differ considerably according to their absorption spectra (Table 13.5). Due to its third absorption band in the visible range (Figure 13.6) riboflavin is of particular interest and will be discussed separately. It is also interesting to note that light exposure increases milk vitamin D content.[40] This reaction was used[41] as a production technique in the United States around 1920.

13.4.1.1　*Milk.*　Exposure of unpackaged pasteurised milk to the cool white fluorescent tube Philips 33 (which has long been recommended for the lighting of storerooms) reduced the contents of vitamin A, B_6 and E. The vitamin A concentration decreased by 50% within 6 hours.[9] This confirms the photosensitivity of vitamin A and its precursor β-carotene in milk already demonstrated by Sinha,[42] Hedrick and Glass,[43] Sattar *et al.*[44] and deMan.[45] According to these authors, the β-carotene in milk fat was destroyed by wavelength lower than 465 nm. The vitamin A content decreased[44] at a wavelength below 415 nm and to a lesser extent at wavelengths between 415–455 nm. It has also been proved that vitamin A added to milk is more sensitive to light than naturally occurring vitamin A.[45,46] Enrichment in β-carotene reduced photodegradation of vitamin A.[47] The photosensitivity of added vitamin A also depends on the holder (e.g. butter oil or coconut oil) and the form in which it is used (e.g. retinyl

Table 13.5 Absorption maxima of some vitamins

Vitamin	Solvent	λ_{max} (nm)	Remarks
A	Isopropanol	326	
	Cyclohexane	328	
β-carotene	Cyclohexane	approx. 456, 484	
D	In alcoholic solution	265	
Tocopherol	In alcoholic solution	292	Minimum at 255 nm
K_1	Cyclohexane	243, 249, 261, 270	Minima at 254 and 285 nm
Ascorbic acid	In very acid medium	approx. 245	
	after neutralization	365	
B_1		200–300	Variable according to pH
	In 0.1 molar HCl	approx. 245	
B_2	In 0.1 molar HCl	223, 267, 374, 444	
B_6	In aqueous solution:		Variable according to pH
	acid medium	291	
	neutral medium	245, 324	
	alkaline medium	245, 309	
B_{12}	In aqueous solution	approx. 278, 361, 550	
Nicotinic acid	In aqueous solution	approx. 261	Variable according to pH
Folic acid	In 0.1 molar NaOH	256, 283, 365	Variable according to pH

Source: reference 39.

palmitate) (Table 13.6). An increase in light intensity led to an increasing loss of vitamin A in milk, however this is not directly proportional.[50]

No loss of vitamin B_1 was found by Ford[51] in sterilised milk exposed to sunlight. Ferretti *et al.*[52] reported a light induced vitamin B_1 decrease of 10% in UHT milk stored for 90 days exposed to light in comparison with milk stored in the dark. In contrast Mohammad *et al.*[53] indicate thiamine losses in untreated milk of up to 40% after 6 hours exposure to fluorescent light or sunlight in the presence of oxygen. Light irradiation of a thiamine solution for 5 days with a mercury-vapour lamp at 254 nm led to the production of perceivable odorous compounds.[54] A review of the photo-sensitivity of vitamin B_1 has been recently published.[23]

Folic acid in homogenised milk underwent no significant changes after 48 hours of exposure to fluorescent light at 2160 lx, regardless of the type of packaging used (cardboard, plastic jug and clear polyethylene bags).[55] In sterilised milk previously purged with carbon dioxide or saturated with air, exposure to light caused a decrease in the folic acid content. These losses are additional to those caused by heat treatment. Sunlight also decreased the vitamin B_6 and B_{12} contents, whereas biotin and nicotinic acid concentrations remained unchanged.[51]

Table 13.6 Effect of light on milk vitamin A concentration

Milk	Vitamin A	Support system	Container	Conditions of light exposure	Loss in %	Reference
Whole milk	Native		Plastic pouches	30 h at 2200 lx	32	45
2% Fat milk	Added				73	
Skimmed milk					96	
Whole milk	Retinyl palmitate		Paperboard	48 h at 1614 lx	43	50
2% Fat milk					47	
Skimmed milk					55	
Skimmed milk	Retinyl palmitate	Butter oil	Glass	96 h at 1614 lx	~40	48
		Coconut oil			~40	
		Corn oil			~80	
		Peanut oil			~70	
2% Fat milk		Butter oil			~30	
		Coconut oil			~30	
		Corn oil			~60	
		Peanut oil			~70	
Milk	Retinyl palmitate	Butter oil	Glass tubes	32 h at 4300 lx	~60	46
	Concentrated retinyl				~95	
Skimmed milk	Oil		Polyethylene	72 h at 1076 lx	58	409
	Aqueous				69	
2% Fat milk	Oil				35	
	Aqueous				26	

There is much published literature on the photolysis of vitamin C in milk. Losses of vitamin C vary considerably according to the light source, the light intensity and the exposure time. It also depends on a possible presence of metals[56] and the packaging materials used (Table 13.7). The presence of oxygen, even at trace levels, accelerated the photodestruction of vitamin C. In pasteurised whole milk, the ascorbic acid content decreased by about 90% within 30 minutes of exposure to daylight. When milk is exposed to fluorescent tubes, the colour of the light may affect the vitamin C content. The 'warmer' the white light, the smaller the loss of vitamin C.[62] The use of an appropriate packaging material with low oxygen permeability may reduce the rate of ascorbic acid degradation in milk exposed to light (Table 13.8). According to Nordlund,[67] the rate of loss of ascorbic acid induced by light is of first-order kinetics and can be described with the following equation:

$$A = A_0 \cdot 10^{(-k \cdot t + l \cdot t^2)}$$

where

A = ascorbic acid content after light exposure; A_0 = original ascorbic acid content; t = light exposure time; k and l = constants.

Pasteurised milk in polyethylene bags without a light barrier (put on the Swiss market since 1990) underwent vitamin C losses in excess of 50% after 12 hours of exposure to cool white light (see 13.3: Test conditions 4). No loss was observed[21] in milk packaged in Tetra-Brik and stored in the same conditions (Figure 13.7). With a lower light intensity and warm white light (see 13.3: Test conditions 5) the ascorbic acid content of unpackaged heat-treated milk decreased by one-third within 20 hours of exposure.[22]

13.4.1.2 *Yoghurt.* Natural yoghurt was exposed to light in five different packagings (180 g jars or cups). Light (cold white, Philips 33) induced losses of vitamin A and B_{12} could only be found in yoghurt packaged with uncoloured polystyrene.[12,13] Assays were not carried out for other vitamins. The results of these tests are shown in Table 13.9.

The effect of light on vitamin A and B_2 was analysed in strawberry, mocha and chocolate yoghurt.[18] The vitamin A content remained unchanged even after 18 days of lighted storage in either polystyrene or brown-coloured glass. The beetroot juice extract used for the coloration of strawberry yoghurt is thought to act as a barrier against light or as an absorbent of light and thus protects the vitamin A. The light absorption of an extract containing 1% of betanin is highest at 530 nm. In the critical spectral area from 366 to 436 nm light absorption fluctuates between 25 and 40% of maximum absorption. The stability of vitamin A in mocha and chocolate yoghurt must be ascribed to the antioxidants naturally occurring in coffee and cacao, which inhibit photooxidation. This hypothesis is

Table 13.7 Effect of light on milk vitamin C concentration

Milk	Packaging	Type of light	Lighting conditions time/temp.	Initial concentration (mg/l)		Loss (% of initial concentration)		Reference
				A	A+D	A	A+D	
p		Direct daylight	4 h		22.0		56	57
	Clear gl	Diffuse daylight	1 h 3 h				78 100	59
p	Clear gl	Fl.t.: 200, 500, 1000, 1500, 2000 lx	6 h				12, 46, 68, 86, 87	60
p	Clear gl Brown gl Tetra	Fl.t: white, warm white, yellow, red UV	6 h			84, 82, 20, 17 18, 16, 10, 7 11, 11, 6, 5 0		61
p. ui	PE	Daylight Daylight	5 h 30d/20°C 30d/37°C 90d/20°C 90d/37°C	17.4	18.6	77	40 24 33 45 51	58 62 52
r		Sunlight 360 W UV	1 h 6 h	19.5 20.4		45 66		63
p	Clear gl Opaque gl Clear pb Opaque pb	40 W fl.t.	6 h 24 h	18.0 12.6 12.8		26 54 8 7 0		64
p	Pb PE	4000 lx	17 h/7°C 9 h/8°C	18 20	15.5 16.5	18 39	9 20	93
p	Clear gl Brown gl	700, 1500, 2300 lx	12 h				40–70 20–40	66

Abbreviations: A = ascorbic acid; D = dehydroascorbic acid; R = raw; p = pasteurised; u = uperised; i = indirect; gl = glass; Pb = paperboard; PE = polyethylene; fl.t. = fluorescent tube.

Table 13.8 Concentrations of ascorbic and riboflavin in milk in different packaging after 24 hours of fluorescent light exposure

Packaging material	Ascorbic acid (mg/l)			Riboflavin (mg/l)		
	Before light exposure	Light exposure 1080 lx	Light exposure 2160 lx	Before light exposure	Light exposure 1080 lx	Light exposure 2160 lx
Clear pouch	12.34	1.12	0.92	1.75	1.36	1.26
Opaque pouch	12.34	10.74	10.22	1.75	1.74	1.73
Paperboard	12.57	9.60	8.68	1.82	1.71	1.65
Plastic jug	12.34	1.70	1.28	1.80	1.60	1.56

Source: reference 32.

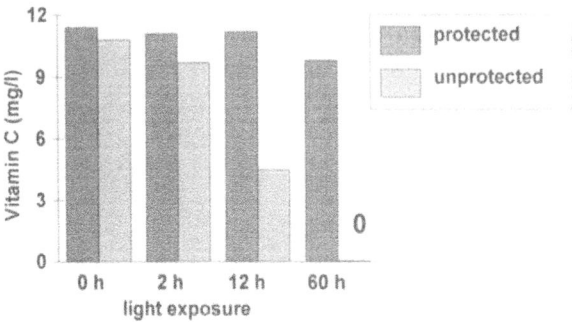

Figure 13.7 Concentration of vitamin C in pasteurised milk (80°C/15 s) after light exposure (display cabinet).

confirmed by the low peroxide concentration found after light exposure of these types of flavoured yoghurt.[16]

13.4.2 Influence of riboflavin and effects of light on riboflavin

As different studies indicate, riboflavin (vitamin B_2) acts as a photosensitiser in milk and dairy products. It is also a highly photosensitive vitamin.[9,35,62,67–75]

13.4.2.1 *Milk.* The light induced loss of riboflavin in milk depends on the wavelength range and intensity of the light source, the duration of exposure, the product temperature and the light transmittance of the packaging materials. This has been shown by several tests using both sunlight or fluorescent light (Tables 13.10 and 13.11). Exposure to sunlight[62] for 2 hours can reduce the riboflavin content of milk in transparent uncoloured glass by more than 50%. Unpackaged pasteurised milk exposed to cool white fluorescent light (Philips 33) for 7 hours undergoes riboflavin losses[9] of over 75%. Very mild lighting conditions (see 13.3: Test conditions 5) scarcely alter the riboflavin content of milk submitted to different types of homogenisation and heat treatment.[24] After 60 hours of light exposure the riboflavin content of milk in polyethylene bags without a light barrier was 20% lower than that of milk in Tetra-Brik cartons.[21] The use of a suitable packaging material is essential for the reduction of riboflavin losses in milk stored under lighted conditions (Table 13.8).

According to Nordlund,[67] the photolysis of milk riboflavin follows zero-order kinetics at light intensities of 40000 to 44000 lx, with an average molar activation energy of 23 kJ mol^{-1} (temperature range: 10–50°C). Other authors[35,50,70,97] consider riboflavin photodegradation a reaction of first-order kinetics at light intensities of 1620–5400 lx (temperature range 2–25°C). After exposure to 1614 lx for 48 hours, whole milk lost less

Table 13.9 Vitamin losses in light-exposed unpackaged milk (20000 lx) (from ref. 9), natural yoghurt (2000 lx) (from refs 12, 13) and flavoured yoghurt (2000 lx) (from ref. 18) in different packagings (artificial cool white light, Philips 33)

Vitamin	Uncoloured glass	Brown–red glass	Uncoloured polystyrene	Brown–red polystyrene	Paperboard wrapped polystyrene
Vitamin A					
Pasteurised milk	Slight decrease				
Natural yoghurt	Strong decrease	Slight decrease	Strong decrease	Increase	Increase
Strawberry yoghurt		Unchanged			Slight decrease
Mocha yoghurt		Unchanged			Unchanged
Chocolate yoghurt		Unchanged			Unchanged
Vitamin E					
Pasteurised milk	Strong decrease				
Vitamin B$_2$					
Pasteurised milk	Strong decrease	Unchanged	Strong decrease	Unchanged	Unchanged
Natural yoghurt	Unchanged	Strong decrease			Unchanged
Strawberry yoghurt		Unchanged			Unchanged
Mocha yoghurt		Unchanged			Unchanged
Chocolate yoghurt					
Vitamin B$_6$					
Pasteurised milk	Slight decrease				
Vitamin B$_{12}$					
Natural yoghurt	Strong decrease	Unchanged	Strong decrease		Unchanged

Table 13.10 Effects of sunlight and diffuse daylight on milk vitamin B_2 concentration

Milk	Packaging material	Conditions of light exposure		Initial vitamin B_2 conc.	Loss	Remarks	Reference
		Time	Temp. (°C)	(mg/l)	(%)		
n.s.	n.s.	5 min 0.5 h	100	n.s.	26 48	Boiling under the effect of light	76
r	White gl	1 h 3 h	16–22	1.90	36 72	Sunlight 4 series	77
r	White gl	2 h	17–21	2.06 1.92	57 61	Sunlight	78
r	White gl	2×1 h		1.66	66	Sunlight	79
	Cl gl	2 h		1.62/	40 35 28	Sunlight	68
r/p/ph		2 h		1.64/1.63	7 4 3	Stored in the dark	
ph	Cl gl/Br gl 4 Pb	6 h		1.78	78 6 6 5 6 7 17	Sunlight	
h	White gl	2 h			37–45	Sunlight	80
ph	White gl Red gl Pb	2 h			60–80 2 3	Sunlight	81
s	Gl	180 d	Room temp.		60	Diffuse daylight	82
p	White gl	6–7 h		1.54–1.86	20 50	February–May June–August	83

Milk	Container	Temp (°C)	Time			Light	Ref.
p	n.s.		2 h	1.78/1.94	93 (sum) 31 (win)	Clear sky	84
				1.82/1.96	46 (sum) 16 (win)	Cloudy	
				1.89/1.89	30 (sum) 10 (win)	Cloudy	
s	White gl	20	14 d	1.03	83	Diffuse light	85
		38	14 d		96	Sunlight + artificial light	
s	White gl	Room temp.	42 d	0.16	100 20	Intense light	86
	Amber gl				50 2	Dim light	
s	PE		8 h	~1.7	~90	Sunlight	51
p	PE		6 h	~2.4	~40	Diffuse daylight	62
ui	n.s.	20 37	30 d		19 16	Sunlight	52
		20 37	90 d		26 27	Indirect light	
					~90		
n.s.	Cl gl		2 h		90	Sunlight	87
Hum	PE		7 h		~50	Daylight	88
Cow	Opaque		6 h	1.76	64 45	1st series with O_2	53
Sheep	container			3.42	66 49	2nd series without O_2	
Goat				1.24	66 49		
Buf				1.32	64 48		
Cow	Black		1 h	1.94	83	Sunlight	89
Buf	varnished			1.59	77	Ref.: Milk stored for 1 h	
	bottle					at room temp.	

Abbreviations: **Milk:** hum = human; buf = buffalo
Milk: r = raw; p = pasteurised; h = homogenised; s = sterilised, ui = UHT indirect; gl = glass; Pb = paperboard; PE = polyethylene;
Cl = clear; Sum = summer; win = winter
n.s. = non-specified

Table 13.11 Effect of fluorescent light on milk vitamin B$_2$ concentration

Milk	Type of artificial light	Packaging material	Storage conditions		Initial vitamin B$_2$ conc. (mg/l)	Loss (%)	Remarks	Reference
			Time	Temp. (°C)				
p	White, warm, white, yellow, red	White gl, Br gl	6 h	n.s.	n.s.	40 26 16 17 24 21 10 14 14 9 8 11		61
p	Cool white	Plastic	5 d	7	2.8	17		104
n.s.	White	Petri dish	3 h			60	Without filter Blue, br, red filters	90
p	1614 lx	Pb Plastic	5,10, 24 h	4	1.66 1.67	20 30 10 3 4 3 5 7 8	n=11	43
n.s.	2200 lx, cool white, cool white + filter, warm white	6 pb + plastic packagings	48 h	4		13–28 17 11 15 19 13 17		33
p	1076 lx	Gl, PC, PE	72 h	7	1.9	27 13 10	Method: HPLC	91
p	20000 lx	Pb	7 h			~85		9
	35 W, cool white	Pb	16 h	1.5		~55		92
p	4000 lx	PE Pb	90 h	7	1.5	31 37	2 tests	93
p	1614 lx	Pb	48 h	4		0 55 64 70	Whole milk 2% fat milk Skimmed milk	50
p	4000 lx	Gl	14 d	4.5	~1.5	~80 ~60	Past. 72°C 82°C	94
p	4300 lx	PE, PE + Alu	72 h	6	0.84 0.79	13 1		95
		Cl gl	12 h					
p Cow	700, 1500, 2300 lx	Opaque container	6 h		1.76	25–40	1st series with O$_2$	96
Sheep					1.42	56 39	2nd series without O$_2$	53
Goat					1.24	59 39		
Buf					1.32	59 39		
Cow		Black varnished bottles	2 h		1.94	58 36	Ref: Milk stored for	89
Buf			1 h		1.59	26 22	1 h at room temp.	

Abbreviations: see Table 13.10

riboflavin than partially skimmed milk (2% fat) or skimmed milk. This might be due to the diffusion of scattering by the fat globules.[50] An extensive review on this subject has been recently published.[24]

13.4.2.2 *Yoghurt.* Brown-coloured polystyrene and glass are capable of partially protecting riboflavin of natural yoghurt against light, since they filter light of the wavelength range absorbed by this vitamin (Figures 13.5 and 13.6). Cardboard packagings offer still better protection, whereas the use of transparent uncoloured polystyrene cups leads to a decrease in the riboflavin content of natural yoghurt by about 90% within 18 days of light exposure.[14] Tagliaferri[18] confirmed this result, however the riboflavin loss was only 55% after 21 days of light exposure.

Strawberry yoghurt lost about half of its riboflavin content within 18 days of light exposure in a transparent uncoloured polystyrene container, whereas no change occurred after storage in transparent brown-stained glass. Unlike vitamin A, riboflavin does not seem to be protected from light induced changes by the beetroot juice extract added for the coloration of strawberry yoghurt. In mocha and chocolate yoghurt stored under light in the same containers, the riboflavin content remained practically unaltered. This might be due to the protective effect of pigments present in chocolate and coffee extracts.[18]

13.4.2.3 *Cheese.* Like other dairy products, cheese is an important source of riboflavin sensitive to the effects of light. In the form of small pieces it is frequently wrapped in transparent films, stored in display cabinets and exposed to fluorescent light for a longer time period than milk. Deger and Ashoor[98] have dealt with riboflavin photodegradation in cheese. In Cheddar cheese samples, 0.6, 2.5 or 5.1 cm thick, exposed to 538, 1614 and 5380 lx at 5 to 10°C for 12 days, the riboflavin content decreased with increasing light intensity. Exposed in a transparent wrapping to 5380 lx, the losses were slightly over 40%, regardless of the thickness of the cheese slices. This might be ascribed to the low penetration depth of light in a compact medium of this kind. In 3 to 4.1 cm thick commercial Cheddar and Colby cheese, an unexplained increase in riboflavin was observed after 14 days of light exposure at 1076 to 1829 lx; in both grated and finely sliced cheese riboflavin losses of 1 to 12% were found at the same light intensities.

13.4.3 *Effects on proteins and free amino acids*

Light can induce photooxidative changes in milk proteins and amino acids.[99] This leads to the photoaggregation of whey proteins. Sunlight and fluorescent light may provoke the hydrolysis of peptides. The principal amino acids subject to photochemical changes are methionine, tryptophan, cysteine, histidine and tyrosine.[12,13,99] Methionine is broken down to methional and other sulphur compounds[100,101] and tryptophan to kynurenine and N-formylkynurenine.[102]

Fluorescent light did not modify free and total amino acid concentrations in milk held in glass, fibreboard, cardboard or plastic containers.[43,103] The amino acid composition of the whey proteins β-lactoglobulin and α-lactalbumin did not change significantly after 72 hours of fluorescent light exposure. However compositional changes occurred in the immunoglobulin fraction.[104] In the presence of riboflavin, a high intensity illumination of the whey proteins β-lactoglobulin and α-lactalbumin led to the formation of high molecular weight protein fractions, the hydrolysis of peptide bonds[105] and the production of superoxide anions.[106]

13.4.4   *Effects on peroxide formation*

13.4.4.1   *Milk*.   The peroxide value is a sensitive indicator of oxidative and photooxidative changes in fats and oils. It permits photometric determination of the degree of oxidation of unsaturated fatty acids. This method is based on the oxidation of Fe^{2+} to Fe^{3+} by the peroxides, the latter having been generated by oxidation of unsaturated fatty acids by oxygen.[107] Peroxide formation is delayed with respect to off-flavours and thus a poorer indicator of flavour alterations. Peroxides are labile substances producing mainly carbonyl compounds of low olfactory and gustative perception thresholds (see next section).

13.4.4.2   *Yoghurt*.   In natural yoghurt the production of peroxides under the effect of light varies substantially according to the type of packaging used (Figure 13.8). With uncoloured materials the peroxide value increased under lighted conditions. This increase was greater with uncoloured transparent polystyrene pervious to gas than with uncoloured glass.[12,13] With uncoloured polystyrene, the peroxide value increased from 0.06 to 0.6 meq O_2/kg after light exposure, whereas no significant change was found after storage in the dark. It also increased in yoghurt exposed to light in brown-stained polystyrene, but not in brown-stained glass. This can again be explained by the difference in oxygen permeability of these materials. In strawberry yoghurt the peroxide value increase was smaller. No significant change occurred in mocha and chocolate yoghurt exposed to light even when they were packaged in uncoloured transparent polystyrene, which has the lowest protective effect. This demonstrates the antioxidant effect of some of their ingredients.[16]

13.4.4.3   *Butter*.   Butter packaged in 10 different materials was exposed to fluorescent light of various intensities and then stored at 5°C for 4 weeks.[108] The peroxide value was higher at the surface of butter than at the interior, except for two samples exposed to the lowest light intensity

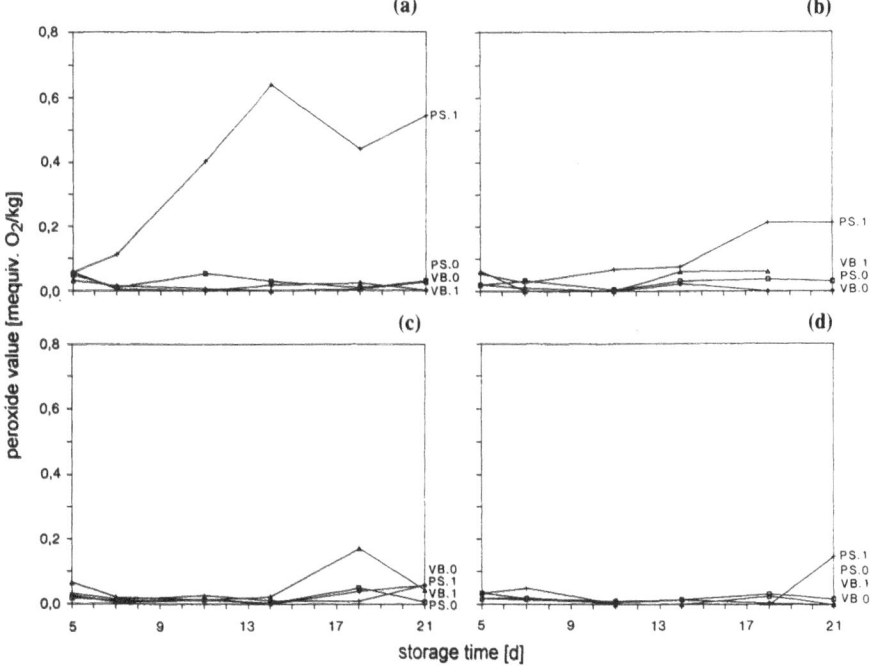

Figure 13.8 Peroxide values of (a) natural, (b) strawberry, (c) mocha and (d) chocolate yoghurt stored in the following conditions:[16] PS.0 = polystyrene/dark; PS.1 = polystyrene/lighted; VB.0 = brown glass/dark; VB.1 = brown glass/lighted.

(Table 13.12). The increase was not proportional to the exposure time. Luby *et al.*[109] also found higher peroxide values at the surface of blocks of butter exposed to light for 20 days. In butter displayed in cabinets at 5°C and exposed to fluorescent light of 648 lx, the peroxide value increased progressively. A linear relationship was found between the peroxide value of butter and light intensity.[110] The peroxide value did not increase in butter packaged in aluminum foils and exposed to fluorescent light for 15 days, as it did with five other packaging materials under identical lighting conditions.[111]

13.4.4.4 *Cheese.* The thiobarbituric acid number (as an indicator equivalent to the peroxide value) was determined at the surface of iron-enriched Cheddar cheese exposed to fluorescent light. It did not differ significantly from the thiobarbituric acid number of the control stored in the dark.[112]

13.4.5 *Light induced formation of undesired volatile flavour compounds*

Light induced alterations in the chemical composition of milk and dairy products affect their olfactory and gustative properties. They are responsi-

Table 13.12 Oxidised flavour intensity, flavour score and peroxide values – differences between the surface and the interior of butter after exposure to different lighting conditions and 4 weeks of storage at 5°C

Packaging material	Oxidised flavour intensity				Flavour score				Peroxide value (meq/kg/feet)			
	400 lx 1 d	400 lx 6 d	2000 lx 2 d	2000 lx 4 d	400 lx 1 d	400 lx 6 d	2000 lx 2 d	2000 lx 4 d	400 lx 1 d	400 lx 6 d	2000 lx 2 d	2000 lx 4 d
Basic material: paper												
A	0.9*	4.6**	5.6**	6.0**	-0.4	-0.12**	-1.5**	-1.7**	1.1**	2.0**	3.6**	5.3**
B	0.8*	4.9**	5.8**	7.0**	-0.4**	-1.2**	-1.5**	-1.9**	0.7**	2.4**	3.1**	4.0**
C	0.2	0.6*	1.3**	2.8**	-0.1	-0.2*	-0.3*	-0.8**	0.2	0.5**	0.7**	1.0**
D	0.9*	1.2*	1.8**	2.6**	-0.4**	-0.4*	0.5**	-0.6**	0.3*	0.5**	1.0**	1.1**
E	0	0	-0.1	0	0.1	0	0	0.1	0	0.2	0.4**	0.3
Basic material: polyethylene												
F	1.1**	5.2**	5.8**	6.0**	-0.3**	-1.5**	-1.5**	-1.6**	0.7**	2.6**	3.7**	4.7**
G	0.1	3.2**	4.3**	5.4**	-0.1	-0.9**	-1.1**	-1.5**	0.4**	1.9**	2.6**	3.8**
H	0.2	3.5**	5.1**	4.9**	-0.2	-0.9**	-1.3**	-1.7**	0.5**	1.7**	2.9**	3.8**
J	0.4	3.5**	4.6**	6.6**	-0.2	-0.9**	-1.3**	-1.7**	0.6**	1.3**	2.6**	2.2**
K	-0.1	3.6**	4.9**	6.4**	-0.1	-0.9**	-1.2**	-1.7**	1.0**	1.9**	2.7**	3.0**

Source: reference 108.
$*p < 0.05$ $**p < 0.01$

ble for the so-called 'sunlight taste', which has been the subject of many extensive investigations over the last decades.[38]

13.4.5.1 *Milk*. The photosensitivity of milk varies considerably from one type of milk to the next. Thus the typical light induced off-flavour appeared to be stronger in UHT skimmed milk than in UHT whole milk or UHT coffee cream.[9] This is most probably due to the differences in light reflection and light scattering of these dairy products.

In raw and pasteurised milk exposed to sunlight for 30 to 120 minutes the contents of the following undesired volatile flavour-active compounds increased by 3 to 7 fold: acetaldehyde, propanal, pentanal and hexanal.[113,114] Additionally, Bassette[115] has studied the behaviour of methyl sulphide, acetone and butanone in light exposed milk. The microbially produced substances acetone and butanone underwent practically no change, whereas the methyl sulphide content increased after storage of milk under fluorescent light. Surprisingly sunlight induced no change in this component. During sunlight exposure, the acetaldehyde concentration increased more in skimmed milk than in pasteurised whole milk.

In UHT whole milk and UHT skimmed milk exposed to fluorescent light for 4 hours, an off-flavour was observed which became stronger with advancing exposure time.[9] After only 2 hours of light exposure, pasteurised milk in polyethylene bags without a light barrier (see 13.3: Test conditions 4) showed clearly perceivable flavour defects described as 'tallowy', 'oxidised', 'sunlight' and 'impure'.[21] Homogenised milk of different fat concentrations (3.25%, 2% or skimmed milk) packaged in plastic containers and exposed to fluorescent light of 2200 lx for 24 hours was judged to have a significantly altered flavour with respect to milk stored in the dark, independently of the unexposed milk's fat content.[116]

In contrast, Olsen and Ashoor[34] found no flavour alterations in milk from grocery stores exposed to light intensities of 129–1076, 215–1076 or 915–4304 lx for 3 to 7 days. This is irrespective of the season, packaging material (synthetic vs. fibreboard), size of the containers or milk fat content.

13.4.5.2 *Yoghurt*. The behaviour of some aldehydes and methyl ketones (from C-3 to C-11) was studied in natural and strawberry yoghurt exposed to light in uncoloured transparent polystyrene and brown-stained glass cups of 180 g volume (see 13.3: Test conditions 3).[15] An increase in the propanal, butanal, hexanal and/or hexanone-2 levels was measured. In natural yoghurt there was also a slight increase in the contents of butanone-2, pentanal and/or pentanone-2 as well as heptanal and/or heptanone-2. No significant light induced changes were found for methional. This might however be due to analytical difficulties. Acetaldehyde and acetone, which are products of microbiological fermentation, were not influenced by light.

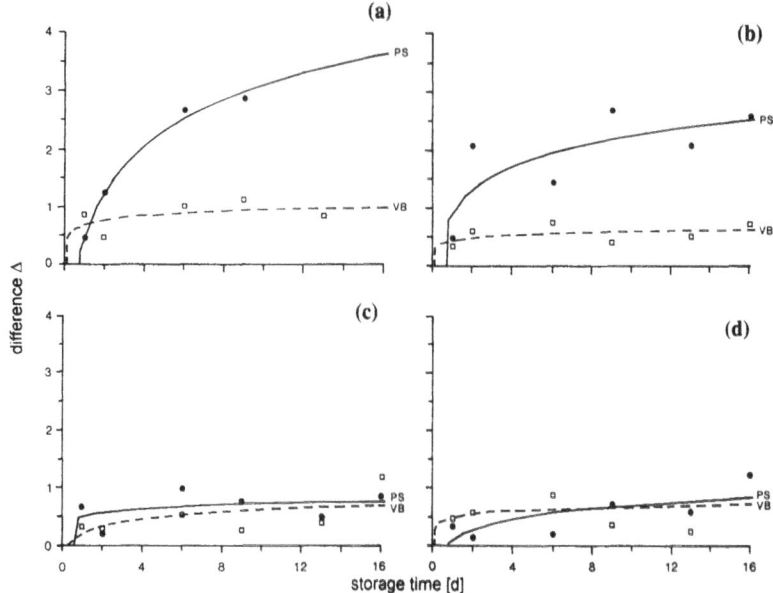

Figure 13.9 Kinetics of flavour changes in (a) natural, (b) strawberry, (c) mocha and (d) chocolate yoghurt stored under light:[19] PS = transparent uncoloured polystyrene; VB = brown glass.

Natural yoghurt was exposed for 4 hours at light intensities of 10 000, 25 000 and 50 000 lx. A typical sunlight taste was produced after only 3 hours at the higher intensities.[117] Triangle and ranking tests showed that light induced flavour changes in natural and to a lesser extent in flavoured yoghurt are dependent upon the type of packaging materials and the exposure time.[19] Figure 13.9 shows the results of the sensory evaluation of samples exposed to light and samples stored in the dark. Natural yoghurt in uncoloured transparent polystyrene jars which were exposed to light suffered the greatest flavour changes. In strawberry yoghurt stored in the same packaging these alterations were less pronounced, but nevertheless significant after only 2 days. The kinetics of flavour changes can be expressed by the following logarithmic regressions:

natural yoghurt: $y = 0.49 + 1.135 \ln x$ (coefficient of correlation $r = 0.996$)
strawberry yoghurt: $y = 0.925 + 0.572 \ln x$ (coefficient of correlation $r = 0.767$)
in which x = storage time in days.

Greater stability was achieved when these yoghurts were stored in brown glass jars.

On the other hand, mocha and chocolate yoghurt underwent no significant flavour changes even in transparent packaging. Only a few panelists assessed mocha yoghurt to be 'not as good' and 'less aromatic' after 9 days and 'clearly not as good' after 13 days. Neither sunlight nor tallowy off-flavours were observed[19] which confirms the hypothesis on the protective effect of the antioxidants in these types of yoghurt.

13.4.5.3 *Butter.* The oxidised taste of butter grew stronger with increasing peroxide values as a result of fluorescent light exposure. This taste defect was more pronounced in the serum than in the fat fraction of melted butter.[110] In the trials made by Emmons *et al.*[108] with 10 different packing materials, the oxidised taste was only slightly noticeable after exposure at 400 lx for 1 day (Table 13.12). However exposure to 400 lx for 6 days or to 2000 lx for either 2 and 4 days produced an oxidised flavour which was stronger at the interior than at the surface of the butter. The sensory flavour evaluation confirmed these results. Among the tested packagings, aluminum foil (packaging E, Table 13.12) transmitted no detectable light. In another trial, these authors surveyed 66 butter samples taken from the top and the front of piles in 6 stores.[118] Seven of the samples were found to have a stronger oxidised taste than butter exposed to light at 2000 lx for 2 days. According to this study[108] a tolerance of up to 1% of light transmittance was fixed for butter packagings in Canada.

The influence of light on butter in different packagings (different sizes and materials) was also studied by our institute.[36] The aim of these studies was to evaluate the possibility of replacing aluminum foils and PVC with ecological sound packaging materials which provide sufficient light and oxygen protection. Packaged butter samples were stored under simulated standard sales conditions (see 13.3: Test conditions 6). The colour, aspect, odour and taste of the samples were assessed by a trained sensory panel. The following conclusions were drawn from this sensory analysis:

- Small butter portions (10–20 g): PVC packaging as used in Switzerland can be replaced by the so-called blends (polystyrene/polyethylene). A thickness of 0.4 to 0.6 mm offers sufficient protection against oxidation and photooxidation (Table 12.13).
- 100 to 200 g butter portions: The currently used aluminum foils were first replaced by aluminum coated paper and then the coating was replaced by using appropriate pigmentation of the paper. However it was also found that the quality of the paper used is important. Butter wrapped in aluminum foils or aluminum free HIFI LS 50 packaging does not change significantly after light exposure. However, when stored in a similar packaging of another manufacturer it appears to have a tallowy, oxidised or sunlight flavour.
- 200 g butter portions: The cardboard wrapped plastic containers (currently in use) were proved to be perfectly adequate. The only weak point with regard to light protection is the junction of the cardboard between the bottom and the walls of the packaging.

Table 13.13 Protection against oxidation of small portion of butter by various containers, wall thickness 0.4 mm (sensory assessment)

Portion in g	Film material	Lid material	Scores		Control	Weight loss %
			Storage in the dark	Storage under light		
20	ABS	PET-alu	10.5 Stale, yellowy	7 Tallowy, yellowy	11.0	4.2
20	ABS	Operflex	10.5 Idem	7 Idem	11.0	4.2
20	PS/PE	PET-alu	11.25	9 Slightly tallowy	11.25	1.4
10	ABS	PET-alu	11.0	10.75	11.25	1.7
10	PS/PE	PET-alu	11.25	11.0	11.25	1.1
10	PS/PE	PET-alu	11.0	10.25	11.0	1.3

Source: reference 36.
Scale of 12 scores: 1 = very bad, 12 = very good
ABS = acrylinitrile-butadiene-styrene
PS/PE = polystyrene/polyethylene monofilm
PET-alu = polyethylene-terephthalate
Operflex = polypropylene

13.4.5.4 *Cheese.* Exposure of cheese to 538, 1614 and 5380 lx fluorescent light for 12 days did not cause any perceptible change in flavour or odour.[98] However, Kristoffersen *et al.*[119] observed a decrease in the flavour quality of Cheddar and Swiss cheese exposed to light, which were nevertheless comparable with the control stored in the dark. The flavour defects were described as follows:

- Cheddar exposed to light: oxidised, metallic, burnt
- Swiss cheese exposed to light: butyric, unclean, oxidised
- Control (Cheddar stored in the dark): acid, fermented, whey taint, utensil
- Control (Swiss cheese stored in the dark): acid, barny, utensil.

13.4.5.5 *Other dairy products.* Butter milk exposed to 2420 lx cool white fluorescent light for 4 days showed no light induced flavour alteration.[120]

Whey obtained from pasteurised milk and neutralised to pH 6.7 proved to be more susceptible to fluorescent light than pasteurised milk or skimmed milk. Dialysed whey is still even more photosensitive. This can be explained by the high light transmission of whey and the high water solubility of riboflavin in whey.[113]

13.4.6 *Effects on colour*

Milk and dairy products change their colour under the influence of light. This change might be due to the photodestruction of some of the brightly coloured constituents such as riboflavin, β-carotene and vitamin A, directly influencing the colour components *a* and *b* according to Hunter. It may also be explained by the modification of the light scattering structure of dairy product (photoagglomeration, photolysis, etc.), which has a direct influence on the colour component *L* (brightness) (see 13.4.3: Effects on proteins and free amino acids) and an indirect one on the colour components *a* and *b*.

13.4.6.1 *Milk and coffee cream.* In addition to riboflavin photodegradation, Toba *et al.*[121] mention light induced degradation of the strongly fluorescent amino acids tyrosine and tryptophan as a cause of colour changes. The penetration depth of the incident light (photons) in the medium is closely related to its turbidity (relationship between transmitted and reflected energy). Thus, the colour components *L*, *a* and *b* undergo greater changes in skimmed milk than in whole milk or coffee cream during storage in lighted conditions. Raw milk is, however, more stable than homogenised milk.[9]

Light exposure of different unpackaged dairy products (skimmed, raw, pasteurised, condensed milk, coffee cream) resulted in a decrease in the colour component L (less brilliant), an algebraic increase in the negative component a (less green) and a decrease in the positive component b (less yellow). The sigmoid course of the colour component curves during light exposure indicates that chain reactions might have taken place. After a certain time, these values become stable.[9]

13.4.6.2 *Yoghurt.* In natural yoghurt exposed to light the colour changes varied from one packaging to the next. In the normal upright position of the cups, the Δ variations of the colour components were as follows (Δ = difference between sample exposed to light and control stored in the dark):

ΔL (decreasing luminance):   C/PS < B/PS = B/G < U/G < U/PS
Δa (decreasing green component):   C/PS = B/PS < B/G < U/G < U/PS
Δb (decreasing yellow component): C/PS = B/PS < B/G < U/G < U/PS.

The degree of light protection afforded by the packagings is in reverse order, i.e. C/PS = cardboard wrapped polystyrene > B/PS = brown-coloured polystyrene > B/G = brown-coloured glass > U/G = uncoloured glass > U/PS = uncoloured polystyrene. With the cups positioned upside down, the light penetrates into the centre of the yoghurt, where its effect is reduced by light scattering.[12,13,17]

The colour component L (luminance), which is naturally weak in mocha and chocolate yoghurt, increased after exposure to 2000 lx for 21 days. This increase was slightly lower in the interior of the yoghurt. The colour components a and b remained unchanged regardless of variations in lighting conditions and packaging materials. However in strawberry yoghurt, there was an overall increase in these components.[17]

13.4.6.3 *Cheese.* Cheese exposed to 1614 lx and more for 12 days underwent a visually perceivable colour change.[98]

13.4.7 *Effects on other components*

In natural yoghurt the following parameters were not influenced by light: pH, partial pressure of carbon dioxide (pCO_2), total carbon dioxide, D- and L-lactic acid concentrations, concentration of 4-carboxylic acids and the concentration of biogenic amines.[12,13] The pH value did not change in any tested yoghurt type.[17]

At the surface of butter exposed to 1500 lx fluorescent light Luby *et al.*[109] detected oxidised cholesterols (7α- and 7β-hydroxycholesterol) by thin layer chromatography. The concentrations of these cholesterol oxidation products was lower in packaged than in unpackaged butter.[111]

Cheddar cheese powder exposed to 1611 lx fluorescent light for 12 weeks also contained different oxidised cholesterols.[122] The contents of 5,6-β-epoxy-5β-cholestan-3β-ol and 7β-hydroxycholesterol remained constant, whereas the concentrations of 5,6-α-epoxy-5α-cholestan-3-β-ol and 7-ketocholesterol were highest during week 3 of exposure and then decreased until week 12.

13.4.8   Comparison of kinetics and limits of detection of light induced changes

During storage under light, the dairy products undergo different chemical, biochemical, physico-chemical and sensorial modifications, which have been extensively discussed in the preceding chapters. In a study on the photosensitivity of natural yoghurt the following changes have been observed chronologically (Figure 13.10): first alterations of taste after only 2 days;[19] the first significant vitamin B_2 losses after about 4 days;[18] change in colour component *b* after approximately 7 days[17] and the peroxide value increase.[16] These alterations seem to follow first-order kinetics.[19] From the order of appearance of the photodegradative changes it may be concluded that the sensory evaluations are more sensitive and yield results earlier than the physico-chemical analyses.[20]

The fact that carbonyl compounds such as propanal, butanal, pentanal, pentanone, hexanal and hexanone[15] appear at the same time as the flavour defects is probably due to a hazardous extrapolation to time zero of the detection of these compounds. The first GC–MS determinations were not actually carried out before day 3 of exposure to light. This hypothesis is confirmed by Bassette's work[115] in milk directly exposed to sunlight as well as by an accelerated test made by Daget.[19] According to Bassette,[115] the concentration of pentanal increased to 150 µg/kg within 20 minutes and then decreased to about 110 µg/kg within a further 20 minutes, whereas the hexanal concentration continued to increase. Daget's accelerated test showed a strong correlation between pentanal production and the appearance of the flavour defect.[19] Both aldehydes are products of degradation of the peroxides, which are themselves products of unsaturated fatty acid oxidation. However, the initial products of this chain of reactions (free radical, peroxides, carbonyl compounds) have not yet been clearly identified. It is presumed that they are responsible for the first sensorial flavour deteriorations.

13.5   Photodegradation mechanisms in milk and dairy products

The photolytic processes occurring in milk and milk products are numerous and complex. Compounds involved in these processes are riboflavin as a photosensitiser, dissolved oxygen as a source of activated oxygen (singlet

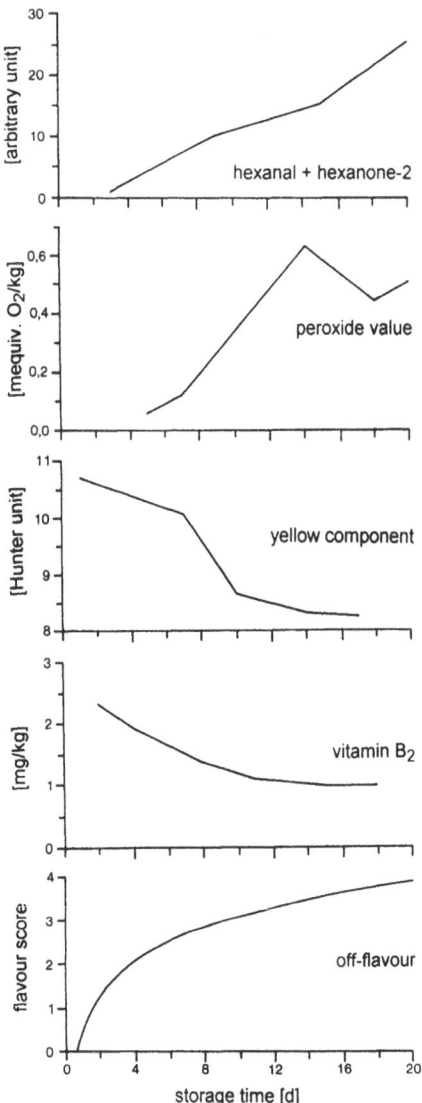

Figure 13.10 Photooxidation indicators in natural yoghurt exposed to light in uncoloured polystyrene cups: a comparison of their response times and their sensitivities.[20]

and triplet oxygen), and both methionine and unsaturated fatty acids as reactants in different chemical processes.

The light induced off-flavours are to be ascribed to two different groups of photodegradation reactions. The reactions of the first group are fast and involve proteins and amino acids. They produce the so-called sunlight flavour (also described as burnt cabbage, cooked cabbage or mushroom). The second group of reactions probably occur later and involve unsaturated fatty acids. They give rise to the oxidised flavour (also described with the terms 'papery', 'cardboardy', 'cappy', 'metallic', 'tallowy' or 'oily').[123]

Riboflavin and vitamin C are highly photosensitive compounds and participate largely in photodegradation. The photodestruction of vitamin C is closely related to the photosensibilisation of riboflavin.

13.5.1 *Light and riboflavin*

According to Richardson and Korycka-Dahl,[124] photolysis of riboflavin is associated with reactions involving oxygen and other substrates (Figure 13.11). The photosensitiser riboflavin is excited by the absorption of photons. The subsequent reactions depend on the relative concentrations of potential reactants. At high oxygen concentrations the reduced ribofla-

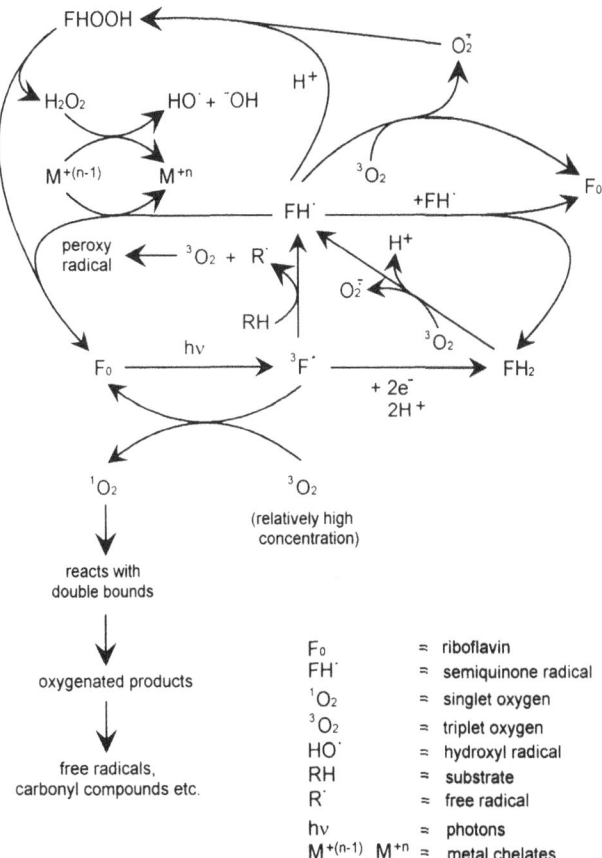

Figure 13.11 Possible photocatalytical reactions involving riboflavin (modified scheme from [124]).

vin is reconverted to its original form producing active oxygen. In a second phase, the active oxygen destroys the riboflavin.

Toyosaki et al.[75,125] have carried out extensive research on the photolysis of riboflavin in milk. They found that riboflavin photolysis is more complex in milk than in purely aqueous solution. There are two phases: The first phase corresponds to the degradation of pure riboflavin in aqueous solution with the generation of active oxygen. In the second phase this active oxygen destroys the riboflavin. The authors described these reactions as follows:[125]

1st phase:
$$\text{Riboflavin} + h\nu \rightarrow \text{riboflavin* (excited)}$$
$$\text{Riboflavin*} + DH_2 \text{ (electron donor)} \rightarrow \text{riboflavin--}H_2 + D$$
$$\text{Riboflavin--}H_2 + O_2 \rightarrow \text{riboflavin} + H^+ + O_2^-$$

2nd phase:
$$\text{Riboflavin--}H_2 + O_2 \rightarrow \text{riboflavin} + H_2O_2$$
$$O_2^- + H_2O_2 \rightarrow OH^- + {}^{\cdot}OH + O_2$$
$$\text{Riboflavin} + O_2^- \text{ (or OH)} \rightarrow \text{riboflavin decomposition}$$

Excited oxygen in milk may be transformed into hydrogen peroxide in the presence of superoxide dismutase. Both excited oxygen and hydrogen peroxide were detected in milk serum exposed to light. At high light intensities the excited oxygen did not seem to participate in the destruction of riboflavin.[75] Riboflavin decomposition was quicker in model systems containing the whey proteins α-lactalbumin or β-lactoglobulin than in a pure riboflavin solution. This indicates that these proteins are involved in the photolysis of riboflavin in milk.[126] The same degradation products were formed in both the model systems and the milk.[127] Analysis by GC–MS identified lumichrome as the major product of riboflavin photolysis in skimmed milk exposed to sunlight.[72]

13.5.2 Methionine as a primary product of sunlight flavour formation

Photodecomposition of methionine generates methional (mercaptomethyl-propionaldehyde) with its secondary products (mercaptans, sulphides, disulphides).[128] These are the compounds principally responsible for the development of the sunlight flavour in milk and dairy products.[114] Patton and Josephson[129] observed a strong flavour defect in methionine enriched skimmed milk exposed to sunlight. Allen and Parks[100] used mass spectrometry to identify methional in skimmed milk exposed to sunlight. Patton[130] detected this compound in milk with concentration of up to 50 μg/kg. Riboflavin is always involved in these reactions (Figure 13.12).

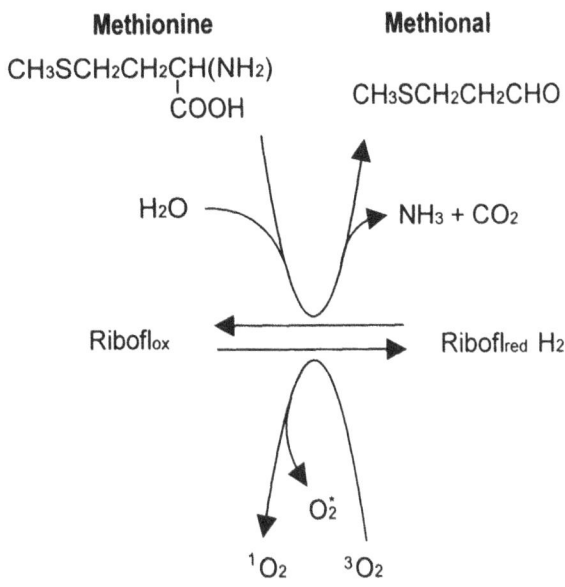

Figure 13.12 Light induced methional formation in the presence of riboflavin (modified after[5]).

13.5.3 *Fat oxidation*

The complex photocatalytic reactions involving riboflavin as a sensitiser produce activated singlet oxygen[131] (Figure 13.11). This oxygen reacts with unsaturated fatty acids, producing peroxides. It is in the initial phase that this photooxidation differs from autooxidation, in which the oxygen reacts with the double bonds in the absence of photons.[132]

13.6   Effects of oxidants and antioxidants

The development of off-flavours due to exposure to light may be influenced by many oxidants and antioxidants present in milk and dairy products, i.e. ascorbic acid, carotene, tocopherol, unsaturated fatty acids, transition metals (e.g. $Cu(II)$), sulfhydryl bonds, superoxide dismutase and particularly dissolved oxygen and riboflavin. Dairy product packagings are not completely oxygen proof and it has been seen that even trace levels of oxygen, which may permeate the packaging, suffice to cause flavour defects.

Aurand *et al.*[131] demonstrated the synergistic effect between different agents and light. They investigated the influence of light on lipid peroxidation in milk in conjunction with singlet oxygen quenchers. Using the thiobarbituric acid test they measured the amount of malonyldialdehyde produced during lipid oxidation. It appeared that in the absence of oxygen light does not alter the thiobarbituric acid level. Superoxide dismutase, a potent inhibitor of reactions involving superoxide anion including singlet oxygen-generating spontaneous dismutation, had no effect[131] (Table 13.14).

Trials carried out by the authors and their colleagues[12,13] using natural yoghurt have shown that in the packaging materials tested the protection offered against light is not identical with that against oxygen (Table 13.15). Light protection seems to be much more important than oxygen protection, particularly during the first two weeks of storage as the active yoghurt flora has a reducing effect on oxygen (redox potential). Protection against oxygen is nevertheless essential because of its synergistic effect with light. Light protection is best afforded by opaque materials (cardboard wrapped polystyrene) or brown–red materials (glass or polystyrene). Glass offers the best protection against oxygen.

Table 13.14 Inhibition of light induced lipid peroxidation by single oxygen quenchers

Sample description	Value obtained by the TBA test ($A \times 10^3$)			
	initial	1st day	3rd day	5th day
Raw milk (control)	21	22	20	24
Raw milk + light	34	42	86	92
Raw milk + light without oxygen	27	29	29	26
Raw milk + light + filter	18	16	18	14
Raw milk (control)	27	28	28	28
Raw milk + light	40	50	62	97
Raw milk + SOD (10^{-8} M)	23	26	26	32
Raw milk + SOD (10^{-8} M) + light	48	58	65	80
Raw milk + SOD (6×10^{-8} M) + light	47	62	69	73

Source: reference 131.
A = Extinction
SOD = Superoxide dismutase
TBA = Thiobarbituric acid

Table 13.15 Protection of natural yoghurt by various packaging materials

Packaging material	Light protection	Oxygen protection	Classification by decreasing overall protection
Brown–red glass	Good	Optimum	1
Paperboard wrapped polystyrene	Very good	Bad	2
Brown–red polystyrene	Good	Medium	3
Uncoloured glass	Medium	Optimum	4
Uncoloured polystyrene	Bad	Medium	5

Source: reference 13.

13.7   Effects of processing on the photosensitivity of milk and dairy products

Before applying a processing technique or defining a technological operation procedure, the specific photosensitivity of a product and the conditions under which it will be exposed to light should be considered. Changes in the contents of reducing and oxidative components (redox system) produced by technological operations, aimed at increasing the photoprotection of the product, may influence the formation of degradation products of light induced reactions.

Since it is presumed that not all pasteurised milk reacts in the same way when exposed to light, the effects of different steps of homogenisation and heat treatment were studied. Raw milk with an aerobic mesophilic germ count of less than 500 000 CFU/ml was homogenised, less than 48 h after milking, at pressures of either 60, 120 or 180 bar and at respective temperatures of 60, 65 or 70°C. It was then pasteurised in a STORK pilot plant, with a capacity of 80 l/h, for 16 s at either 75, 82 or 89°C and bottled in 1 litre clear glass bottles with a headspace of 250 ml of air. The bottles were exposed to 750 lx of warm white light for 20 hours (see 13.3: Test conditions 5).

13.7.1   *Effects of homogenisation pressure and temperature*

Increasing the homogenisation pressure from 60 to 180 bar caused a reduction in the size of the fat globules. The pressure at which milk was homogenised appeared to have little effect on the photosensitivity of milk, whereas the exposure time proved to be very important. An exposure time of 12 h provoked marked taste defects such as sunlight, tallowy and oxidised (Figure 13.13), particularly at the homogenisation pressure of 120 bar.

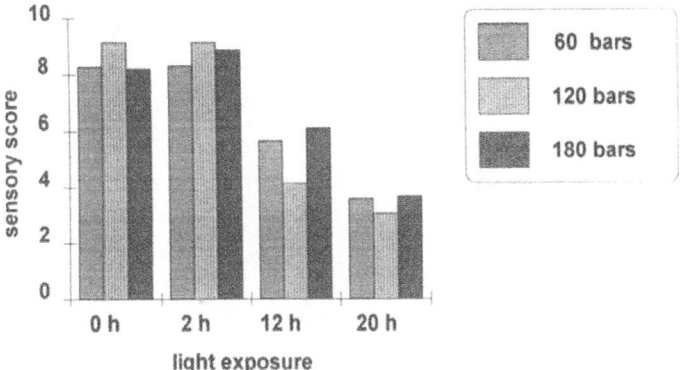

Figure 13.13 Effect of homogenisation pressure on the photosensibility of milk (sensorial evaluation).

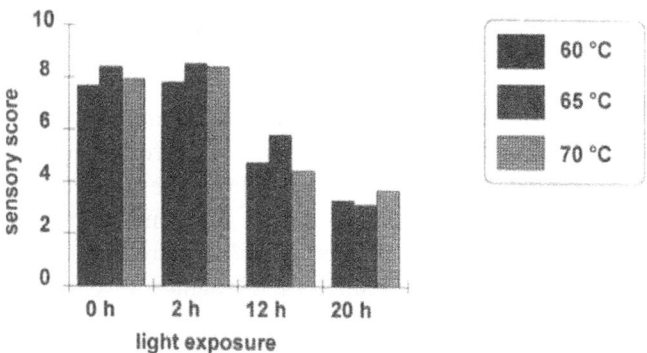

Figure 13.14 Effect of homogenisation temperature on the photosensibility of milk (sensorial evaluation).

The influence of homogenisation temperature on photodegradation of milk was investigated. Milk was subjected to 2-stage homogenisation at pressures of 120 bar and 30 bar and at temperatures between 60 and 70°C. The milk was then pasteurised at 82°C for 16 s. As in the aforementioned tests, light exposure of the milk variants led to marked flavour changes within 12 h. However, no relationship between deterioration of the quality and homogenisation temperature could be established (Figure 13.14). This is probably due to the effect of subsequent pasteurisation.

13.7.2   Effects of pasteurisation temperature

Pasteurisation temperature ranging from 75 to 89°C aggravated whey protein denaturation as well as decreasing the germ count. Figure 13.15 shows the influence of the pasteurisation temperature on the photosensitiv-

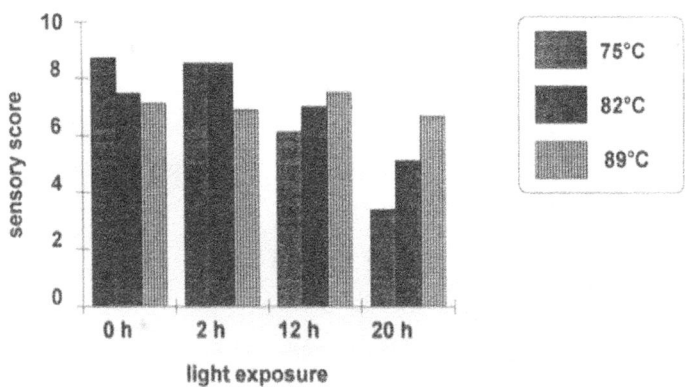

Figure 13.15 Effect of pasteurisation temperature on the photosensibility of milk (sensorial evaluation).

ity of milk. Decreasing pasteurisation temperature led to better scores in sensory analysis for both milk in the absence of light and milk exposed to light for less than 2 h, the evaluation criterion being the cooking taste. The best milk sample was considered to be that heated at the lowest temperature (75°C). However, with increasing exposure time (>12 h), the classification was inverted. The milk sample heated at the highest temperature (89°C) was considered to be the best and was the only sample acceptable for consumption. The samples heated at lower temperatures were rejected because of their taste defects (sunlight taste, oxidised, tallowy). It may be concluded there is an antagonistic relationship between the effects due to heat treatment and those due to exposure to light. Heat treatment liberates sulfhydryl groups from the disulphides of the whey proteins and thus creates a reducing medium. Light exposure generates an oxidative state.[133]

This interpretation has been confirmed by the following test results: UHT milk in glass bottles impervious to oxygen was irradiated at 600 and 4000 lx for 17 days; the initial dissolved oxygen content of 6.5 mg/l fell below the level of detection after 8 days of light exposure. No oxidised or photooxidised taste was noted. It may therefore be presumed that the oxygen remaining after UHT treatment was entirely consumed in the oxidation of the sulfhydryl groups liberated during heating (e.g. conversion of cystine into cysteine) as well as for the oxidation of other reducing agents such as ascorbic acid. In the same conditions but with a headspace volume of 8% (excess oxygen), Schröder[65] recognised an oxidised or sunlight taste after exposure to either 600 lx for 10 days or 4000 lx for 2 days. Ford[51] had already shown in 1967 that, with decreasing ascorbic acid, the partial pressure of dissolved oxygen (pO_2) decreased in pasteurised milk from 187 to 13 mm Hg within 30 min of exposure to sunlight. However, after a quick nitrogen purge, the milk showed no ascorbic acid loss. These observations were explained by another of Schröder's test.[65] UHT milk in glass bottles impervious to oxygen was exposed to 600 lx for 24 days; the acceptability of its flavour did not decrease. In polyethylene bottles permeable to oxygen an oxidised taste was noted after only 48 hours of light exposure. In direct-heated UHT milk stored for 12 weeks in brown-stained reusable glass bottles under natural diffuse light (500 to 1600 lx) or fluorescent light (500 lx) no significant sensorial difference was found on comparison with control milk stored in the dark.[134]

13.8 Practical conclusions

This study shows the importance to be attached by dairy industry to the protection of milk and dairy products against the influence of light. The effects of light are actually detrimental to the flavour and nutritional properties of foodstuffs. Riboflavin (vitamin B_2) and dissolved oxygen are key components involved in undesired photodegradative reactions. The principal substrates of these reactions are the proteins, free amino acids and unsaturated fatty acids.

The numerous factors capable of increasing or reducing the photosensitivity of milk and dairy products can be classified into two main groups. The *intrinsic* factors are linked to the product's composition and constitution. Therefore it is difficult to influence them without modifying the nature of the product. They include the composition of the product (oxidants and antioxidants concentration), pH, redox potential, oxygen content, level of light transmission or turbidity (light scattering) of the product, as well as the application of technological treatments such as homogenisation, heat treatment with formation of reducing sulfhydryl groups at the most elevated temperatures. The *extrinsic* factors are those due to the product environment. The most important are:

- the spectrum and intensity of the light source as well as the exposure time;
- the level of light transmittance and oxygen permeability of the packaging material;
- the storage temperature of the product.

In most cases the appropriate extrinsic factors can be selected to ensure optimum light protection.

13.8.1 *Intrinsic factors*

The specific photosensitivity of each dairy product must be considered before choosing the type of packaging to be used.

Raw milk or milk pasteurised at relatively low temperatures (72–75°C) is more sensitive to photooxidation than milk subjected to higher temperatures. The latter is protected by its increased content of reducing agents and the antioxidant effect of the sulfhydryl groups liberated by heat. The photosensitivity of milk depends therefore to a great extent on the temperature applied during its processing.[10,135] Today's milk pasteurisation technology aims at a mild treatment of the milk in order to preserve its natural properties. Since this increases the photosensitivity of pasteurised milk and limits its shelf life (approx. 5 days), it is particularly important to protect the milk against light. Ecological considerations should not therefore override the need for effective light protection when choosing the packaging for pasteurised milk.

Active flora of yoghurt has a considerable redox potential. This dairy product is therefore better protected against oxygen, which intensifies the photolytic influence of light. However, in comparison with pasteurised milk, yoghurt has certain disadvantages that aggravate photodegradation, i.e. longer shelf-life, more unfavourable surface/volume ratio, higher free amino acid content, the fact that thin-walled packagings are used in order to lower costs and weight. Such packagings, particularly the frequently used polystyrene, have a higher level of light transmittance and oxygen permeability. The more natural and attractive the yoghurt appears, the greater the risk of it undergoing photodegradation. Certain yoghurt types (e.g. mocha and chocolate) are better protected against photooxidation by the pigments and antioxidative compounds they contain.

Cheese is less sensitive to the effects of light than yoghurt. It has a more compact structure reducing the penetration depth of light. It contains less dissolved oxygen because of its increased dry matter and the reducing effect of its microbial flora. The riboflavin content is also less, since it is water-soluble and therefore partially eliminated with whey drainage. However, one should not forget that cheese is stored for long periods of time and the form in which it is displayed (particularly grated or sliced) makes it more susceptible to photodegradation, as well as the fact that the transparent plastic films used for prepacking offer little light protection.

Butter is particularly susceptible to photodegradation because it is rich in fatty acids especially unsaturated fatty acids. It contains enough riboflavin and water to dissolve the oxygen and thus to generate excited oxygen. Further factors add to the risk of photodegradation, namely low storage temperatures, which increase the solubility of oxygen, long storage times (several months) and deep penetration of light (less light scattering potential than cheese).

13.8.2 Extrinsic factors

13.8.2.1 *Packaging materials.* The ideal packaging material should be either opaque or strongly light scattering, in order to limit the transmission of light (energy) into the product. These packaging materials have a two-fold disadvantage: Since they are not transparent the product appears less attractive to the consumer. They are significantly most expensive (e.g. aluminum foil lining for UHT milk or '3K' yoghurt cups). If transparency is desired, the packaging should be tinted red–brown. This is the complementary colour of blue–green and minimises the light absorption by riboflavin, which acts as a photosensitiser.

When choosing the most appropriate material,[136] it is also essential to consider the processing characteristics of the materials, their mechanical and functional properties as well as ecological aspects (recycling, elimination) (Figure 13.1). The actual risks to which each food product is exposed, especially during distribution to the consumer, have to be carefully evaluated in order to avoid overestimation of their light protection requirements. There are numerous and often conflicting factors to be taken into consideration and tests carried out in either normal or accelerated conditions give the best results and the answer as to what is the lowest protection level necessary.[20] The author's trials have shown that the results obtained for natural yoghurt cannot be directly extrapolated to other yoghurt types. Mocha and chocolate yoghurt proved to be less sensitive to light and hence require less light protection.

To avoid light induced changes in milk, a light barrier can be incorporated into the packaging material instead of using a brown–red colouration. This solution was adopted in Switzerland for the light protection of the polyethylene bags used for pasteurised milk.[21]

13.8.2.2 *Light source.* The light source used during storage should be poor in blue–green components (350–550 nm) in order to prevent emission in the spectral region which corresponds to the third absorption band of riboflavin (λ_{max} at approx. 444 nm). Therefore warm white fluorescent tubes with predominantly yellow–red components (e.g. Philips TL 58, TL 82, Osram 36 or Thorn NX) should be used in preference to cool white tubes with predominantly blue–green components (e.g. Philips 33).

The light intensity should be as low as possible to reduce to a minimum photolytic and photocatalytic reactions. To diminish the effects of light additional measures are recommended, i.e. storing the products as far away from the light source as possible (as the light intensity is inversely proportional to the square of the distance), appropriate position of the product, stacking of the product, using all possible optical obstacles for the product transport (crates, baskets, etc.).[34] The exposure time should be limited, e.g. by storage in the dark or the use of timers in display cabinets.[12,13]

13.8.2.3 *Temperature.* The storage temperature should be as low as possible in order to limit energy transmission and to lower the rate of photodegradation reactions. This is also a basic requirement for the microbiological preservation of milk and dairy products. However, low temperatures increase the solubility of oxygen. As far as possible one should compensate for this negative factor by choosing a packaging material more impermeable to oxygen. It has also been shown that the conversion of light energy into thermal energy may raise the temperature in dark-coloured packages such as brown-stained glass bottles exposed to sunlight.[59]

Acknowledgements

The authors wish to thank their colleagues at Nestlé in Lausanne and La Tour-de-Peilz, N. Daget, C. Desarzens, A. Dieffenbacher and E. Tagliaferri, as well as their colleagues from the Swiss Federal Dairy

Research Institute, Liebefeld-Bern, in particular P. Eberhard, H. Eyer and R. Gauch for their interest and help (assays, investigations, availability of tables and figures), with their research work upon which the literature summary in this review is based.

References

1. Bekbölet, M. (1990) *J. Food Protect.*, **53**, 430–440.
2. Bojkow, E. (1970) *Oest. Milchwirt.*, **25**, 449–454, 472–478.
3. Bojkow, E. (1984) *Dt. Molk. Ztg.*, **105**, 1592–1598.
4. Bradley, R.L. (1980) *J. Food Protect.*, **43**, 314–320.
5. Dimick, P.S. (1982) *Can. Inst. Food Sci. Technol. J.*, **15**, 247–256.
6. Janda, J.M. (1990) *Agric. Rev.*, **11**, 94–96.
7. Sattar, A. and deMan, J.M. (1975) *CRC Crit. Rev. Food Sci. Nutr.*, **7**, 13–37.
8. Stull, J.W. (1953) *J. Dairy Sci.*, **36**, 1153–1164.
9. Desarzens, C., Bosset, J.O. and Blanc, B. (1983) *Lebensm. -Wiss. u. -Technol.*, **17**, 241–247.
10. Bosset, J.O., Desarzens, C. and Blanc, B. (1983) *Lebensm. -Wiss. u. -Technol.*, **17**, 248–253.
11. Bosset, J.O. and Flückiger, E. (1985) *Schweiz. Milchztg.*, **111**, 440.
12. Bosset, J.O., Flückiger, E., Lavanchy, P., Nick, B., Pauchard, J.-P., Daget, N., Desarzens, C., Dieffenbacher, A. and Tagliaferri, E. (1986) *Lebensm. -Wiss. u. - Technol.*, **19**, 104–106.
13. Bosset, J.O., Daget, N., Desarzens, C., Dieffenbacher, A., Flückiger, E. Lavanchy, P., Nick, B., Pauchard, J.-P. and Tagliaferri, E. (1986) In: *Food Packaging and Preservation. Theory and practice.* Ed. Mathlouthi, M., Elsevier App. Sci. Publ., London, New York, 235–270.
14. Bosset, J.O. and Flückiger, E. (1986) *Dt. Milchwirt.*, **37**, 908–914.
15. Bosset, J.O. and Gauch, R. (1988) *Trav. chim. aliment. hyg.*, **79**, 165–174.
16. Dieffenbacher, A. and Trisconi, M.-J. (1988) *Trav. chim. aliment. hyg.*, **79**, 371–377.
17. Desarzens, C. (1988) *Trav. chim. aliment. hyg.*, **79**, 378–391.
18. Tagliaferri, E. (1989) *Trav. chim. aliment. hyg.*, **80**, 77–86.
19. Daget, N. (1989) *Trav. chim. aliment. hyg.*, **80**, 87–99.
20. Bosset, J.O. and Flückiger, E. (1989) *Lebensm. -Wiss. u. -Technol.*, **22**, 292–300.
21. Eberhard, P. and Gallmann, P.U. (1991) *Schweiz. Milchztg.*, **117**(26), 3.
22. Bosset, J.O., Eberhard, P., Bütikofer, U., Sieber, R. and Tagliaferri, E. (1991) *Trav. chim. aliment. hyg.*, **82**, 433–456.
23. Tagliaferri, E., Bosset, J.O., Bütikofer, U., Eberhard, P. and Sieber, R. (1992) *Mitt. Gebiete Lebensm. Hyg.*, **83**, 435–452.
24. Tagliaferri, E., Sieber, R., Bütikofer, U., Eberhard, P. and Bosset, J.O. (1992) *Mitt. Gebiete Lebensm. Hyg.*, **83**, 467–491.
25. Sattar, A., Durrani, M.J., Khan, R.N. and Hussain, B.H. (1989) *Z. Lebensm. Unters. Forsch.*, **188**, 430–433.
26. Kamimura, M. and Kaneda, H. (1992) *Developments Food Science*, **28**, 433–472.
27. Chen, A.O., Tsai, Y.S. and Chiu, W.T.F. (1992) *Developments Food Science*, **28**, 375–410.
28. Maujean, A. and Seguin, N. (1983) *Sci. Aliment*, **3**, 603–613.
29. Ohba, T. and Akiyama, H. (1992) *Developments Food Science*, **28**, 473–483.
30. Neumann, M. and Garcia, N.A. (1992) *J. Agr. Food Chem.*, **40**, 957–960.
31. NN (technical notice). *Solar simulation for research & industry.* Oriel, Stamford.
32. Sattar, A. and deMan, J.M. (1973) *J. Inst. Can. Sci. Technol.*, **6**, 170–174.
33. deMan, J.M. (1978) *Can. Inst. Food Sci. Technol. J.*, **11**, 152–154.
34. Olsen, J.R. and Ashoor, S.H. (1987) *J. Dairy Sci.*, **70**, 1362–1370.
35. Sattar, A., deMan, J.M. and Alexander, J.C. (1977) *Can. Inst. Food Sci. Technol. J.*, **10**, 61–64.

36. Eyer, H. (1992), personal communication.
37. Dunkley, W.L., Franklin, J.D. and Pangborn, R.M. (1962) *Food Technol.*, **16**, 112–118.
38. Shipe, W.F., Bassette, R., Deane, D.D., Dunkley, W.L., Hammond, E.G., Harper, W.J., Kleyn, D.H., Morgan, M.E., Nelson, J.H. and Scanlan, R.A. (1978) *J. Dairy Sci.*, **61**, 855–869.
39. NN (1980) *Vitamin-Compendium*. Hoffmann-La Roche, Basel 2.Aufl.
40. Kon, S.K. and Watson, M.B. (1936) *Biochem. J.*, **30**, 2273–2290.
41. Burton, H. (1951) *Dairy Sci. Abstr.*, **13**, 229–245.
42. Sinha, S.P. (1963) *Int. J. Vit. Res.*, **33**, 262–268.
43. Hedrick, T.I. and Glass, L. (1975) *J. Milk Food Technol.*, **38**, 129–131.
44. Sattar, A., deMan, J.M. and Alexander, J.C. (1977) *Can. Inst. Food Sci. Technol. J.*, **10**, 56–60.
45. deMan, J.M. (1981) *J. Dairy Sci.*, **64**, 2031–2032.
46. Bartholomew, B.P. and Ogden, L.V. (1990) *J. Dairy Sci.*, **73**, 1485–1488.
47. Zahar, M., Smith, D.E. and Warthesen, J.J. (1992) *Int. Dairy J.*, **2**, 363–371.
48. Zahar, M., Smith, D.E. and Warthesen, J.J. (1986) *J. Dairy Sci.*, **69**, 2038–2044.
49. Fellman, R.L., Dimick, P.S. and Hollender, R. (1991) *J. Food Protect.*, **54**, 113–116.
50. Gaylord, A.M., Warthesen, J.J. and Smith, D.E. (1986) *J. Dairy Sci.*, **69**, 2779–2784.
51. Ford, J.E. (1977) *J. Dairy Sci.*, **34**, 239–247.
52. Ferretti, L., Lelli, M.E., Miuccio, C. and Ragni, C. (1970) *Quad. Nutr.*, **30**, 124–133.
53. Mohammad, K.S., Al-Thalib, N.A. and Al-Kashab, L.A. (1990) *Egypt. J. Dairy Sci.*, **8**, 37–44.
54. van Dort, H.M., van der Linde, L.M. and de Rijke, D. (1984) *J. Agric. Food Chem.*, **32**, 454–457.
55. Hoppner, K. and Lampi, B. (1985) *Can. Inst. Food Sci. Technol. J.*, **18**, 266–267.
56. Sattar, A., deMan, J.M. and Alexander, J.C. (1977) *Can. Inst. Food Sci. Technol. J.*, **10**, 65–68.
57. Andersen, K.P. (1959) *XV. Int. Milchw. Kongr.*, **3**, 1746–1753.
58. Wodsak, W. (1960) *Nahrung*, **4**, 209–224.
59. Hendrickx, H. and de Moor, H. (1962) *Rev. Agric.*, **15**, 723–738.
60. Radema, L. (1962) *XVI. Int. Milchw. Kongr.*, A, 561–568.
61. Somogyi, J.C. and Ott, E. (1962) *Intern. Z. Vitaminforsch.*, **32**, 493–498.
62. Kiermeier, F. and Waiblinger, W. (1969) *Z. Lebensm. -Unters. -Forsch.*, **141**, 320–331.
63. Renner, E. and Baier, D. (1971) *Dt. Molk. Ztg.*, **92**, 541–543.
64. Berlage-Weinig, L. (1983) *Untersuchungen zur sensorischen Qualität und zur Vitamin-wertigkeit von UHT-Milch und pasteurisierter Milch*. Dissertation Justus-Liebig-Universität Giessen 1–152.
65. Schröder, M.J.A. (1983) *J. Soc. Dairy Technol.*, **36**, 8–12.
66. Renner, E., Renz-Schauen, A., Drathen, M. and Jelen, S. (1989) *Dt. Molk. Ztg.*, **110**, 1006–1008.
67. Nordlund, J. (1984) *Finn. J. Dairy Sci.*, **42**, 49–51.
68. Stamberg, O.E. and Theophilus, D.R. (1945) *J. Dairy Sci.*, **28**, 269–275.
69. Singleton, J.A., Aurand, L.W. and Lancaster, F.W. (1963) *J. Dairy Sci.*, **46**, 1050–1053.
70. Singh, R.P., Heldman, D.R. and Kirk, J.R. (1975) *J. Food Sci.*, **40**, 164–167.
71. Maniere, F.Y. and Dimick, P.S. (1976) *J. Dairy Sci.*, **59**, 2019–2023.
72. Parks, O.W. and Allen, C. (1977) *J. Dairy Sci.*, **60**, 1038–1041.
73. Allen, C. and Parks, O.W. (1979) *J. Dairy Sci.*, **62**, 1377–1379.
74. Toyosaki, T., Yamamoto, A. and Mineshita, T. (1984) *Agric. Biol. Chem.*, **48**, 2919–2922.
75. Toyosaki, T., Yamamoto, A. and Mineshita, T. (1988) *Milchwissenschaft*, **43**, 143–146.
76. Williams, R.R. and Cheldelin, V.H. (1942) *Science*, **96**, 22–23.
77. Peterson, W.J., Haig, F.M. and Shaw, A.O. (1944) *J. Am. Chem. Soc.*, **66**, 662–663.
78. Ziegler, J.A. (1944) *J. Am. Chem. Soc.*, **66**, 1039–1044.
79. Holmes, A.D. and Jones, C.P. (1945) *J. Nutr.*, **29**, 201–209.
80. Josephson, D.V., Burgwald, L.H. and Stoltz, R.B. (1946) *J. Dairy Sci.*, **29**, 273–284.
81. Herreid, E.O., Ruskin, B., Clark, G. and Parks, T.B. (1952) *J. Dairy Sci.*, **35**, 772–778.

82. van der Mijll Dekker, L.P. and Engel, C. (1952) *Neth. Milk Dairy J.*, **6**, 104–108.
83. Kon, S.K. and Thompson, S.Y. (1953) *XIII. Int. Dairy Congr.*, **2**, 363–367.
84. Funai, V. (1956) *Shikoku Acta Med.*, **9**, 78–88.
85. Causeret, J., Hugot, D., Goulas-Scholler, C. and Mocquot, G. (1961) *Ann. Technol. Agric.*, **10**, 289–300.
86. Hugot, D., Lhuissier, M. and Causeret, J. (1962) *Ann. Technol. Agric.*, **11**, 145–151.
87. Paik, J.J. and Kim, H. (1976) *Korean J. Nutr.*, **9**, 164–168, cited after *Dairy Sci. Abstr.*, **39**, 462 (1977).
88. Bates, C.J., Liu, D.-S., Fuller, N.J. and Lucas, A. (1985) *Acta Paed. Scand.*, **74**, 40–44.
89. Sikka, P., Narayan, R. and Atheya, U.K. (1990) *Indian J. Dairy Sci.*, **43**, 598–600, cited after *Dairy Sci. Abstr.*, **53**, 887 (1991).
90. Fukumoto, J. and Nakashima, K. (1975) *J. Jap. Soc. Food Nutr.*, **28**, 257–261, cited after *Dairy Sci. Abstr.*, **38**, 791 (1976).
91. Hoskin, J.C. and Dimick, P.S. (1979) *J. Food Protect.*, **42**, 105–109.
92. Fanelli, A.J., Burlew, J.V. and Gabriel, M.K. (1985) *J. Food Protect.*, **48**, 112–117.
93. Schröder, M.J.A., Scott, K.J., Bland, M.A. and Bishop, D.R. (1985) *J. Soc. Dairy Technol.*, **38**, 48–52.
94. Ford, J.E., Schröder, M.J.A., Bland, M.A., Blease, K.S. and Scott, K.J. (1986) *J. Dairy Res.*, **53**, 391–406.
95. Hoskin, J.C. (1988) *J. Food Protect.*, **51**, 19–23.
96. Renner, E., Renz-Schauen, A. and Drathen, M. (1988) *Dt. Molk. Ztg.*, **20**, 609–612.
97. Palanuk, S.L., Warthesen, J.J. and Smith, D.E. (1988) *J. Food Sci.*, **53**, 436–438.
98. Deger, D. and Ashoor, S.H. (1987) *J. Dairy Sci.*, **70**, 1371–1376.
99. Dimick, P.S. and Kilara, A. (1983) *Kieler Milchwirt. Forschungsber.*, **35**, 289–299.
100. Allen, C. and Parks, O.W. (1975) *J. Dairy Sci.*, **58**, 1609–1611.
101. Samuelsson, E.-G. and Harper, J.W. (1961) *Milchwissenschaft*, **16**, 344–347.
102. Kanner, J.D. and Fennema, O. (1987) *J. Agr. Food Chem.*, **35**, 71–76.
103. Dimick, P.S. (1973) *J. Milk Food Technol.*, **36**, 383–387.
104. Dimick, P.S. (1976) *J. Dairy Sci.*, **59**, 305–308.
105. Gilmore, T.M. and Dimick, P.S. (1979) *J. Dairy Sci.*, **62**, 189–194.
106. Korycka-Dahl, M. and Richardson, T. (1979) *J. Dairy Sci.*, **62**, 183–188.
107. Dieffenbacher, A. and Lüthi, B. (1986) *Mitt. Gebiete Lebensm. Hyg.*, **77**, 544–553.
108. Emmons, D.B., Paquette, G.J., Froehlich, D.A., Beckett, D.C., Modler, H.W., Butler, G., Brackenridge, P. and Daniels, G. (1986) *J. Dairy Sci.*, **69**, 2437–2450.
109. Luby, J.M., Gray, J.I., Harte, B.R. and Ryan, T.C. (1986) *J. Food Sci.*, **51**, 904–907.
110. Foley, J., O'Donov, D. and Cooney, C. (1971) *J. Soc. Dairy Technol.*, **24**, 38–45.
111. Luby, J.M., Gray, J.I. and Harte, B.R. (1986) *J. Food Sci.*, **51**, 908–911.
112. Zhang, D. and Mahoney, A.W. (1990) *J. Dairy Sci.*, **73**, 2252–2258.
113. Jenq, W., Bassette, R. and Crang, R.E. (1988) *J. Dairy Sci.*, **71**, 2366–2372.
114. Gaafar, A.M. and Gaber, F.L. (1992) *Egypt. J. Dairy Sci.*, **20**, 111–115.
115. Bassette, R. (1976) *J. Milk Food Technol.*, **39**, 10–12.
116. deMan, J.M. (1980) *Milchwissenschaft*, **35**, 725–726.
117. de Moor, H. and Hendrickx, H. (1970) *Rev. Agric.*, **23**, 1647–1654.
118. Emmons, D.B., Froehlich, D.A., Paquette, G.J., Butler, G., Beckett, D.C., Modler, H.W., Brackenridge, P. and Daniels, G. (1986) *J. Dairy Sci.*, **69**, 2248–2267.
119. Kristoffersen, T., Stüssi, D.B. and Gould, I.A. (1964) *J. Dairy Sci.*, **47**, 496–501.
120. Hoskin, J.C. (1989) *Cult. Dairy Prod. J.*, **24**(1), 14–15.
121. Toba, T., Adachi, S. and Arai, I. (1980) *J. Dairy Sci.*, **63**, 1796–1801.
122. Sander, B.D., Smith, D.E., Addis, P.B. and Park, S.W. (1989) *J. Food Sci.*, **54**, 874–879.
123. Azzara, D. and Campbell, L.B. (1992) *Developments Food Science*, **28**, 329–374.
124. Richardson, T. and Korycka-Dahl, M. (1983) *Developments in Dairy Chemistry*, **2**, 241–363.
125. Toyosaki, T., Yamamoto, A. and Mineshita, T. (1987) *Milchwissenschaft*, **42**, 364–367.
126. Toyosaki, T. and Mineshita, T. (1989) *Milchwissenschaft*, **44**, 292–294.
127. Toyosaki, T. and Mineshita, T. (1990) *Milchwissenschaft*, **45**, 80–82.
128. Samuelsson, E.-G. (1962) *Milchwissenschaft*, **17**, 401–405.
129. Patton, S. and Josephson, D.V. (1953) *Science*, **118**, 211.

130. Patton, S. (1954) *J. Dairy Sci.*, **37**, 446–452.
131. Aurand, L.W., Boone, N.H. and Giddings, G.G. (1977) *J. Dairy Sci.*, **60**, 363–369.
132. Mörsel, J.-T. (1990) *Nahrung*, **34**, 3–12.
133. Eberhard, P. (1992), personal communication.
134. Biewendt, H.-G., Manasterny, K. and Moltzen, B. (1991) *Kieler Milchwirt. Forschungsber.*, **43**, 307–316.
135. Bürki, C. (1977) *Verhalten der Sulfhydril- und Disulfidgruppen in proteinhaltigen wässerigen Systemen, dargestellt am Beispiel der Milchproteine.* Thesis Nr. 5924, ETH Zürich.
136. Fink, P. (1990) *Mitt. Gebiete Lebensm. Hyg.*, **81**, 10–22.

Index

The manufacturer's authorised representative in the EU is Springer
Nature Customer Service Centre GmbH, Europaplatz 3, 69115 Heidelberg,
Germany. If you have any concerns regarding our products, please
contact ProductSafety@springernature.com

Printed and bound by CPI Group (UK) Ltd, Croydon, CR0 4YY
23/04/2026
02095624-0002